AF679248

Archaeometric and Archaeological Approaches to Ceramics

Papers presented at EMAC '05, 8th European Meeting on Ancient Ceramics, Lyon 2005

Edited by

S. Y. Waksman

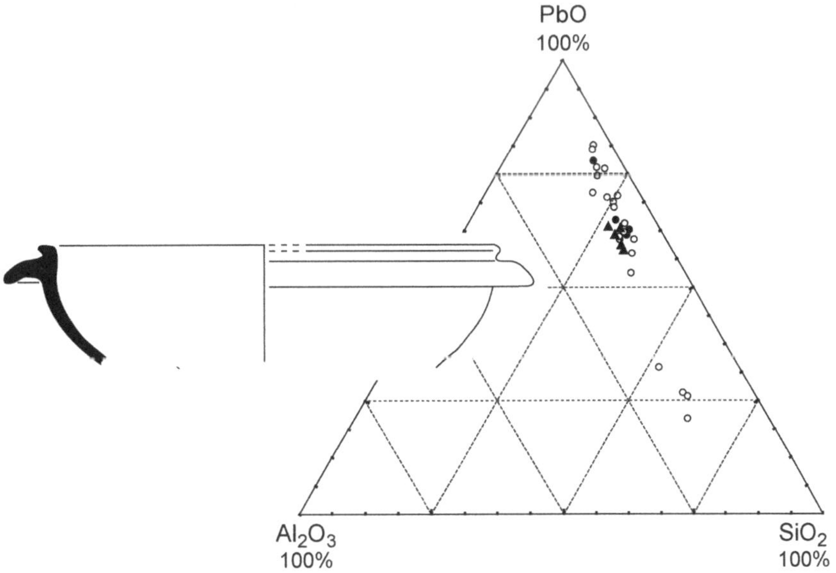

BAR International Series 1691
2007

Published in 2016 by
BAR Publishing, Oxford

BAR International Series 1691

Archaeometric and Archaeological Approaches to Ceramics

ISBN 978 1 4073 0129 7

© The editors and contributors severally and the Publisher 2007

COVER IMAGE *(left):*
Late Roman glazed mortar found in Saint-Blaise excavations, possibly from northern Italy. [After C.A.T.H.M.A., Importations de céramiques communes méditerranéennes dans le midi de la Gaule (Ve - VIIe s.), in A cerâmica medieval no Mediterrâneo ocidental, *1991, Mertola, p. 39, fig. 28]*

The authors' moral rights under the 1988 UK Copyright,
Designs and Patents Act are hereby expressly asserted.

All rights reserved. No part of this work may be copied, reproduced, stored, sold, distributed, scanned, saved in any form of digital format or transmitted in any form digitally, without the written permission of the Publisher.

BAR Publishing is the trading name of British Archaeological Reports (Oxford) Ltd. British Archaeological Reports was first incorporated in 1974 to publish the BAR Series, International and British. In 1992 Hadrian Books Ltd became part of the BAR group. This volume was originally published by Archaeopress in conjunction with British Archaeological Reports (Oxford) Ltd / Hadrian Books Ltd, the Series principal publisher, in 2007. This present volume is published by BAR Publishing, 2016.

Printed in England

BAR
PUBLISHING

BAR titles are available from:

BAR Publishing
122 Banbury Rd, Oxford, OX2 7BP, UK
EMAIL info@barpublishing.com
PHONE +44 (0)1865 310431
FAX +44 (0)1865 316916
www.barpublishing.com

TABLE OF CONTENTS

PREFACE

This volume presents a selection of papers delivered at the 8th European Meeting on Ancient Ceramics (EMAC) which took place in Lyon (France) in 2005. The EMAC series of conferences, initiated in Rome in 1991, meets every two years in a European city and brings together specialists carrying out research on ancient ceramics using archaeological sciences. EMAC provides the opportunity to present and debate recent advances in this field of research, from methodological aspects to archaeological studies with fully integrated laboratory approaches.

The theme sessions covered by the 2005 meeting included both "traditional" EMAC subjects (Methodological developments / Production, diffusion, commercialization / Dating ceramics / Ceramics as containers / Ceramics as building material / Technical ceramics) and new subjects proposed on a tentative basis (Ceramics in conservation and restoration / Slip, gloss, glaze). The dynamism of some of the sessions, including "Slip, gloss, glaze", "Technical ceramics" and "Production, diffusion, commercialization" was noticeable and may reflect the main trends of research in the past years.

The conference was organized in Lyon by the *Laboratoire de Céramologie* (UMR 5138 *Archéométrie et Archéologie: Origine, datation et technologie des matériaux*, CNRS – Universities Lyon 2 and Lyon 1), through the initiative of its director, A. Schmitt. The *Laboratoire de Céramologie,* particularly under the impetus of its former director M. Picon, has undertaken many projects covering a wide spectrum of ceramics studies over the last 30 years. It is thus not by chance that the first French meeting of EMAC took place in Lyon.

Papers presented in this volume were peer-reviewed by an international editorial committee, whose work is gratefully acknowledged. Many thanks are due to H. Hatcher for the English revision of the texts and to E. Capet for the page setting and the homogenization of the volume.

EMAC'05 benefited from the financial and logistical support of the following partners: *Université Lumière Lyon 2*; *Ministère de l'Éducation Nationale, de l'Enseignement Supérieur et de la Recherche*; *Centre National de la Recherche Scientifique* (CNRS); *Conseil général* of the Rhône department; *IUFM* of Lyon; *Mairie de Lyon*; *Maison de l'Orient et de la Méditerranée*. We are especially grateful to the *Conseil régional* of the region Rhône-Alpes and the *Association des Amis de la Maison de l'Orient* (AAMO) who provided support specifically for the publication of this volume.

The editor
S.Y. Waksman

ASSESSMENT OF ANCIENT VESSEL DESIGN WITH THE FINITE ELEMENT METHOD (FEM)

ANNO HEIN, VASSILIS KILIKOGLOU

Institute of Materials Science, N.C.S.R. "Demokritos", 15310 Aghia Paraskevi, Greece

1. INTRODUCTION

Ceramics have always been, and still were up to the middle of the 20th century, the most common material for a large variety of applications in daily life as they fulfilled miscellaneous functions. Pottery vessels have been used mainly for storage, transport and cooking. Fine ceramics have been used as tableware but also for purely decorative purposes. Furthermore, ceramics have been used as tools in technological applications, such as metallurgy or glass production, and construction. Ceramics became so popular not only due to the wide availability of raw materials, but due to some basic properties which were advantageous in all uses. First, due to its plasticity the unfired clay paste is easy to form and practically any possible shape can be fabricated. Second, with firing, ceramic material becomes hard and extremely resistant to environment and use. In this context, the most important properties induced with firing are strength, toughness, impermeability and heat resistance.

Diverse functions, however, required a variety of properties for different ceramic products. Whereas, for example, cooking pots had to withstand primarily thermal stresses, the use of vessels as transport containers required high fracture strength. Furthermore, the contents of storage or transport jars demand particular properties. For example, liquids require impermeable walls but at the same time, water jars with high porosity in order to cool down the water. Suitable material properties could be achieved with modification of clay paste and with appropriate firing technology (Steponaitis 1984, Bronitsky and Hamer 1986). However, the clay paste itself was only one component of the actual ceramic structure, with the others being non-plastic inclusions and voids or pores.

Apart from the nature of the clay paste components and the firing conditions, another parameter that affects the mechanical and thermal performance of the ceramic vessels is its shape and geometric characteristics. Important parameters are the wall thickness, the vessel size, the curvature and angles in its form. Furthermore, its performance is dependent on the points of load, *i.e.* the areas at which mechanical or thermal loads are actually applied to the vessel. It is, therefore, apparent that the design of a vessel is an important parameter for the assessment of technology and in this context a continuous diachronic change in the design can be observed (Braun 1983). Experimental testing of the mechanical properties of the material and the effect of shape on the performance of actual archaeological vessels is out of the question as long as these measurements are linked with destruction of the object. Nevertheless, considering particular material properties, which can be determined in the laboratory, in combination with shape, a computer model of the vessel can be generated. By using the Finite Element Method (FEM) mechanical or thermal loads can be applied to the model in a computer simulation and it is possible to assess the performance of the vessel and to predict consequently critical loads, which could cause failure of the vessel. The present paper will introduce the approach and illustrate its potential with three case studies.

2. MATERIAL PROPERTIES OF CERAMICS

2.1. Mechanical properties

When a solid object is exposed to a tensile force or stress it is deformed. In the beginning this deformation is elastic and therefore reversible, but for stresses beyond a critical value the change becomes irreversible. Principally, two different modes of behaviour are possible: plastic deformation, which is typical of metals, and brittle behaviour, more typical of ceramics (Kilikoglou *et al.* 1998). Depending on microstructure and parameters, such as temper or porosity, earthenware ceramic materials show in many cases a reaction which is a combination of the above. After the initial fracture the ceramic body is not totally destroyed, but cracks develop and additional energy is dissipated until the body finally breaks apart. In this case the ceramics exhibit a stable fracture unlike the unstable brittle behaviour of ceramics containing very low amounts of tempering material and having low porosity. The fracture mode of a specific ceramic material can be examined with bending experiments, in which the total fracture energy or toughness can be determined as the sum of the intrinsic fracture energy or strength plus the energy dissipated until the ultimate failure of the ceramic. Strength and toughness allow for estimation of the critical strain and therefore for failure prediction.

For the elastic deformation, Young's modulus and Poisson ratio of the ceramic material also need to be used in the failure prediction modeling. Taking into account the material's density, FEM determines the deformation by simulating different modes and values of loading. From these deformations the critical loads, which would cause fracture, can finally be estimated.

2.2. Thermal properties

In general, ceramics are considered as heat resistant materials exhibiting high thermal capacity. Nevertheless, they are affected by temperature changes because of their thermal expansion. Temperature changes and temperature differences cause thermal stress in the ceramic body, which can lead to cracks. When the temperature is increased in a particular part of the ceramic body, for example by exposing it to a heat source, heat is transferred through the body depending on the specific heat conductivity of the material. At the same time the material expands due to the increased temperature, causing compressive stress in the surrounding area.

Apart from the assessment of resistance to thermal stress and thermal shock, FEM can be used to simulate the temperature development in the ceramics by considering thermal conductivity and heat capacity of the specific material. In this way, operating conditions can be simulated and the functionality of the ceramics can be examined with respect to heat input.

3. FINITE ELEMENT METHOD (FEM)

The Finite Element Method is a numerical approach originating from engineering sciences, which was implemented in order to estimate performance of a model in terms of *e.g.* deformation, stress and temperature. An object, or rather its model, is divided into sub-domains, the so-called finite elements, to which the respective material properties are attributed. These elements are connected and form a mesh with constraints, which can be defined. When loads are applied to the model, the behaviour of the object can be estimated by a set of linear equations based on the performance of the finite elements, providing an approximate solution for the partial differential equations, which describe structural deformation or heat transfer. For a good approximation the size of the elements has to be small. With FEM both two-dimensional and three-dimensional models can be examined. However it has, to be borne in mind that the number of elements, *i.e.* equations, is considerably higher in a three-dimensional model and therefore the time for computer calculations is increased.

When applying FEM to an archaeological ceramic object the first step is to digitize its shape, which can be done for instance by processing with a CAD program. Because vessel bodies are commonly axially symmetric a three-dimensional model can be constructed from a two-dimensional section. Material properties, based on actual measurements of ceramic specimens coming from the vessel under examination or based on estimates taking into account the fabric, are then attributed to the model. At the next stage an element is selected, which must be small enough in order to describe the details of the vessel shape. The next step is to define the constraints, which are in the case of *e.g.* structural analysis of a vessel for example the domains which cannot be moved, such as its base. In the final step, the points of load have to be defined and the particular loads have to be quantified. For the solution of the problem either static analysis can be selected for purely structural problems, or transient analysis, which considers the development over time, for example in heat transfer problems. The status of the object, displayed by the FEM solution, can be finally evaluated in view of critical strains and stresses, which could cause the failure of the ceramic object.

4. CASE STUDIES

4.1. Myrtos piriform jars

One of the first applications of FEM with regard to archaeological ceramics was a study on Minoan jars as attempt to develop a model for failure prediction (Kilikoglou and Vekinis 2002). The ceramic material was characterized as calcareous and extensively vitrified, with an estimated firing temperature of 1000-1080 °C. The mechanical properties of this type of ceramic were then determined in the laboratory following the methodology described by Kilikoglou *et al.* (1998). The fracture strain of the material was estimated as 0.09% and the Young's modulus was estimated as 12.5 GPa.

The failure prediction method developed was tested with fracture tests performed on model jars, which were constructed from controlled material. Afterwards FEM model of a piriform jar was designed. The volume of the modeled water jar was approximately 40 litres with an empty weight of approximately 10 kg, resulting in a total weight of the filled vessel of approximately 50 kg. Based on the model different loading conditions were examined. Lifting the filled vessel by the two handles, for example, resulted in a maximum strain of 0.058% at the joints of the handles, which appeared to be critical. Therefore a ring support for the filled vessel was suggested.

4.2. Hellenistic transport amphorae

A recent project concerns the study of the changes that took place in the shape of Hellenistic transport amphorae. In antiquity amphorae constituted the most important transport containers for a large variety of liquid and dry products (Twede 2002). During their use they had to withstand considerable mechanical loads and stresses. These functional ceramics featured a high level of standardization concerning quality, size and particularly shape, which appeared to be related to the specific production place and its content.

In a geochemical study of amphorae from Halasarna on Cos (Greece) a certain continuity in ceramic production could be observed for the entire period examined from the 5th to the 1st century BC (Hein *et al.* forthcoming). Apparently there was no change in raw material selection, clay paste preparation or firing technology and therefore the basic material properties of the ceramics remained the same. Laboratory tests of selected body fragments indicated the high mechanical strength of the ceramics, clearly higher than in the example of the Minoan piriform jars. However, the ceramics presented an unstable fracture, a typical behaviour for containers of liquids.

On the other hand, during the same period a change in the vessel shape of Coan amphorae can be observed (Georgopoulou 2006). Fig. 1 shows two digitalized amphorae from the 4th and the 1st century BC, respectively. On the basis of the models, a volume of 32 litres was estimated for the earlier amphora type and

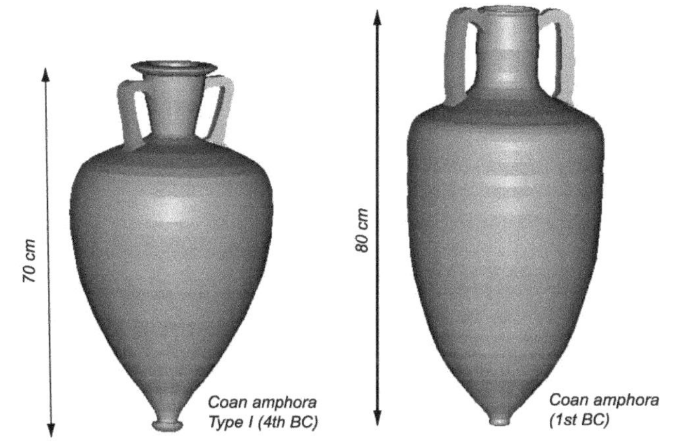

Fig. 1 – Two digital models of Coan amphorae from the 4th century BC and from the 1st century BC, respectively.

41 litres for the later, whereas the weight of the vessels remained roughly the same (approximately 9.5 kg for the earlier amphora and approximately 10.2 kg for the later amphora). The two models were examined with FEM simulating four point loads of 1000 N on their shoulders (Fig. 2). Assuming stowage and transport in a cargo ship with amphorae piled up in layers (Twede 2002), this load would correspond to the load on the bottom layer of nine layers of amphorae with a total weight of 50 kg each. The simulated deformation was higher in the model of the later

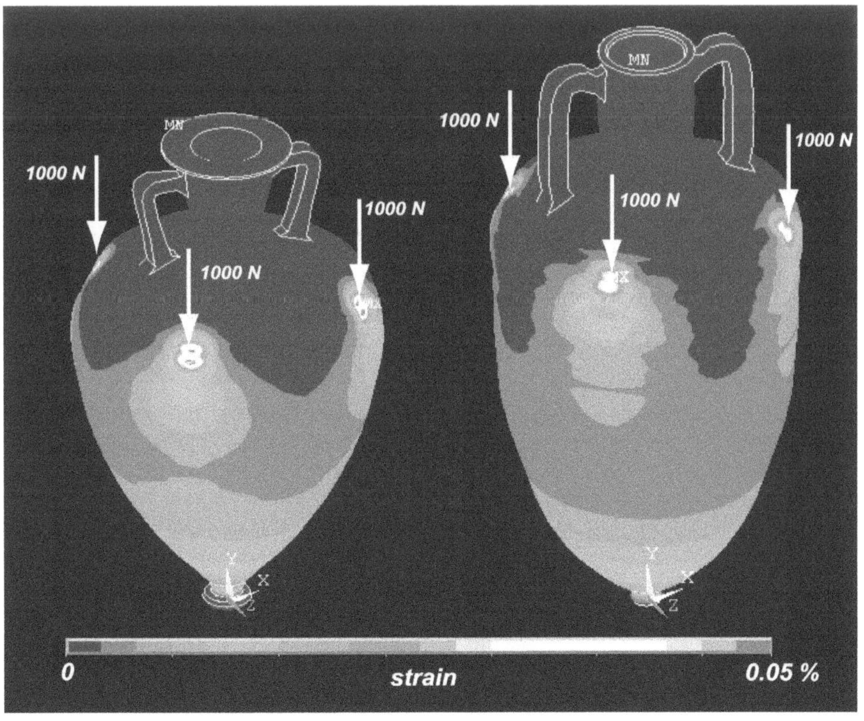

Fig. 2 – Strain as calculated by the FEM software: in this case loads of 1000 N at four points on the shoulders were simulated, corresponding to an amphora in the bottom layer of nine layers of amphorae piled up. Maximum strains in this simulation emerged at the points of load and close to the amphora feet, which is indicated by colour variation.

amphora. In contrast, the resulting maximum strain in this simulation was higher in the model of the earlier amphora with 0.07% compared to 0.05% in the later amphora. Both values, however, were below the fracture strain assumed for the ceramics from Cos.

These first results indicated that the later type was produced with approximately the same amount of clay paste, which resulted in the same vessel weight. But the content, *i.e.* the wine the amphora contained, was clearly increased by at least 25%. At the same time the mechanical properties with respect to stresses and strains during use did not degrade.

4.3. Heat transfer in Bronze Age copper smelting furnaces

An example of FEM study of metallurgical ceramics was implemented with regard to thermal properties and heat transfer (Hein and Kilikoglou in press). On the basis of Bronze Age copper smelting furnaces from Cyprus (Hein *et al.* 2007) a model was designed and the heat transfer during the smelting process was simulated. The FEM model considered as well as heat conduction in the actual furnace, heat transfer into the environment, *i.e.* soil and air, respectively. In combination with temperature estimations, based on observations of the degree of vitrification in particular furnace fragments, the operating conditions of the furnaces were assessed. Furthermore, points of maximum thermal stress could be located, which, however, emerged as rather uncritical due to the thick furnace walls.

The FEM approach provided some advantages compared to the numerical solution of the heat transfer equation, which was applied in earlier studies of ancient metallurgical ceramics (Kingery and Gourdin 1976, Tite *et al.* 1982). It was for example possible to examine 2-dimensional or even 3-dimensional furnace models. Furthermore heat convection, *i.e.* the heat transfer to the air, was considered. This was particularly important in view of free standing furnace walls in contrast to furnace linings which were examined in the earlier studies.

5. CONCLUSIONS

Computer simulations help to understand functionality of ancient vessels and other ceramic objects. For this reason the basic mechanical and thermal properties of the specific ceramic material have to be determined at a first stage. This can be done in the laboratory by testing small brick specimens. These specimens can be cut from the actual ceramics in question or they can be produced from a similar material, concerning base clay, firing temperature, porosity and percentage, type and size of inclusions. Knowing the shape and material properties, the entire object can be modeled on the computer and tested for its assumed performance, simulating different kinds of loads.

REFERENCES

Braun D.P. 1983, "Pots as tools", *in* A. Keene and J. Moore (eds.), *Archaeological hammers and theories*, Academic Press, New York, p. 107-134.

Bronitsky G. and Hamer R. 1986, "Experiments in ceramic technology: The effects of various tempering materials on impact and thermal-shock resistance", *American antiquity* 51, p. 89-101.

Georgopoulou V. 2006, *Coan transport amphorae: Typology, chronology and dissemination in the Eastern Mediterranean and the Black Sea*, Ph.D., University of Athens.

Hein A., Kilikoglou V. and Kassianidou V. 2007, "Chemical and mineralogical examination of metallurgical ceramics from a Late Bronze Age copper smelting site in Cyprus", *Journal of archaeological science* 34, p. 141-154.

Hein A. and Kilikoglou V., "Finite element analysis (FEA) of metallurgical ceramics: Assessment of their thermal behaviour", *in* I. Tzachili (ed.), *Aegean metallurgy in the Bronze Age*, Cretan University Press, in press.

Hein A., Georgopoulou V., Nodarou E. and Kilikoglou V., "Koan amphorae from Halasarna – Investigations in a Hellenistic amphorae production center", forthcoming.

Kilikoglou V., Vekinis G., Maniatis Y. and Day P.M. 1998, "Mechanical performance of quartz-tempered ceramics: Part I: Strength and toughness", *Archaeometry* 40, p. 261-279.

Kilikoglou V. and Vekinis G. 2002, "Failure prediction and function determination of archaeological pottery by Finite Element Analysis", *Journal of archaeological science* 29, p. 1317-1325.

Kingery W.D. and Gourdin W.H. 1976, "Examination of furnace linings from Rothenberg Site #590 in Wadi Zaghra", *Journal of field archaeology* 3, p. 351-353.

Steponaitis V. 1984, "Technological studies of prehistoric pottery from Alabama: Physical properties and vessel function", *in* S.E. van der Leeuw and A.C. Pritchard (eds.), *The many dimensions of pottery: Ceramics in archaeology and anthropology*, University of Amsterdam, Amsterdam, p. 79-127.

Tite M.S., Maniatis Y., Meeks N.D., Bimson M., Hughes M.J. and Leppard S.C. 1982, "Technological studies of ancient ceramics from the Near East, Aegean, and Southeast Europe", *in* T.A. Wertime and S.F. Wertime (eds.), *Early pyrotechnology*, Smithsonian Institution Press, Washington D.C., p. 61-71.

Twede D., 2002, "The packaging technology and science of ancient transport amphoras", *Packaging technology and science* 15, p. 181-195.

CONTRIBUTION FOR A MINERALOGICAL THERMOMETER TO BE APPLIED TO LOW FIRED AND/OR NON-CARBONATE CERAMICS

P. RICCIARDI[1], L. NODARI[2], B. FABBRI[1], S. GUALTIERI[1], U. RUSSO[2]

[1] CNR-ISTEC, Institute of Science and Technology for Ceramics, Faenza
[2] Chemistry Dept., University of Padova, INSTM

ABSTRACT

This research project was suggested by the need to define the ceramic firing temperature in some particular situations, such as non-carbonate ceramic pastes and low fired bodies (about 650 °C). This last condition is very frequent in the case of prehistoric artefacts, while non-carbonate pastes are frequently encountered in classic and medieval ceramics. The firing temperature is usually evaluated through the mineralogical transformations which occur in carbonate mixtures. But this method cannot be used in our case because no distinct crystalline phases were formed during firing, so the samples were analyzed by the usual techniques for chemical (XRF) and mineralogical (XRD) characterization and through innovative techniques such as micro-Raman and Mössbauer spectroscopy. The Raman technique allowed us to identify individual phase changes through the study of the evolution of vibrational spectra of titanium compounds, in particular, the transformation anatase/rutile. The use of Mössbauer spectroscopy is related to the presence of iron oxides and oxy-hydroxides. So, it was possible to verify the deviation of the magnetic properties of these compounds according to the effects of temperature on particle size and/or isomorphous substitution of iron by other elements that affect the Mössbauer spectra. In order to obtain reference data, two sets of laboratory specimens, made with an illitic and a kaolinitic clay, were prepared appropriately and characterized.

KEY WORDS: FIRING TEMPERATURE, TITANIUM DIOXIDE, RAMAN SPECTROSCOPY, MÖSSBAUER SPECTROSCOPY, SIMULATED FIRING

1. INTRODUCTION

Firing temperature determination is among the most difficult tasks in ancient pottery studies. The knowledge of the thermal treatment provides the experts with precious information about social, economic and cultural characteristics of the society that produced them. But firing conditions in ancient times were liable to great variation even within the same production, and also not all materials carry a distinct trace of the thermal treatment they underwent during pottery making. In particular, low fired and/or non-carbonate ceramic pastes do not show the new Ca-phases which are often effectively used as temperature markers.

A detailed bibliographic review shows that several different techniques have so far been used in archaeometric studies in order to evaluate firing temperatures, and more generally the firing conditions of archaeological pottery artefacts (Fabbri 1998). Amongst them, most notably are X-ray diffraction for the analysis of crystalline phases (Cultrone *et al.* 2001; Maniatis *et al.* 2002), thermal analyses (Moropoulou *et al.* 1995), SEM observations (Wolf 2002), colour measures (Mirti and Davit 2004), FT-IR analysis (Artioli *et al.* 2000; De Benedetto *et al.* 2002) and Mössbauer spectroscopy (Wagner and Wagner 2004).

Phase analysis is probably the main method for studying ceramic technology; our work aims at adding different perspectives to previous well-known studies on widely used kinds of clays (Maggetti 1982, and references therein) by applying different techniques. In trying to develop a mineralogical thermometer for low fired and/or non-carbonate materials, we focused on the study of the transition between two of the polymorphs of titanium dioxide, anatase and rutile. This transition has so far been reported to take place over a range of temperatures as great as 600 °C (Gennari and Pasquevich 1998; Ghosh *et al.* 2001; Rodriguez-Talavera 1997), in studies generally not dedicated to pottery analysis. The same transition has nonetheless been used as a firing temperature marker (Liem *et al.* 2000; Lofrumento *et al.* 2005). We used Raman spectroscopy for the detection of TiO_2, because this technique has long proved its great effectiveness in the detection of titanium compounds (Murad 1997, 2003). We also chose to use Mössbauer spectroscopy to follow in detail the evolution of iron compounds. This methodology has often been effectively used for detailed analysis of iron compound modifications in raw and fired clays (Murad 1998; Häusler 2004; Nodari *et al.* 2004).

2. MATERIALS AND METHODS

2.1. Sample preparation

Two different raw materials (one illitic clay named "Sala10" and one kaolin named "C1641") were characterized and used to prepare sets of laboratory specimens. From the mineralogical point of view, the illitic clay is mainly made up of quartz and a certain quantity of illite; accessory phases, present in small quantities, are plagioclase, K-feldspar, chlorite, kaolinite, illite-smectite mixed layers and hematite. The kaolin is mainly composed of kaolinite, with small quantities of quartz and illite. Their chemical composition is reported in Table 1.

Both materials were hand shaped in small disks about 3 cm in diameter and 0.4 cm thick. Sets of three such disks were kiln-fired in an oxidising atmosphere, at temperatures between 600 °C and 1100 °C for the illitic clay, and between 800 °C and 1100 °C for the kaolin, in steps of 50 °C. Powders were obtained by crushing in an agate mortar a mixture of fragments taken from each of the three pieces fired at each temperature; each sample obtained was then characterized by means of XRD, Raman spectroscopy and Mössbauer analyses at 298K. Mössbauer analyses were performed only on the illitic clay samples, because kaolinitic ones do not contain a significant amount of iron.

Illitic clay samples fired at the highest temperatures (≥ 1050 °C) developed an extensive "black core", associated with an obvious bulge (Emiliani and Corbara 1999). In these cases, powders for Mössbauer analysis were prepared by discarding the black zones, where iron is present in a highly reduced form, because they do not represent the oxidising firing conditions.

Samples of the illitic clay were also pit-fired, in a first attempt to simulate ancient pottery production conditions. In order to evaluate the maximum firing temperature reached in the pit, some of these samples had been previously kiln-fired respectively at 650 °C, 700 °C and 750 °C. If the temperature reached in the pit exceeded that of the previous firing, some transformations should be observed. XRD and Raman analyses were performed on all samples.

2.2. Experimental setup

X-ray diffraction analyses were performed with a 300 W power Miniflex instrument with Cu anode. We scanned an angular interval from 6° to 55° 2 θ, whose central part (26° to 30° 2 θ), containing the main peak of quartz, has been eliminated in Fig. 1 and Fig. 4 to allow easier identification of minor peaks. Measurements were made, with a step of 0.02° and an acquisition time of 3 sec at each point. The identification of the crystalline phases was carried out by comparison with JCPDS data sheets.

Raman spectra were collected using a Renishaw RM1000 spectrometer with a green Ar^+ laser (λ = 514.5 nm) and maximum power of 3.5 mW. A small quantity of powder from each sample was optically examined through the microscope coupled to the instrument and then analyzed in 10 to 20 spots. Only a few significant spectra were collected and analysed for each sample.

Mössbauer data were collected at 298K by a conventional instrument with constant acceleration velocity and a ^{57}Co source in a Rh matrix, using a proportional counter as a γ-ray detector. The spectra were fitted using a procedure which interpolates experimental data by means of Lorentzian functions. A small quantity of powder (about 80 mg) from each sample was mixed with vaseline and inserted into the sample holder. Each measurement takes one to two days; therefore only a limited number of samples can be studied by this method.

3. RESULTS AND DISCUSSION

3.1. Illitic clay

A comparative plot of X-ray diffractograms of the illitic clay (Fig. 1) shows that quartz is only slightly affected by the temperature rise, and that clay minerals disappear by 900 °C, while cristobalite starts forming above 950 °C and mullite around 1050 °C. The traces of hematite which were present in the raw material become more evident as the temperature (T) increases, most notably above 800 °C. The spinel-type phase which is weakly visible by XRD at the highest firing temperatures most likely corresponds to hercynite, whose presence has to be related to the "black core". The evolution of Fe-compounds can be followed in greater detail by examining the Mössbauer spectra shown in Fig. 2; the one corresponding to the raw material shows a sextet due to a well crystallized oxide (hematite), together with paramagnetic absorption peaks which have been fitted with one Fe(III) and one Fe(II) sixfold site. At 600 °C the complete oxidation of Fe(II) promotes the formation of a new Fe(III) site which disappears above 850 °C, when there is a rearrangement of the silicate framework. In the spectra shown, this phenomenon is first

	SiO_2	Al_2O_3	TiO_2	Fe_2O_3	MgO	MnO	CaO	Na_2O	K_2O	P_2O_5	L.O.I.
Sala10	62.04	22.76	1.02	6.99	1.97	0.10	1.38	0.88	2.67	0.19	12.77
C1641	53.68	43.09	1.09	1.29	0.14	0.01	0.16	0.08	0.41	0.05	13.54

Table 1 – Chemical composition of the illitic clay "Sala10" and of the kaolin "C1641" (oxide wt. % normalized without L.O.I.).

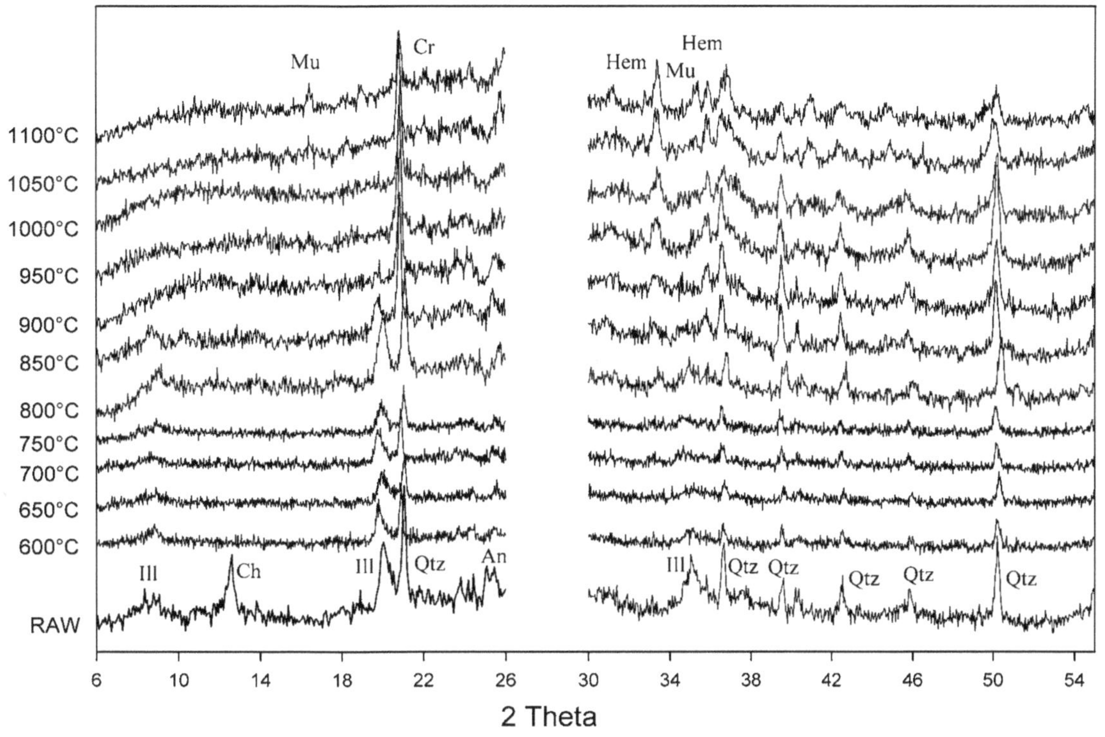

Fig. 1 – Comparative plot of XRD analyses on the illitic clay (Qtz: quartz, Ill: illite, Ch: chlorite, An: anatase, Hem: hematite, Mu: mullite, Cr: cristobalite).

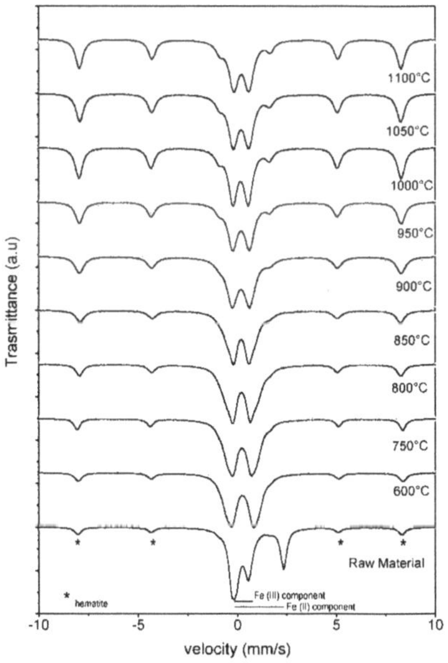

Fig. 2 – Mössbauer spectra of the illitic clay.

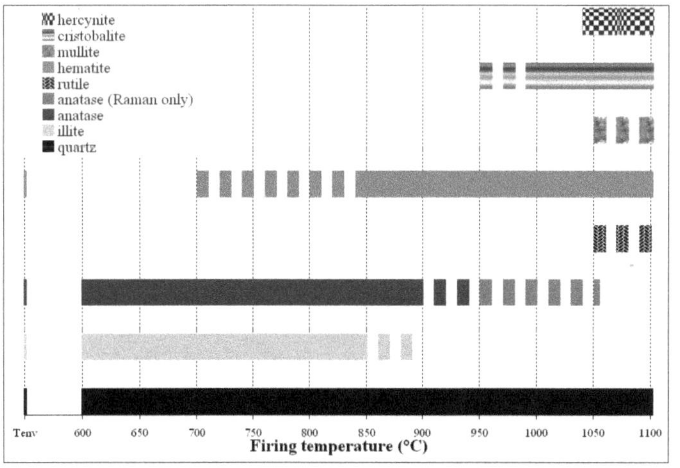

Fig. 3 – Synthesis of experimental data on the illitic clay.

indicated by the widening of the central doublet between these two temperatures. The amount of hematite grows clearly between 800 °C and 1000 °C.

A synthesis of the experimental data on the illitic clay, including those from Raman spectra, is presented in Fig 3; the results mostly confirm literature data about the mineralogical changes occurring in clays during firing (Cultrone *et al.* 2001; Riccardi *et al.* 1999).

The combined use of XRD and Raman results shows the transition between anatase and rutile taking place around 1050 °C. This is too high a temperature if one wants to use the transition as a mineralogical thermometer for prehistoric/low fired artefacts.

Diffractograms of the pit-fired samples show that the firing temperature in the pit cannot have exceeded 700 °C. The Raman spectra are very different from those of the kiln-fired samples, in that they only show a high fluorescence background, which masks any traces of peaks and often saturates the detector. This phenomenon is clearly due to an interaction between ceramics and soil. In support of this statement, we observed the same feature in the Raman spectra of a few ceramic sherds belonging to prehistoric (6th-4th millennium BC) archaeological contexts in Italy and Romania. Further investigations are necessary to better specify the nature of such an interaction.

3.2. Kaolin

A comparative plot of X-ray diffractograms of the fired kaolin samples (Fig. 4) shows the presence of a high background signal between 15° and 35° 2 θ, which is due to the great quantity of amorphous material deriving from the destruction of kaolinite well below 800 °C. Two large peaks are observable above 950 °C, which correspond to γ-alumina that forms, in a scarcely crystalline structure, following the collapse of metakaolin. Quartz is hardly affected by the temperature increase, while illite disappears by 950 °C and mullite starts forming above 1000 °C; these results confirm data in literature about the evolution of kaolinitic clays during firing (Lee *et al.* 1999). Anatase is present in all the fired samples without significant variations.

Raman analyses show the coexistence of anatase and rutile in the whole temperature range, as well as in the raw material (Fig. 5); thus this is another case in which the transition between the two polymorphs cannot be used as a temperature marker. Fig. 6 shows a synthesis of the experimental data on the kaolin.

4. CONCLUSIONS

The anatase/rutile transition was observed only in the illitic clay, both with Raman spectroscopy and XRD, at a temperature of about 1050 °C. In the kaolin, both polymorphs coexist in the raw material and in the whole range of temperatures (up to 1100 °C), and no significant changes can be observed in their relative amounts.

Experimental data therefore confirm that this transition is strongly dependent on the composition of the clay, and that it can hardly be used as a mineralogical thermometer in the archaeometric study of pottery. Some applications could perhaps be found in studies on ceramic typologies fired at temperatures above 1000 °C. It still seems interesting to investigate more deeply the dependence of this transition on the composition of the raw material in a systematic manner. From a methodological point of view, the use of a macro- rather than a micro-Raman approach seems to be advisable in order to obtain quantitative data.

Mössbauer spectroscopy has once more proved to be a useful tool to follow the evolution of iron oxides during firing, in a temperature range of interest for ancient pottery studies. Data collected at 298K suggest that a deeper understanding of the evolution of iron compounds, particularly the formation and size increase of iron oxide crystals, is needed. This is going to be achieved by undertaking Mössbauer measures at low temperatures (80 and if necessary 10K). Such data, together with further

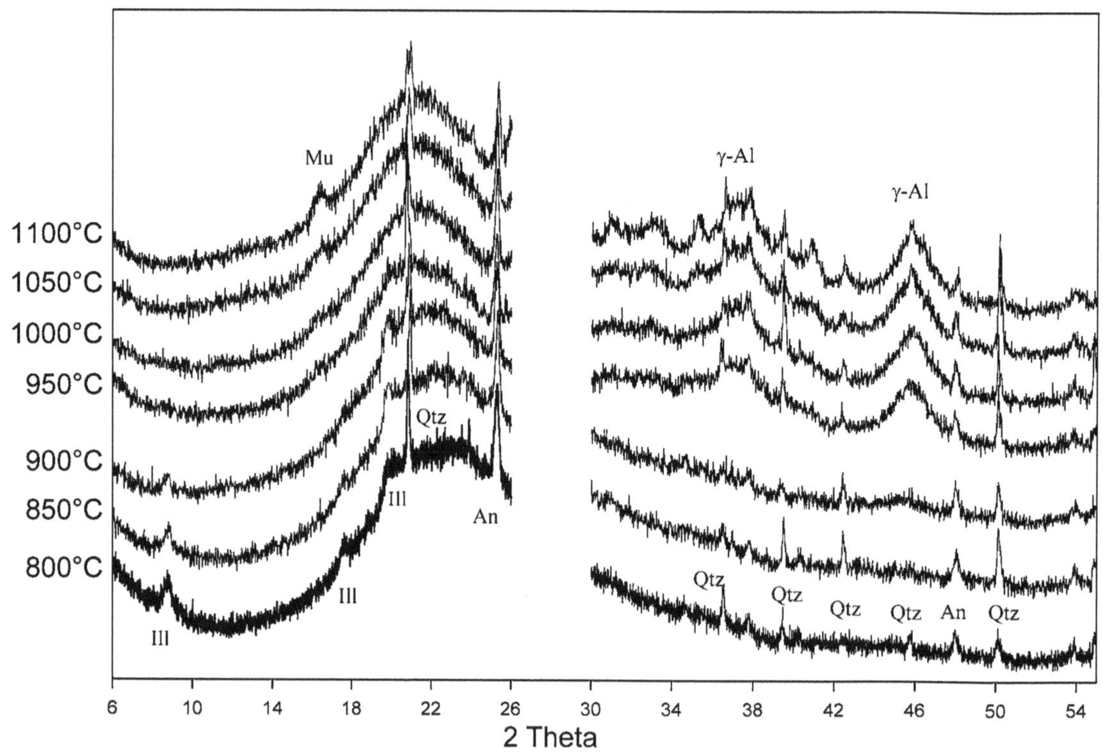

Fig. 4 – Comparative plot of XRD analyses on the kaolin (Qtz: quartz, Ill: illite, An: anatase, Mu: mullite, γ-Al: γ-alumina).

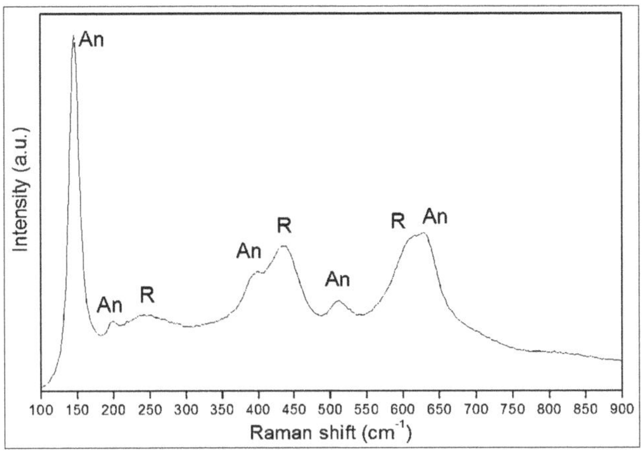

Fig. 5 – Raman spectrum of kaolin fired at 1100 °C (An: anatase, R: rutile).

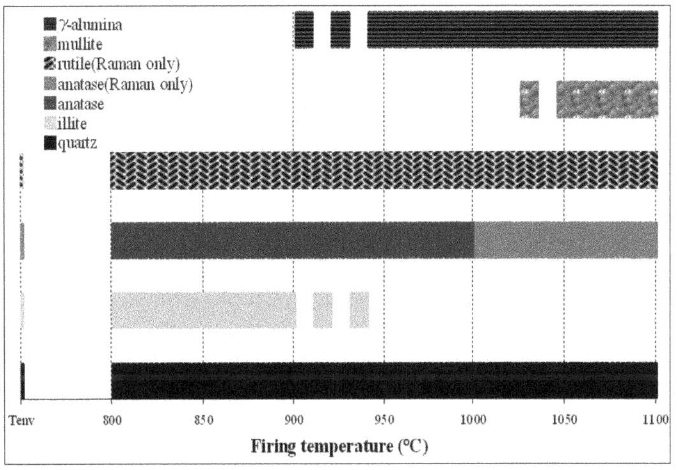

Fig. 6 – Synthesis of experimental data on the kaolin.

analyses on materials with different oxide concentrations and particle sizes, might allow a "calibration curve" for a fast comparison with the ancient pottery spectra to be developed.

ACKNOWLEDGMENTS

The authors wish to thank Prof. G. Di Lonardo and Dr. F. Ospitali from the Industrial Chemistry Dept. of the University of Bologna for performing the Raman analyses and for fruitful discussions.

REFERENCES

ARTIOLI G., BAGNASCO GIANNI G., BRUNI S., CARIATI F., FERMO P., MORIN S., RUSSO U. 2000, "Studio spettroscopico della tecnologia di cottura di ceramiche etrusche dagli scavi di Tarquinia", *in* M. Martini (ed.), *Atti del I congresso nazionale di archeometria*, p. 335-349.

CULTRONE G., RODRIGUEZ-NAVARRO C., SEBASTIAN E., CAZALLA O., DE LA TORRE M. J. 2001, "Carbonate and silicate phase reactions during ceramic firing", *European journal of mineralogy* 13, p. 621-634.

DE BENEDETTO G.E., LAVIANO R., SABBATINI L., ZAMBONIN P.G. 2002, "Infrared spectroscopy characterization of ancient pottery", *Journal of cultural heritage* 3, p. 177-186.

EMILIANI G.P., CORBARA F. 1999, *Tecnologia ceramica 2. La lavorazione*, Faenza.

FABBRI B. 1998, "The problem of defining the firing temperature of ceramic artefacts", *in* C. Ariao, A. Bietti, L. Castelletti and C. Peretto (eds.), *Proceedings of the XIII UISPP Congress Volume I*, p. 207-214.

GENNARI F.C., PASQUEVICH D.M. 1998, "Kinetics of the anatase-rutile transformation in TiO_2 in the presence of Fe_2O_3", *Journal of materials science* 25, p. 1571-1578.

GHOSH S.K., VASUDEVAN A.K., PRABHAKAR RAO P., WARRIER K.G.K. 2001, "Influence of different additives on anatase-rutile transformation in titania system", *British ceramic transactions* 100 (4), p. 151-154.

HÄUSLER W. 2004, "Firing of clays studied by X-ray diffraction and Mössbauer spectroscopy", *Hyperfine interactions* 154, p. 121-141.

LEE S., KIM Y.J., MOON H. 1999, "Phase transformation sequence from kaolinite to mullite investigated by an energy-filtering transmission electron microscope", *Journal of the American ceramic society* 82 (10), p. 2841-2848.

LIEM N.Q., SAGON G., QUANG V.X., TAN H.V., COLOMBAN P. 2000, "Raman study of the microstructure, composition and processing of ancient Vietnamese (proto)porcelains and celadons (13-16th centuries)", *Journal of Raman spectroscopy* 31, p. 933-942.

LOFRUMENTO C., ZOPPI A., CASTELLUCCI E.M. 2005, "La spettroscopia micro-Raman: un termometro mineralogico nello studio delle ceramiche archeologiche", *in* B. Fabbri, G. Volpe and S. Gualtieri (eds.), *Tecnologia di lavorazione e impieghi dei manufatti, Atti della 7ª Giornata di archeometria della ceramica*, p. 21-28.

MAGGETTI M. 1982, "Phase analysis and its significance for technology and origin", *in* J. Olin and A.D. Franklin (eds.), *Archaeological ceramics*, Smithsonian Institution Press, Washington D.C., p. 121-133.

MANIATIS Y., FACORELLIS Y., PILLALI A., PAPANTHIMOU-PAPAEFTHIMIOU A. 2002, "Firing temperature determination of low fired clay structures", *in* V. Kililoglou, A. Hein and Y. Maniatis (eds.), *Modern trends in scientific studies on ancient ceramics*, BAR International Series 1001, p. 59-68.

MIRTI P. and DAVIT P. 2004, "New developments in the study of ancient pottery by colour measurement", *Journal of archaeological science* 31, p. 741-751.

MOROPOULOU A., BAKOLAS A., BISBIKOU K. 1995, "Thermal analysis as a method of characterizing ancient ceramic technologies", *Thermochimica acta* 2570, p. 743-753.

Murad E. 1997, "Identification of minor amounts of anatase in kaolins by Raman spectroscopy", *The American mineralogist* 82, p. 203-206.

Murad E. 1998, "Clays and clay minerals: what can Mössbauer spectroscopy do to help understand them?", *Hyperfine interactions* 117, p. 39-70.

Murad E. 2003, "Raman and X-ray diffraction data on anatase in fired kaolins", *Clays and clay minerals* 51, p. 689-692.

Nodari L., Maritan L., Mazzoli C., Russo U. 2004, "Sandwich structures in the Etruscan-Padan type pottery", *Applied clay science* 27, p. 119-128.

Riccardi M.P., Messiga B., Duminuco P. 1991, "An approach to the dynamics of clay firing", *Applied clay science* 15, p. 393-409.

Rodrìguez-Talavera R. 1997, "Modification of the phase transition temperatures in titania doped with various cations", *Journal of materials research* 12 (2), p. 439-443.

Wagner F.E., Wagner U. 2004, "Mössbauer spectra of clays and ceramics", *Hyperfine interactions* 154, p. 35-82.

Wolf S. 2002, "Estimation of the production parameters of very large medieval bricks from St. Urban, Switzerland", *Archaeometry* 44 (1), p. 37-65.

INVESTIGATING THE SUBSTRATE-GLAZE INTERFACE OF CERAMICS WITH SEM-EDS AND RAMAN SPECTROSCOPY

C. PACHECO, R. CHAPOULIE, F. DANIEL

CRP2A, IRAMAT UMR 5060 CNRS – Université Michel-de-Montaigne Bordeaux 3
Maison de l'Archéologie, Esplanade des Antilles, 33607 Pessac, France

INTRODUCTION

While studying some prestigious gilded ceramics from Central Asia (14th-15th c.), we have been examining the problem of the interfaces between different materials: first a gold/glaze interface, then a glaze/ceramic body one. This paper deals with the latter.

One of the main issues concerning the production process of ancient glazed ceramics can be summed up as: how many firings did these glazed ceramics undergo?

As a matter of fact, after shaping and drying the clay, the ceramist can choose between two processes to glaze the pottery:
— a single firing process, consisting in putting the glazing mixture directly on the dry clay and then firing the object;
— a double firing process, consisting in firing the dry clay object, then putting the glazing mixture on the biscuit-fired ceramic, and finally firing the glazed object.

The different processes directly affect the interaction between the clay or ceramic body and the glazing mixture during the firings, creating a ceramic/glass interface with different mechanical properties. The firing process has a direct effect on the adhesion of the glazes to the ceramic substrate.

The lead glaze/ceramic body interface has already been studied by different research teams and the first thing to point out is that this interface is characterized by some devitrification crystals. In archaeological Hispano-Moresque ceramics, they were identified as sanidine crystals $(K_{0.85}Pb_{0.12}Ca_{0.03})Al_{1.08}Si_{2.92}O_8$ (Molera *et al.* 1993). Laboratory-made samples showed discrimination between the two processes was possible, as there were more devitrification crystals in the single-fired samples than in the double-fired ones (Tite *et al.* 1998; Chapoulie *et al.* 2005). Moreover, the thickness of the interface ranged from 30 to 40 μm for the former samples and only from 5 to 10 μm for the latter (Molera *et al.* 1997). Thus, the following interaction process was suggested (Molera *et al.* 2001):
— decomposition of the phases composing the ceramic or clay body;
— inter-diffusion of the elements present in the glaze and the body;
— formation of an interface zone (as lead containing devitrification crystals).

In the light of this knowledge, laboratory samples have been made with well known raw materials and following precise firing processes. The innovative feature of the study under discussion lies in the combination of the results of SEM-EDS and Raman analyses on the very same areas and crystals at the interface, in search of discriminative features of the two types of firings.

CHARACTERISATION OF THE RAW MATERIALS

The chosen clay was kaolinite-enriched, as the chemical composition of this type of fired clay enables comparisons to be made with other laboratory-made samples whose studies are reported in the literature (Tite *et al.* 1998; Molera *et al.* 1997). X-ray diffraction (XRD), using a powder diffractometer (Siemens D500), enabled us to identify the crystalline phases present in the clay as kaolinite, quartz, dolomite and calcite.

The glaze was made from an industrial lead monosilicate frit ($PbO{\cdot}SiO_2$) whose melting point is *ca.* 750-800 °C.

THE EXPERIMENTAL FIRING PROCESSES

- single firing process:
  - rising temperature rate: 100 °C/h
  - 10 minute soak at 800 °C
  - decreasing temperature: kiln inertia

- double firing process:
  - firing of the clay:
    - rising temperature rate: 100 °C/h
    - 10 minute soak at 900 °C
    - decreasing temperature: kiln inertia
  - firing of the glaze
    - rising temperature rate: 100 °C/h
    - 10 minute soak at 800 °C
    - decreasing temperature: kiln inertia

CHARACTERISATION OF THE LABORATORY MADE SAMPLES

The crystalline composition of the ceramic bodies was determined by XRD. Both ceramic substrates contain quartz and calcium carbonates.

The chemical compositions (Table 1) of the ceramic bodies and the glazes were measured on polished cross-sections by EDS coupled to a SEM (JEOL JSM 6460LV). From the XRD and SEM-EDS results, no clear distinction can be made between the two firing processes, although a higher Al content may be noted in the glaze of the single-fired sample, due to its greater diffusion from the body to the glaze (Tite *et al.* 1998). Then, the research focussed on the glaze/ceramic substrate interface.

DOUBLE-FIRED SAMPLE

Fig. 1 shows a back-scattered electron image of the glaze/ceramic body interface. The presence of two types of devitrification crystals should be noted: some are dark grey and others, light grey. The chemical composition of such crystals, determined by SEM-EDS on 1.5 x 1.5 μm^2 areas, is presented in Fig. 2 in atomic weight percentage.

On the same zone, the Raman profile of the interface was measured and is presented in Fig. 3. The spectra were obtained with a 633 nm laser and a Renishaw RM 2000 spectrometer. The four previous zones, described as glaze, dark grey crystals, light grey crystals and ceramic body have characteristic Raman signatures. The quartz peak at 463 cm^{-1} is present in both the light grey zone and the ceramic body. The four peaks at 323, 387, 665 and 1014 cm^{-1} are characteristic of a silicate belonging to the pyroxene family, whose formula $XY(Si,Al)_2O_6$ agrees with the EDS results.

Moreover, the polymerization index of the glaze (zone 1, Fig. 1) was calculated as the ratio of the areas of the SiO_2 stretching band (*ca.* 500 cm^{-1}) out of its bending bands (*ca.* 1000 cm^{-1}) according to the literature (Colomban 2003; Colomban and Paulsen. 2005). The value of 0.1 is compatible with the results obtained on Islamic lead-containing glazes whose polymerization indexes are less than 0.3 (Colomban and Paulsen 2005). The dark grey (zone 2, Fig. 1) and the light grey (zone 3, Fig. 1) crystals present similar polymerization indexes.

No specific spectra corresponding to lead-enriched pyroxenes have been found in literature. Nevertheless, according to Wang *et al.* (2001), the Raman spectral peak positions of these pyroxenes are the key to determining pyroxene cation mole fractions. The study of these positions and the influence of lead cations on them will be the subject of further study.

SINGLE-FIRED SAMPLE

Fig. 4 shows a back-scattered electron image of the glaze/ceramic body interface. The presence of two types of devitrification crystals is notable: some are white and others grey. The chemical composition of such crystals, determined on 1.5 x 1.5 μm^2 areas, is presented in atomic weight percentage in Fig. 5. The white crystals contain a high concentration of lead and the grey crystals have chemical compositions close to those of the dark and light grey crystals in the double-fired sample.

The Raman profile was measured on the very same region at the interface (Fig. 6). The four zones, described as glaze, white crystals, grey crystals and ceramic body have characteristic Raman signatures. The white crystals present characteristic peaks of feldspars $XAlSi_3O_8$ / $YAl_2Si_2O_8$ at 480, 513 and 558 cm^{-1}. The grey crystals respond to the laser excitation not only with the characteristic peaks of pyroxene, as in the grey crystals in the double-fired sample, but also with the characteristic peaks of feldspar. According to Freeman *et al.* (2003), using Raman peaks in three characteristic regions of Raman spectra of feldspars can distinguish seven types of feldspars. As for the previous sample, the influence of the lead cation is still to be investigated.

	Ceramic substrate		Glaze	
weight%	Single-fired sample	Double-fired sample	Single-fired sample	Double-fired sample
MgO	5.8 ± 0.5	5.7 ± 0.4	0.5 ± 0.1	0.2 ± 0.1
Al_2O_3	18.9 ± 0.8	19.4 ± 0.5	3.0 ± 0.3	1.1 ± 0.2
SiO_2	58.2 ± 2.4	58.8 ± 1.2	25.8 ± 0.3	24.7 ± 0.3
K_2O	0.4 ± 0.1	0.3 ± 0.1	–	–
CaO	14.5 ± 1.5	13.8 ± 0.9	0.9 ± 0.2	0.2 ± 0.1
TiO_2	1.2 ± 0.1	1.2 ± 0.1	–	–
Fe_2O_3	1.1 ± 0.1	0.8 ± 0.2	–	–
PbO	–	–	69.7 ± 1.0	73.9 ± 0.4

Table 1 – Chemical composition of the clay substrates and the glazes of the single and double-fired samples. The results are presented in oxide weight percentage (average of ten measures made on 320 x 320 μm^2 areas the ceramic substrates and on 30 x 30 μm^2 areas for the glazes); standard deviation = 2σ.

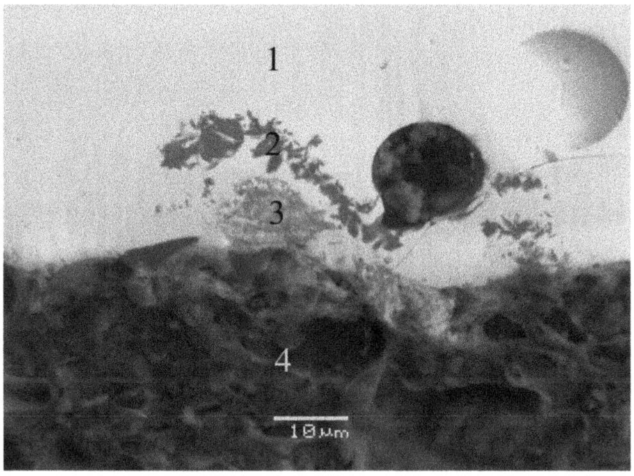

Fig. 1 – Back-scattered electron image of the glaze/ceramic body interface of the double-fired sample. 1: glaze; 2: dark grey crystals; 3: light grey crystals; 4: ceramic substrate.

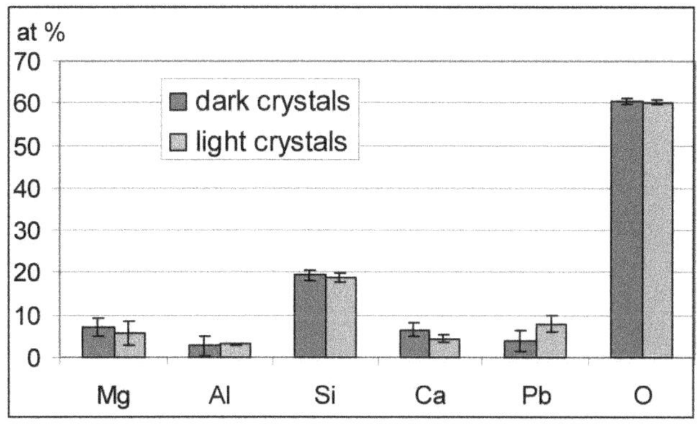

Fig. 2 – Chemical composition (atomic weight percentage) of the two types of crystals present at the glaze/ceramic body interface, determined by SEM-EDS.

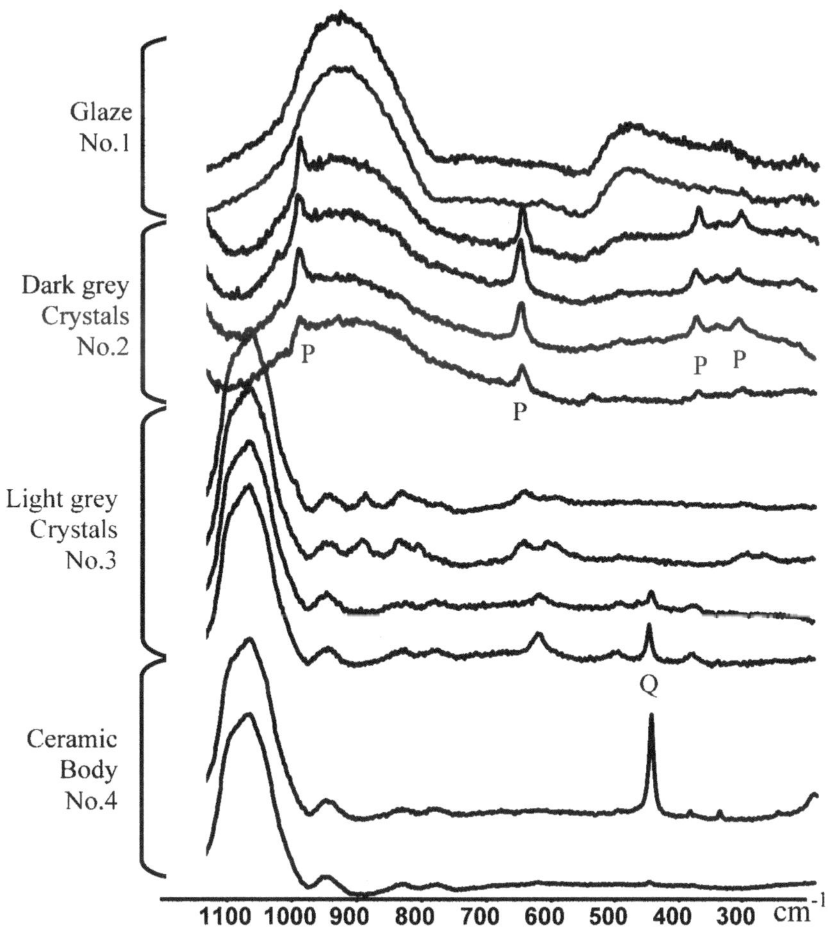

Fig. 3 – Characteristic Raman spectra (633 nm laser) of the different zones shown in Fig. 1 at the glaze/ceramic substrate interface in the double-fired sample. The quartz peak, labelled Q, is identified in both the grey zone and in the ceramic body. The four peaks labelled P, at 323, 387, 665 and 1014 cm^{-1} are characteristic of pyroxene ($XY(Si,Al)_2O_6$).

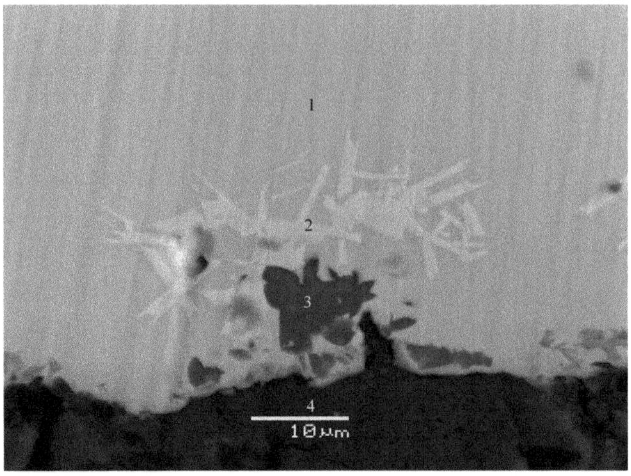

Fig. 4 – Back-scattered electron image of the glaze/ceramic body interface in the single-fired sample. 1: glaze; 2: white crystals; 3: grey crystals; 4: ceramic substrate.

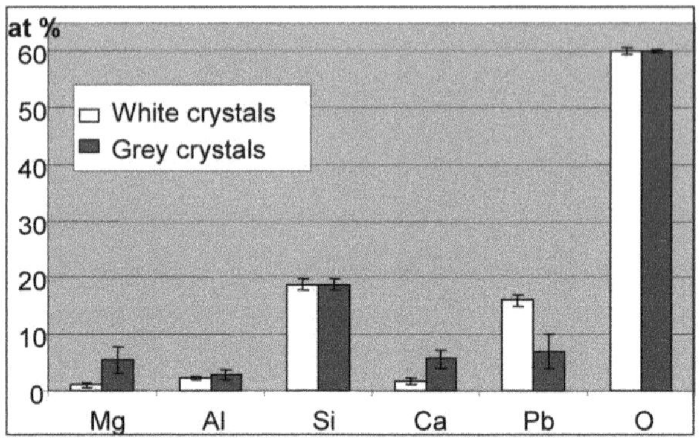

Fig. 5 – Chemical composition (atomic weight percentage) of the two types of crystals present at the glaze/ceramic body interface, determined by SEM-EDS.

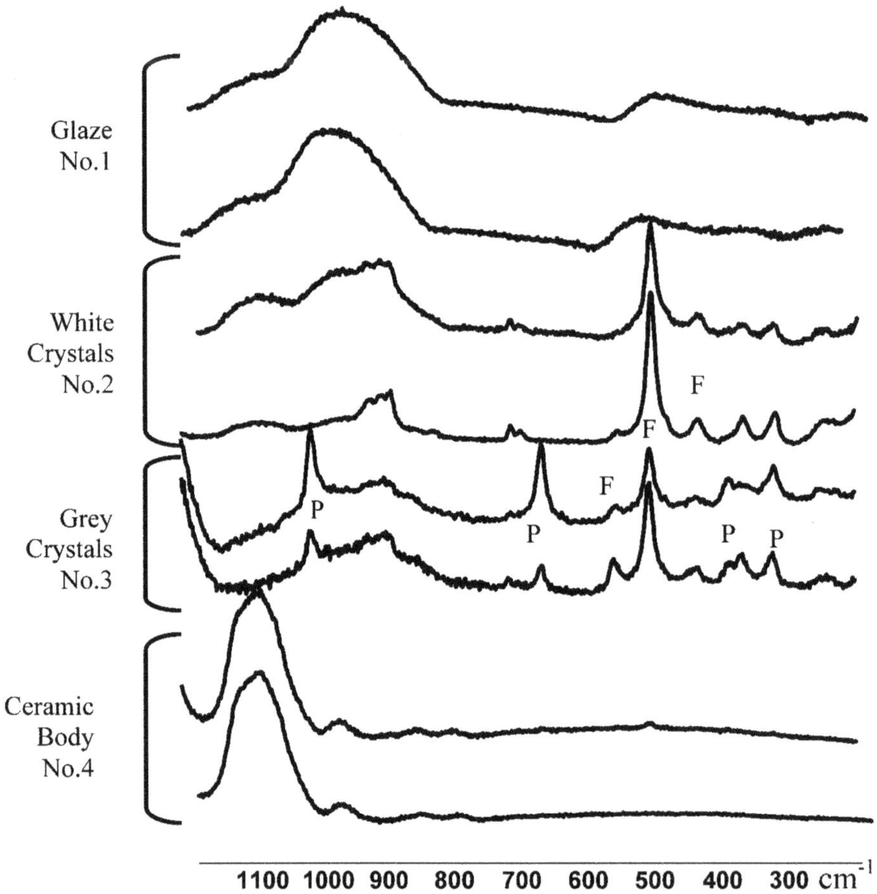

Fig. 6 – Characteristic Raman spectra (633 nm laser) of the different zones shown in Fig. 4 at the glaze/ceramic substrate interface in the single-fired sample. The four peaks labelled P at 323, 387, 665 and 1014 cm^{-1} are characteristic of pyroxene ($XY(Si, Al)_2O_6$). The three peaks labelled F at 480, 513 and 558 cm^{-1} are characteristic of feldspars ($XAlSi_3O_8$ / $YAl_2Si_2O_8$).

Moreover, if the glaze (zone 1, Fig. 4) presents the same polymerization index of 0.1 as for the double-fired sample, the values of this index vary greatly from crystal to crystal: the polymerization index of the white crystals fluctuates between 0.2 and 0.4, and the index for the grey crystals ranges between 0.2 and 0.6.

CONCLUSION AND PERSPECTIVES

Analysing the devitrification crystals present at the glaze-ceramic body interface of double and single-fired samples enabled us to discriminate between the firing processes. In the doubled fired sample, only one type of silicate crystals is found, which may correspond to a lead-containing pyroxene. In the single-fired sample, two types of crystals are observed: high lead containing white crystals identified as feldspars and grey crystals with characteristic peaks of both feldspar and pyroxene. Moreover, the polymerization index of the glaze and the grey crystals is constant for the double-fired sample, but fluctuates greatly for the white and grey crystals in the single-fired sample. This point should be further investigated in order to understand what it means from a physico-chemical point of view, as well as the influence of the lead cations on the Raman bands of feldspar and pyroxene.

The next step in this study is to make a μ-XRD profile under X-ray synchrotron radiation on the identical zone as it would obviously be of great interest to compare the Raman and the μ-XRD profiles, the latter providing the crystallographic properties of the different types of crystals revealed in this paper.

Moreover, the study of other raw materials may be of interest, especially the wide mineralogical clay family of smectites, taking into account their chemical compositions and physical properties.

REFERENCES

COLOMBAN P. 2003, "Polymerization degree and Raman identification of ancient glasses used for jewelry, ceramic enamels and mosaics", *Journal of non-crystalline solids* 323, p. 180-187.

COLOMBAN P., PAULSEN O. 2005, "Non-destructive determination of the structure and composition of glazes by Raman spectroscopy", *Journal of the American ceramic society* 88 (2), p. 390-395.

CHAPOULIE R., DÉLÉRY C., DANIEL F., VENDRELL-SAZ M. 2005, "Cuerda Seca ceramics from al-Andalus, Islamic Spain and Portugal (10th-12th c. AD) — Investigation with SEM-EDX and cathodoluminescence", *Archaeometry* 47, 3, p. 519-534.

FREEMAN J., WANG A., KUEBLER K.E., HASKIN L.A. 2003, "Raman spectroscopic characterization of the feldspars – Implications for in situ surface mineral characterization in planetary exploration", Abstract #1676, 34th LPSc.

MOLERA J., PRADELL T., MARTINEZ-MANENT S., VENDRELL SAZ M. 1993, "The growth of sanidine crystals in the lead of glazes of Hispano-Moresque pottery", *Applied clay science* 7, p. 483-491.

MOLERA J., PRADELL T, MERINO L., GARCIA VALLÉS M., GARCIA ORELLANA J., SALVADO N., VENDRELL SAZ M., 1997, "La tecnologia de la ceramica islamica y Mudejar", *Caesaraugusta* 73, p. 15-41.

MOLERA J., PRADELL T, Salvado N., VENDRELL SAZ M. 2001, "Interactions between clay bodies and lead glazes", *Journal of the American ceramic society* 84 (5), p. 1120-1128.

TITE M.S., FREESTONE I., MASON R., MOLERA J., VENDRELL SAZ M., WOOD N. 1998, "Lead glazes in antiquity – Methods of production and reasons for use", *Archaeometry* 40 (2), p. 241-260.

WANG A., JOLLIFF B.L., HASKIN L.A., KUEBLER K.E., VISKUPIC K.M. 2001, "Characterization and comparison of structural and compositional features of planetary quadrilateral pyroxenes by Raman spectroscopy", *American mineralogists* 86, p. 790 806.

CERAMIC SEQUENCE OF 7000 YEARS: ARCHAEOMETRICAL STUDY OF POTTERY FINDS FROM VÖRS, MÁRIAASSZONYSZIGET (SW HUNGARY)

KATALIN T. BIRÓ[1], KATALIN GHERDÁN[2], GYÖRGY SZAKMÁNY[2]

[1] Hungarian National Museum, Budapest (tbk@ace.hu)
[2] ELTE University, Dept. of Petrology and Geochemistry, Budapest

ABSTRACT

The site considered in this study is a multi-period archaeological site, spanning from the Early Neolithic (around 5500 BC) to the Early Mediaeval Conquest period. The village lies in Southwest Hungary, near Lake Balaton. The exceptionally favourable environmental endowments of the territory offered an ideal setting for habitation. Sites and finds from almost all periods of prehistory were found here, rich till the historical ages. The pottery assemblage of the locality at Máriaasszonysziget, a sandy peninsula protruding into the former lake, provided a good possibility for a diachronic study of changes in raw material selection and pottery manufacture.

There were at least 8 distinct periods of habitation separated on the basis of traditional archaeological methods. 15 to 20 samples from each period were chosen through macroscopic investigation of fabric and form for archaeometrical analyses. In addition, coeval samples from nearby sites from all periods were examined.

Selected samples were thin sectioned and subjected to petrographic analysis. Comparative study of the fabric and the composition of the non plastic material revealed the existence of a pottery group – characterised by serial fabric and sand sizc non plastic inclusions, dominantly quartz, feldspars, micas and accessories – that can be found in all periods, and is presumed to be local. It also became clear that over the 7000 years there were remarkable changes in raw material selection. Early Neolithic Starčevo pottery was tempered with organic material, while potters of the Copper Age (Chalcolithic) and the Bronze Age used grog for the same purpose. The non-plastic material of Late Iron Age and Early Roman Age pottery is more diverse: limestone and quartzite are typical, while in one of the samples metaultrabasic rock fragments were detected in great quantity, suggesting a non-local origin.

Geochemical analysis (XRF, INAA) of bulk samples was used to compare pottery of the different periods and the coeval samples from nearby sites. It was also considered if the petrographically uniform pottery that can be found in all periods can be used to establish an internal reference group, representing local material.

The experiences of Vörs pottery analysis serve as a starting point for ceramic archaeometry studies for archaeological assemblages over a wide area and an essential time span.

KEYWORDS: NEOLITHIC, COPPER AGE, BRONZE AGE, POTTERY, DIACHRONIC STUDY, PETROGRAPHY, GEOCHEMISTRY

1. INTRODUCTION

The aim of the work presented here was to investigate prehistoric pottery by its raw material along different chronological, temporal dimensions for the assessment of technology, origin of raw material(s) and potential transportation of the vessels. The raw material of pottery, as seen by petrographical/geochemical investigation, is the result of numerous factors including the variety of raw materials, production technology and taphonomic factors. To determine possible movements of prehistoric people by the investigation of the raw material of their vessels, we should be capable of distinguishing between the effects of these factors. Therefore we started our investigations on a multi-period archaeological site, offering a variety of chronological horizons with practically the same environment and selected control material of comparable age, near to and further away from the site investigated. The basic method of research was petrographical microscopy, coupled with geochemical analysis on a selected set of samples intended to represent all the macroscopical varieties within the pottery types.

2. VÖRS-MÁRIAASSZONYSZIGET: THE ARCHAEOLOGICAL SITE

The site selected for analysis is a settlement with exceptionally favourable environmental endowments. The confines of the village Vörs, lying in the fringes of the marshland of Kis-Balaton near Lake Balaton abound in important archaeological sites from the Early Neolithic period onwards (Költő and Vándor eds. 1996). On the Máriaasszonysziget itself, remains of eight different archaeological periods and cultures came to light, in course of several seasons of excavation (Kalicz *et al.* 2002). The periods encountered embrace temporally the archaeological periods from the VIth mill. BC to the beginning of the IInd mill. AD. The interdisciplinary elaboration of the site and the finds is the subject of an on-going research program of the Hungarian National Science Foundation (OTKA T-046297). The archaeological investigation of the pottery finds is carried out by several specialists: Zs. M. Virág for Neolithic and Copper Age pottery, V. Kiss for Bronze Age Kisapostag culture, K. Tankó working on the Celtic period and Cs. Aradi on Hungarian Conquest period finds.

3. THE ARCHAEOMETRICAL ANALYSIS OF POTTERY

When planning the project, we wanted to explore both the temporal and spatial information coded in the finds. Therefore we selected from each chronological horizon some representative samples—at least five pieces, but from the more significant periods possibly all the macroscopic and functional varieties. For each period, control samples are going to be taken (for some horizons it has already been done) from nearby coeval settlements, coming from recent excavations. The samples selected for analysis were described macroscopically according to type/function and the physical qualities of the raw material using Munsell colour charts and binocular microscope. Petrographic thin sections were prepared, and geochemical analyses were performed by XRF and INAA on selected samples. Chemical composition of the pottery was compared to the geochemical composition of the Post Archaean Australian Shale (PAAS), with commonly used normalisation values for argillaceous sediments (Taylor and McLennan, 1995). The samples were grouped according to the analysis results, compared for both the temporal and the spatial variables.

4. RESULTS AND DISCUSSION – COMPARISON THROUGH TIME AND SPACE

The earliest remains found at the archaeological site belong to *Starčevo culture* (Early Neolithic, 5600-5400 BC). Through macroscopic investigations, 25 samples were selected and thin sectioned. For comparison five control samples were taken from Szentgyörgyvölgy, a contemporaneous archaeological site, lying at a distance of a few scores of km (Bánffy 2004).

Petrographic analysis revealed that the characteristic Starčevo sample has serial texture with mainly quartz and feldspars as non-plastic constituents and also contains vegetal temper (probably chaff) (Fig. 1). The examined sherds differ only in quantity of the constituents. Only one sherd is different from the general pattern. This sample contains very small amounts of mineral fragments and it is tempered with grog. No signs of vegetal tempering could be found, neither in the sherd itself, nor in the tempering grog fragments. Control samples were found to be similar to the majority of the samples both in texture and mineral composition (Gherdán *et al.* 2005).

Geochemical analyses (XRF analysis of 25 samples from Vörs, and 5 samples from Szentgyörgyvölgy from which 11 sherds from Vörs and 2 from Szentgyörgyvölgy were subjected to INAA) showed that the chemical composition of the petrographically characteristic sherds is uniform concerning both major and trace elements (Table 1). Multi-element diagrams show that SiO_2, TiO_2

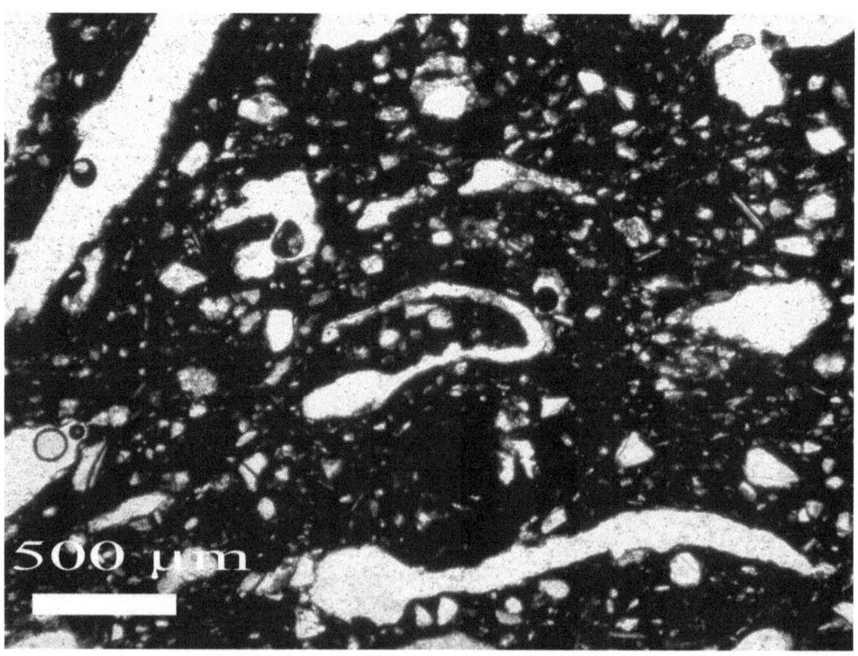

Fig. 1 – Polarisation microscopic photomicrograph of a characteristic Starčevo sample with serial fabric. Non-plastic constituents are dominantly quartz and feldspar grains. Vegetal tempering (probably chaff) is characteristic. PPL.

	Starčevo mean 12 samples	σ	outlier 53/02	Lengyel mean 14 samples	σ	Stroked mean 5 samples	σ	Kostolac mean 6 samples	σ	Kisapostag mean 20 samples	σ	Celtic mean	σ 3 samples	outlier 51/05
SiO_2	64,95	2,37	68,23	63,71	4,57	63,63	0,39	66,35	2,54	65,60	2,00	68,19	0,66	67,53
TiO_2	0,89	0,05	0,82	0,95	0,09	0,85	0,04	0,81	0,07	0,79	0,05	0,87	0,06	1,01
Al_2O_3	18,63	1,15	17,02	19,81	2,72	19,94	0,67	17,86	1,21	17,79	1,73	16,51	0,43	21,76
Fe_2O_3	6,18	0,79	6,41	6,97	1,34	6,81	0,10	5,39	0,54	6,16	0,63	6,52	0,29	4,05
MnO	0,06	0,02	0,08	0,04	0,02	0,04	0,00	0,05	0,02	0,10	0,05	0,09	0,04	0,02
MgO	1,66	0,29	1,96	1,70	0,42	1,87	0,17	1,85	0,40	2,34	0,28	1,86	0,03	1,11
CaO	1,89	0,32	1,18	1,72	0,24	2,68	0,30	1,76	0,20	2,86	1,21	1,69	0,45	1,00
Na_2O	0,71	0,18	0,97	0,56	0,19	0,46	0,02	0,56	0,06	0,64	0,15	1,08	0,05	0,42
K_2O	3,02	0,59	2,71	2,63	0,44	2,83	0,22	2,96	0,41	3,12	0,43	2,59	0,17	2,97
P_2O_5	2,01	1,10	0,61	1,91	0,71	0,89	0,31	2,43	1,42	0,62	0,38	0,60	0,26	0,15
Rb	148,37	18,27	133,90	123,66	24,23	154,64	9,47	156,47	24,54	161,70	21,11	127,83	14,15	169,80
Th	15,58	1,62	7,20			13,30	0,61	12,03	1,25	11,87	0,40	12,30	1,98	20,60
U	2,85	0,46	1,10			2,12	0,20	2,83	0,93	2,22	0,16	2,22	0,47	6,68
Ba	1005,53	164,84	637,60	936,46	146,61	840,98	64,17	1171,82	235,77	753,60	31,69	742,23	87,17	413,20
Sr	194,26	33,57	130,30	172,17	29,40	190,70	36,71	213,55	40,90	284,70	28,88	157,30	26,76	99,20
Nb	16,59	1,42	17,80	17,56	2,35	18,08	0,71	16,60	2,37	15,87	2,59	17,07	1,33	17,80
Hf	6,05	0,70	1,70			4,22	0,27	4,48	0,76	4,10	0,97	5,92	0,21	4,10
Zr	173,92	20,40	192,90	167,67	21,37	140,56	10,27	158,58	23,17	144,40	29,15	189,67	10,63	131,60
Y	31,96	3,18	41,30	35,54	5,26	29,76	1,01	31,53	3,22	26,97	4,40	39,57	7,41	35,30
Sc	15,88	1,25	7,00			15,73	0,40	14,63	0,96	14,83	1,07	15,40	0,85	23,40
Co	16,58	3,44	20,20	13,83	4,91	16,78	0,70	15,07	1,97	19,73	1,24	20,83	5,26	13,60
Cr	111,52	11,48	116,20	125,61	18,03	139,46	9,24	116,50	7,36	130,27	5,28	102,27	9,36	139,10
Ni	72,41	32,61	58,30	83,69	21,50	72,88	7,57	41,87	10,12	56,27	7,54	55,73	11,61	39,20

Table 1 – Chemical composition of pottery samples. Mean composition and standard deviation for each culture and the individual compositions of the outliers (53/02 from Starčevo culture 51/05 from Celtic/Early Roman culture).

and Al_2O_3 concentrations of the sherds do not vary much, normalised values are around 1 (Fig. 5, Fig. 6). Fe_2O_3 and MnO values vary in a wider range: there are both positive and negative anomalies. For CaO slightly positive, for Na_2O and K_2O negative anomalies can be detected. For P_2O_5 strong positive values were found. With regard to trace elements it was found that for mobile elements – such as Rb, U, Ba, Sr, – normalised values are distributed throughout a wider range; positive and negative anomalies can also be found. For Nb, Zr, and Co slightly negative anomalies were detected. Hf has slightly positive and negative anomalies, while Y has slightly positive anomalies. Th, Sc and Cr in general have slightly positive anomalies. Ni behaves differently; it can have either positive or negative anomalies. Control samples differ in both major and trace element composition. With regard to major elements, stronger negative anomalies of MgO and CaO are characteristic. In the case of trace elements, all except for Rb, Ba, and Sr have positive anomalies. The unusual grog tempered sherd (sample 53/02) has different geochemistry. Strong negative anomalies of Th, U, Hf and Sc are very characteristic (Gherdán *et al.* 2005).

The next period represented at the site is the Early Copper Age (4500-4300 BC). 22 pieces of pottery belonging to *Lengyel III culture* were thin sectioned and analysed. (Control samples have not been selected yet.) Petrographic analysis showed that pottery of that time have either serial texture comprising mainly quartz and feldspars as non-plastic components (resembling characteristic *Starčevo* samples without the vegetal temper), or have hiatal fabric, containing grog temper and mineral fragments (mainly quartz and feldspars) (Fig. 2). Without the use of temper the fabric and composition of the grog tempered varieties would be similar to the characteristic *Starčevo* samples.

The geochemical composition (XRF analysis of 14 sherds) of all the examined pottery is very uniform in both major and trace element composition (Table 1). The pattern of the multi-element diagram for major and trace elements follows the trend of that of Starčevo samples (Fig. 5, Fig. 6).

Balaton-Lasinja culture (MCA, ~ 4000 BC): this period is represented at Vörs-Máriaasszonysziget only by sporadic finds. No analysis was made on the pottery.

For the Middle Copper Age (~ 3800 BC) 5 samples belonging to the *Stroked pottery culture* were chosen for analysis. (Control samples have not been selected yet.) Petrographic investigations showed that the pottery has hiatal fabric and it is tempered with grog. Other non-plastic constituents are mineral fragments, dominantly quartz and feldspars. Without grog the fabric would look like that of the characteristic Starčevo samples without the vegetal temper.

Chemical composition of the examined sherds (XRF analysis of 5 samples, from which 3 were chosen for INAA) in regard to both major and trace elements is uniform and follows the pattern of the basic Starčevo group samples (Table 1, Fig. 5, Fig. 6).

The Late Copper Age (~ 2800 BC) is represented by the *Kostolac culture.* 24 sherds of this culture were thin sectioned and examined petrographically. (Control samples

Fig. 2 – Polarisation microscopic photomicrograph of a grog tempered pottery from Lengyel III culture. Without grog fabric would look like that of Starčevo samples without vegetal material. PPL.

have not been selected yet.) All of the samples have hiatal fabric and they are tempered with grog. Without grog, the texture looks like Starčevo fabric without vegetal temper; that is serial texture with non-plastic elements mainly quartz and feldspars.

For chemical analyses 6 sherds were chosen (6 samples for XRF, and 3 samples for INAA from the sherds analysed by XRF). Major and trace element composition of these pieces was found to be uniform (Table 1). Major and trace element patterns are also in agreement with that of the basic Starčevo group samples (Fig. 5, Fig. 6).

16 sherds of the Early Bronze Age *Kisapostag culture* (~ 2000 BC), accompanied by 9 contemporaneous control samples from Vörs, Tótok-dombja were selected for petrographic analysis. Two main petrographic groups were found. One with the usual serial fabric, with quartz and feldspars as dominant non-plastic constituents, and the other with hiatal fabric and grog temper. Besides grog mineral fragments (mainly feldspars and accessories) are present as non-plastic constituents.

15 samples from Vörs, Máriaasszonysziget and 5 samples from Vörs, Tótok-dombja were subjected to XRF analysis. Among these sherds 3 were analysed by INAA. It was found that the geochemical character of both major and trace elements is similar for all including the control (Table 1, Fig. 5, Fig. 6).

The Celtic/Early Roman period (300 BC- 100 AD) is represented by 6 pottery samples. (Control samples have not been selected yet.) In spite of the small number of the samples, the diversity of pottery fabric is the greatest in this period. The basic petrographic group is characterised by serial fabric and quartz and feldspars as dominant non-plastics can also be found in this horizon. In addition three other fabric types were found. One with hiatal fabric and large rounded and worn calcareous clasts (mineral fragments as additional non-plastics), another with hiatal fabric and large rounded polycrystalline quartz grains (and a very small amount of other non-plastic constituents—quartz) (Fig. 3), and a third with hiatal fabric and large metaultrabasic rock clasts and rock forming minerals (without other types of non-plastics) (Fig. 4).

For chemical analysis 4 sherds were subjected to XRF analysis and then 3 of them to INAA. The analysis revealed that the major and trace element composition of the majority of the samples is uniform. The geochemical composition of the petrographically unusual pottery sample containing metaultrabasic rock fragments as non-plastic inclusions is different (sample 51/05, Table 1). Strong positive anomalies of Th, U and Sc are characteristic (Fig. 5, Fig. 6).

Comparison of the ceramic material from the site through time and space revealed that there is a basic pottery group that can be found in all periods. It can be characterised by serial fabric – containing quartz and feldspars as non-plastic constituents – and uniform chemical composition. This chemical composition can be described best by normalising the pottery composition to the geochemical composition of Post Archaean Australian Shale (PAAS).

There are two geochemically outstanding samples, the grog tempered sherd from the *Starčevo culture* and the metaultrabasic rock tempered sherd from the *Celtic/Early Roman period.*

Fig. 3 – Polarisation microscopic photomicrograph of the fabric of the polyquartz-containing Celtic/Early Roman pottery sample. Crossed nicols (XPL).

5. CONCLUSIONS

It is probable that there was a common raw material at the site, available and used at all the examined periods. This raw material is represented by the fabric and chemical composition of the basic pottery group. In certain cases this material was tempered in different ways:

— in *Starčevo culture* with vegetal material;

— in *Lengyel, Kostolac* and *Kisapostag* cultures with grog;

— and in *Celtic/Early Roman* times with carbonatic sand.

In all these cases a local origin is very probable. The Celtic sample tempered with metaultrabasic rock fragments is an exception. In this case—taking into account the geology of the nearby and wider surroundings—a non-local origin is proved. The position of the grog tempered sherd from *Starčevo culture* and the polycrystalline quartz containing sherd from the *Celtic/Early Roman* period is still questionable.

As a final conclusion we can state that petrographic groups characterise distinct levels of the settlement and that diachronic petrographic and geochemical studies of pottery finds of a multi-period archaeological site can help to distinguish the use of different recipes in pottery manufacture and also to identify non-local material.

ACKNOWLEDGEMENTS

The kind help of Zsuzsanna Virág is acknowledged here. The authors are grateful to Dr. Heinrich Taubald for chemical analysis of the samples and Sándor Józsa, Farkas Pintér and Tibor Árgyelán for sample preparation. Participation in EMAC ' 05 conference was supported by the MÖB-DAAD project "Archaeometrical analysis of Neolithic pottery and comparison to potential sources of raw materials in their immediate environment". Analytical work on Vörs-Máriaasszonysziget is supported by the Hungarian National Scientific Grant (OTKA T-046297). The neutron activation analyses were carried out by Ms. Márta Balla at Budapest Polytechnical University.

REFERENCES

Bánffy E. 2004, *The 6th millennium BC boundary in Western Transdanubia and its role in the Central European Neolithic transition (The Szentgyörgyvölgy-Pityerdomb settlement)*, VAH Varia Archaeologica Hungarica 15.

Gherdán K., Biró K.T., Szakmány Gy., Tóth M. 2005, "Technological investigation of Early-Neolithic pottery from Vörs, South-West Hungary", *in Understanding people through their pottery, Proceedings of the 7th European Meeting on Ancient Ceramics*, Lisbon, p. 111-118.

Kalicz N., Biró K.T., Virág Zs. 2002, "Vörs, Máriaasszony-sziget", *in Régészeti Kutatások Magyarországon/ Archaeological investigations in Hungary, 1999*, Budapest, p. 15-26.

Költő L., Vándor L. (eds.) 1996, *Évezredek üzenete a láp világából (Régészeti kutatások a Kis–Balaton területén 1979-1992)*, Kaposvár – Zalaegerszeg.

Taylor S.R., McLennan S.M. 1995, "The geochemical evolution of the continental crust", *Reviews of geophysics* 33, p. 241-265.

Fig. 4 – Polarisation microscopic photomicrograph of the metaultrabasic rock tempered Celtic/Early Roman pottery sample. PPL.

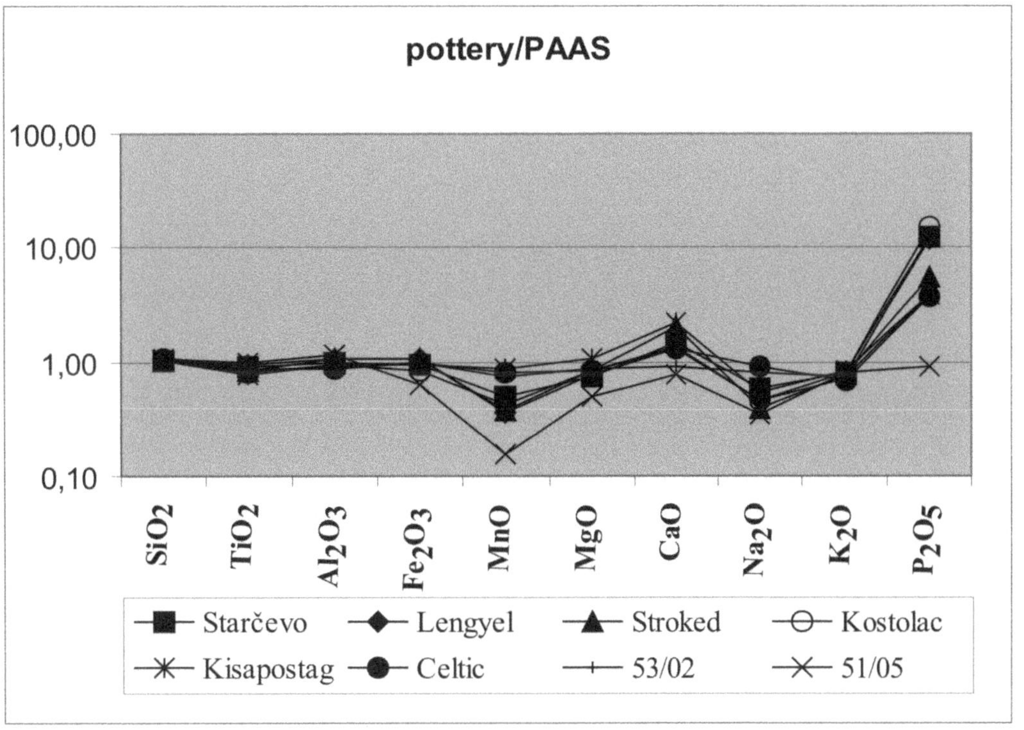

Fig. 5 – Multi-element diagram of major elements – mean values of pottery samples of each period, and the outliers: 53/02 from Starčevo culture, and 51/05 from Celtic/Early Roman pottery culture. PAAS stands for Post Archaean Australian Shale.

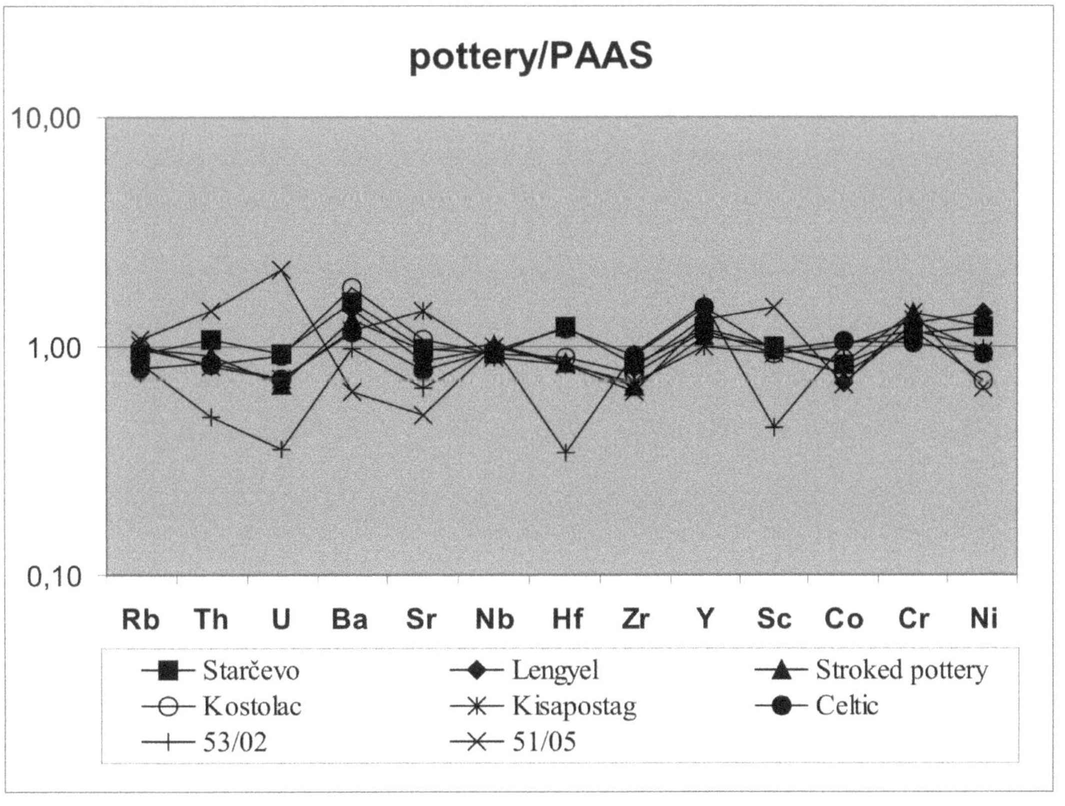

Fig. 6 – Multi-element diagram of trace elements – mean values of pottery samples of each period, and the outliers: 53/02 from Starčevo culture, and 51/05 from Celtic/Early Roman pottery culture. PAAS stands for Post Archaean Australian Shale.

PRODUCTION AND USE: TEMPER AS A MARKER OF DOMESTIC PRODUCTION THE CASE OF TWO MIDDLE NEOLITHIC VILLAGES IN CONCISE (VD, CH)

ELENA BURRI

Section d'archéologie cantonale, 10 pl. de la Riponne, CH-1014 Lausanne; elena.burri@vd.ch

1. A SHORT PRESENTATION OF THE SETTLEMENT

The existence of the lake-side dwellings of Concise, on the northern shore of lake Neuchâtel, has been known since 1855, when work on the railway line resulted in the partial destruction of the site in the bay of Concise and the discovery of many prehistoric objects. Almost 150 years later, work on the « Rail 2000 » project between Yverdon and Neuchâtel necessitated the excavation of a large part of the lake-shore site, involving a surface of almost 4700 m^2. This excavation took place between November 1995 and February 2000 under the direction of C. Wolf, who was mandated by the cantonal archaeologist D. Weidmann.

At the end of the excavation, about 25 villages, including 7949 piles, dating from the Middle Neolithic to the Early Bronze Age, between 4300 BC and 1570 BC, had been brought to light. All the villages, apart from the earliest and those from the Horgen culture, produced an abundance of well stratified material precisely dated by dendrochronology[1] (Winiger to be published, Winiger 2003). The strata were correlated by A. Winiger and dated using the horizontal timbers contained in the archaeological levels.

2. THE MIDDLE NEOLITHIC REMAINS AND THE POTTERY

The Middle Neolithic II is well represented with 6 villages, which are either superimposed or built at other locations within the bay of Concise, and dated between 3868 BC and 3516 BC. The pottery of three of the villages is consistent with the pottery styles of the region and period, being from the well known Cortaillod culture of the Swiss Plateau. The situation is however exceptional for three other villages, in which the NMB (Néolithique Moyen Bourguignon) and the Cortaillod styles coexist. We will focus particularly on two of these villages; the first one dendrochronologically dated to 3713-3675 BC, with two phases of rebuilding and the second one dated to 3645-3635 BC.

3. THE REGIONAL FRAMEWORK

During the Middle Neolithic we have the well known Cortaillod pottery which has been regularly found at lake-shore sites (Carnes 1997; Gauthier 1985; Hafner and Suter 2000; Kaenel 1976; Ramseyer 2000; Schifferdecker 1982; Schwab 1999; Seppey 1991; Stoeckli 1981a and b) on the Swiss Plateau (Fig. 1). It is mostly characterised by its pots with a S profile and knob handles situated near to the lip. Its evolution is also known, with the profile becoming progressively more closed and the proportion of pots in relation to that of dishes increasing from the Classic Cortaillod (approximately 3900-3750 BC) to the Cortaillod Port-Conty (approximately 3500-3350 BC), with the Middle and Late Cortaillod in between. The tempers in all cases include quartz from alpine pebbles or sand, and in some cases they also include broken shells (Fig. 2). It is also possible to use limestone from the Jura, but the only limestone tempers that exist on the Swiss Plateau belong to NMB style pottery and these are considered imports (Pétrequin 1984). They represent less than 30 pots for all the Cortaillod settlements of the Swiss Plateau, for which there are thousands of published pots (Burri in press).

The NMB is usually geographically separated from the Cortaillod (Fig. 1). It exists normally in Burgundy and Franche-Comté on the far side of the Jura. The style, which was defined at the time of the colloquium of Beffia (Pétrequin and Gallay, 1984), is characterized by pots with shoulders underlined by pairs of knob handles. The situation is a little different from that of the Cortaillod, because most of the sites are hilltop settlements or caves and not particularly well dated (Pétrequin and Gallay

1 – The dendrochronological studies and dates are supplied by the LRD (Laboratoire Romand de Dendrochronologie) at Moudon, Switzerland.

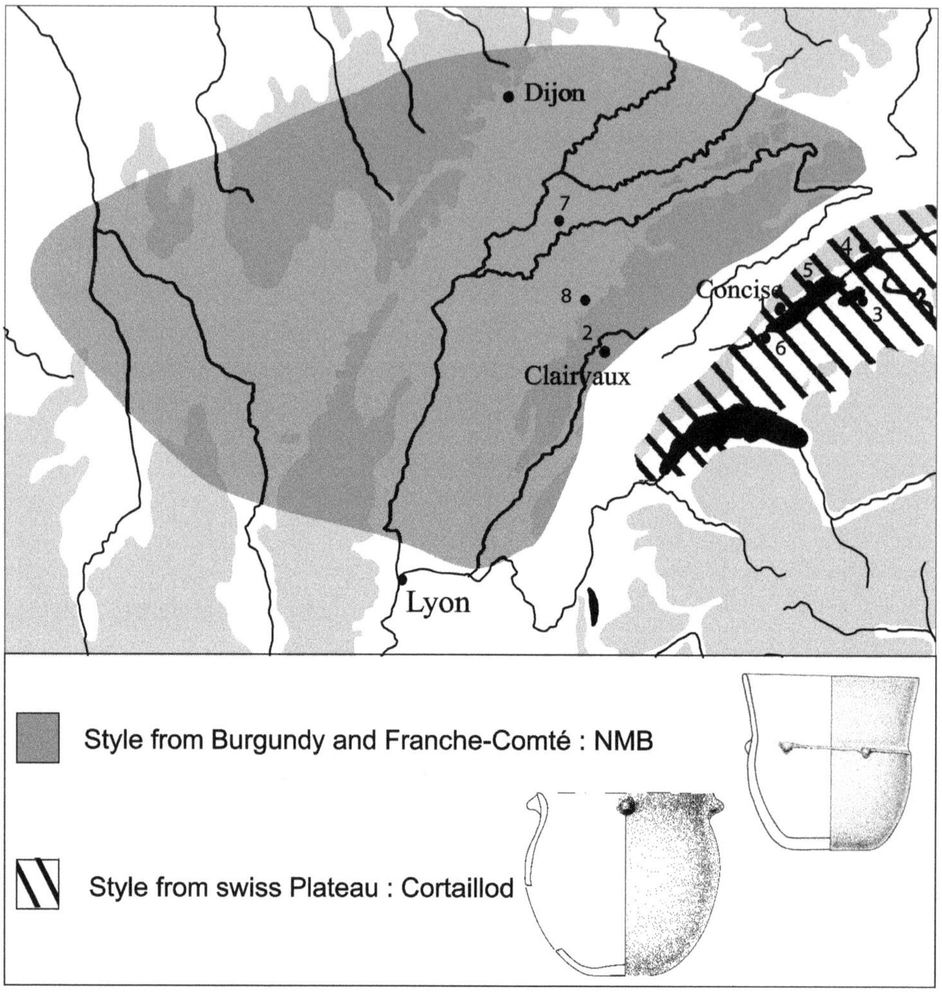

Fig. 1 – The regional framework of Concise and the establishment of the Cortaillod and the NMB cultures. 1. Concise (VD, CH), 2. Clairvaux (Jura, F), 3. Muntelier (FR, CH), 4. Twann (BE, CH), 5. Auvernier (NE, CH), 6. Yverdon (VD, CH), 7. Lavans-les-Dole-Moulin-Rouge (Jura, F), 8. Montmorot (Jura, F).

1984; Lepage 1992; Dufay-Galan 1996; Liégard *et al.* 2000; Rialland 1991; Wernli 1995). The lake-shore dwellings of Clairvaux XIV, Clairvaux V La-Motte-aux-Magnins and, since this summer, Clairvaux VII, are the exceptions, but they are not dated by dendrochronology (Pétrequin and Pétrequin 1989; 2005; Templer 2006). The chronotypology is more difficult to establish, but we can suggest that it follows the evolution of the Cortaillod in that the proportions of pots in relation to that of the dishes increases as well, whilst the shoulder migrates down to the maximum diameter of the belly (Templer 2006). The tempers are limestone in most cases, sometimes with crushed shell. The only exception is the hilltop settlement of Moulin-Rouge where there are tempers including quartz, but representing only 24% of the tempers. One of the explanations for this situation is that the Rhône glacier did not cover the Jura, therefore there are no alpine moraines with quartz-bearing pebbles and the only granite massif of the region is the Massif de la Serre near Moulin-Rouge (Jaccottey *et al.* to be published). The fact that there are only a few pots containing crystal tempers at Moulin-Rouge would indicate that this is a cultural choice (Fig. 2).

We have seen all the pottery from Auvernier, Twann, Yverdon and Muntelier on the Swiss Plateau as well as the pottery from Clairvaux XIV, Clairvaux V La-Motte-aux-Magnins, and Montmorot in Franche-Comté, whilst P. Pétrequin has made determinations for all the pottery from Clairvaux VII, XIV and La-Motte-aux-Magnins, Montmorot and Moulin-Rouge (Pétrequin and Pétrequin 2005). As for Concise, we have made macroscopic determinations of the tempers and these determinations were considered sufficient because the limestone and the crystalline tempers are easy to differentiate. The limestone tempers are abundant, angular and well calibrated, they cannot scratch steel and some of the temper occasionally disappears during the firing of the pots. On the other hand, tempers which contain quartz-rich rocks are often badly

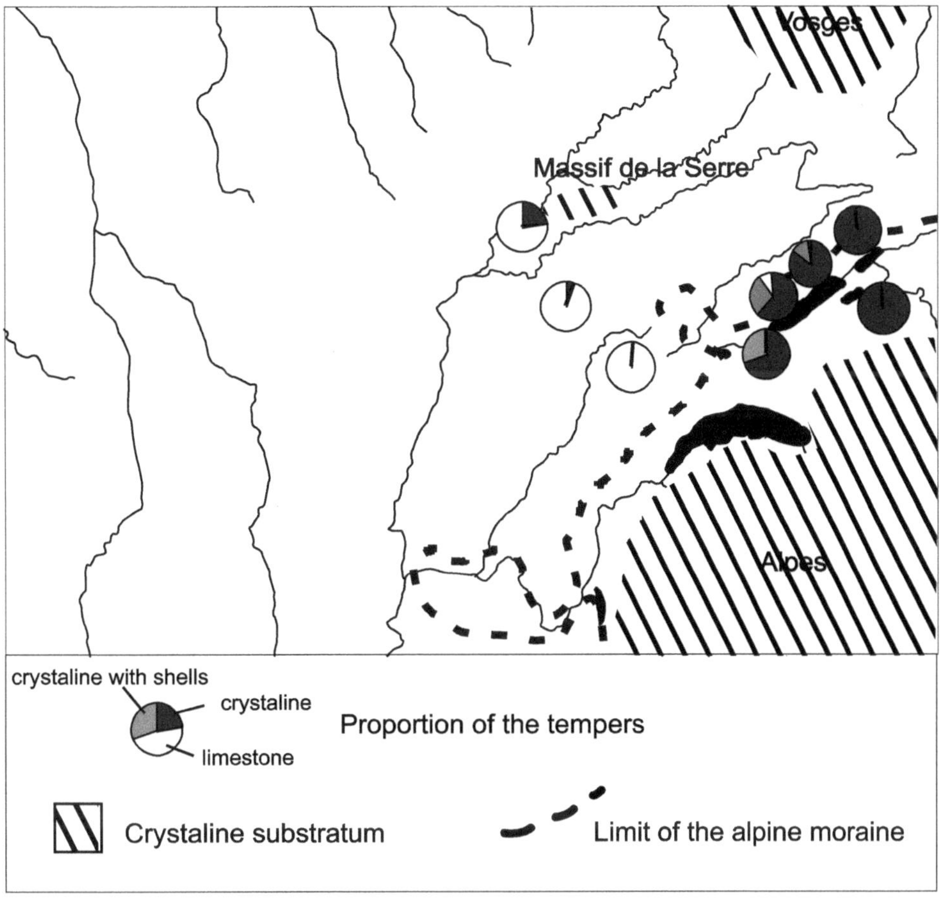

Fig. 2 – The geological substratum and the composition of the tempers.

sorted, less abundant and round; they can also scratch steel and they withstand the firing process very well. We can observe the presence of crushed shell fragments, which may or may not be fossil, because they are often larger than the rest of the temper.

At Concise, it is notable that the two traditions, NMB and Cortaillod, coexisted in at least three villages. One of these, dated between 3570 BC and 3517 BC is strongly eroded and cannot be observed from a planimetric point of view. By contrast, the two others, as well as their pottery, are very well preserved.

4. ZOOM ON TWO VILLAGES OF CONCISE

We will now discuss only the case of the two well preserved villages in which the two pottery styles coexist; the first one dated between 3713 BC and 3675 BC (E2) with two rebuilding phases, and the second one between 3645 BC and 3635 BC (E4A). The assemblage is presented in Fig. 3.

From the village dating from 3713-3675 BC, we see that more than half of the corpus is NMB and that the total number of NMB forms is substantially greater than that which previously existed on all of the Swiss Plateau. For the tempers, the situation is more normal. This demonstrates that most of the "NMB" pots are not imported from Franche-Comté but are rather manufactured in or near Concise. For the second village, the contrast between the style of ceramics and the composition of tempers is more striking. Almost all the pots are manufactured locally.

Apart from the tempers and the hybrid forms, none of the pottery shapes are exceptional from either village, with styles from the normal Cortaillod and NMB typochronologies. What is remarkable is the coexistence and production of both styles in the same village.

E2: 3713-3675 BC	E4A: 3645-3635 BC
226 kg	215,3 kg
362 archaeological profiles	252 archaeological profiles
NMB style: 50%	NMB style: 51%
Cortaillod style: 48%	Cortaillod style 46%
Limestone tempers: 13%	Limestone tempers: 9%

Fig. 3 – The corpus of pottery from the two villages of Concise.

5. THE PLANS OF THE VILLAGES

We carried out a spatial analysis to continue the interpretation. The first stage was to create the plans of both villages and in particular the layout of the houses and rubbish dumps, with a view to assigning the pottery to individual houses. It was not a simple matter, despite the piles, the dendrochronology, and the excellent stratification. Many piles are not dated, there are many postholes and the superimposition of the villages makes for difficult interpretation of architectural structures (Winiger to be published). We therefore had to find an alternative way to reconstruct the layout of the lake-shore dwellings, which could be compared to the architectural plan. To achieve this, we applied a method based on the ethnoarchaeological model of A.-M. and P. Pétrequin (1984). We will not enter into the details of this reconstruction, which we have described elsewhere (Burri to be published). All we need to know at this stage is that such a reconstruction was possible and that the comparison between our model and the plans proposed by of A. Winiger are very convincing (Winiger to be published). We need to stress that our proposals do not come from the layout of the houses, but are plans of the consumption units reconstructed, based on the distribution of the pottery. The final plans will only become available after the integration of the other results, particularly those relating to the architectural structures.

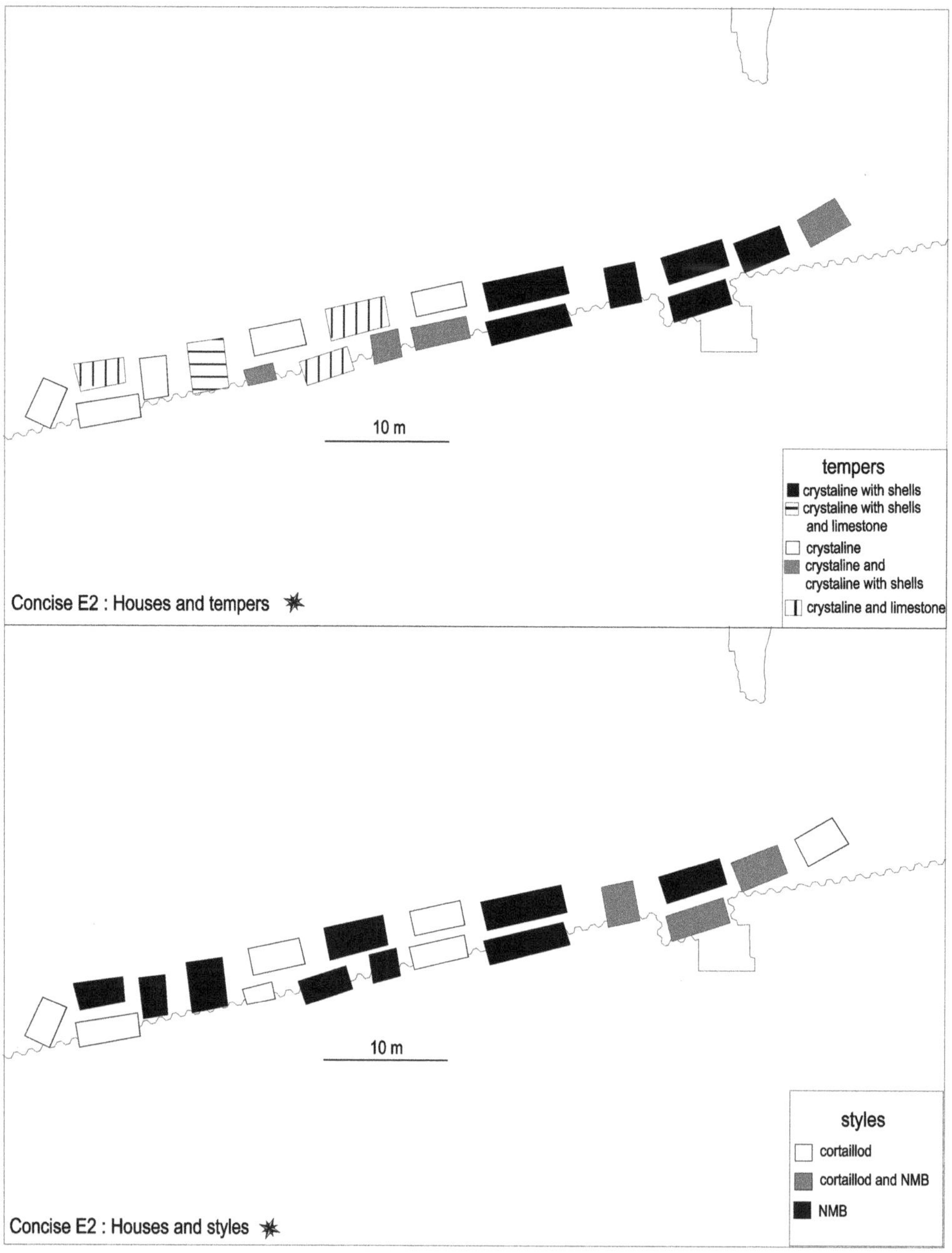

Fig. 4 – The distribution of tempers and styles in the village E2 dated 3713-3675 BC.

6. SPATIAL DISTRIBUTION OF THE POTTERY AND ITS IMPLICATIONS

We will now discuss the distribution of tempers and styles in the villages and between the houses. We have observed that there are obvious groups of tempers which create quarters within each village, but that the styles are equally distributed throughout the villages (Fig. 4 and 5). When we analyse the distribution between the individual houses, we can attribute one or two sorts of temper to each house. The same is true for the styles, with each house having its own style, whether it be Cortaillod or NMB. Just as the distribution of the two parameters is very different between the houses, when the two parameters are combined, it appears that virtually every house has a different combination of temper/style. This implies that the pottery used in each house is also produced there; there is no common workshop nor are there potters who make the pottery for all the village. We can summarise by stating that "the consumption units are production units", or in other words, "the production of pottery is a domestic activity". On the other hand, the collection of the raw materials can be common to some neighbouring houses.

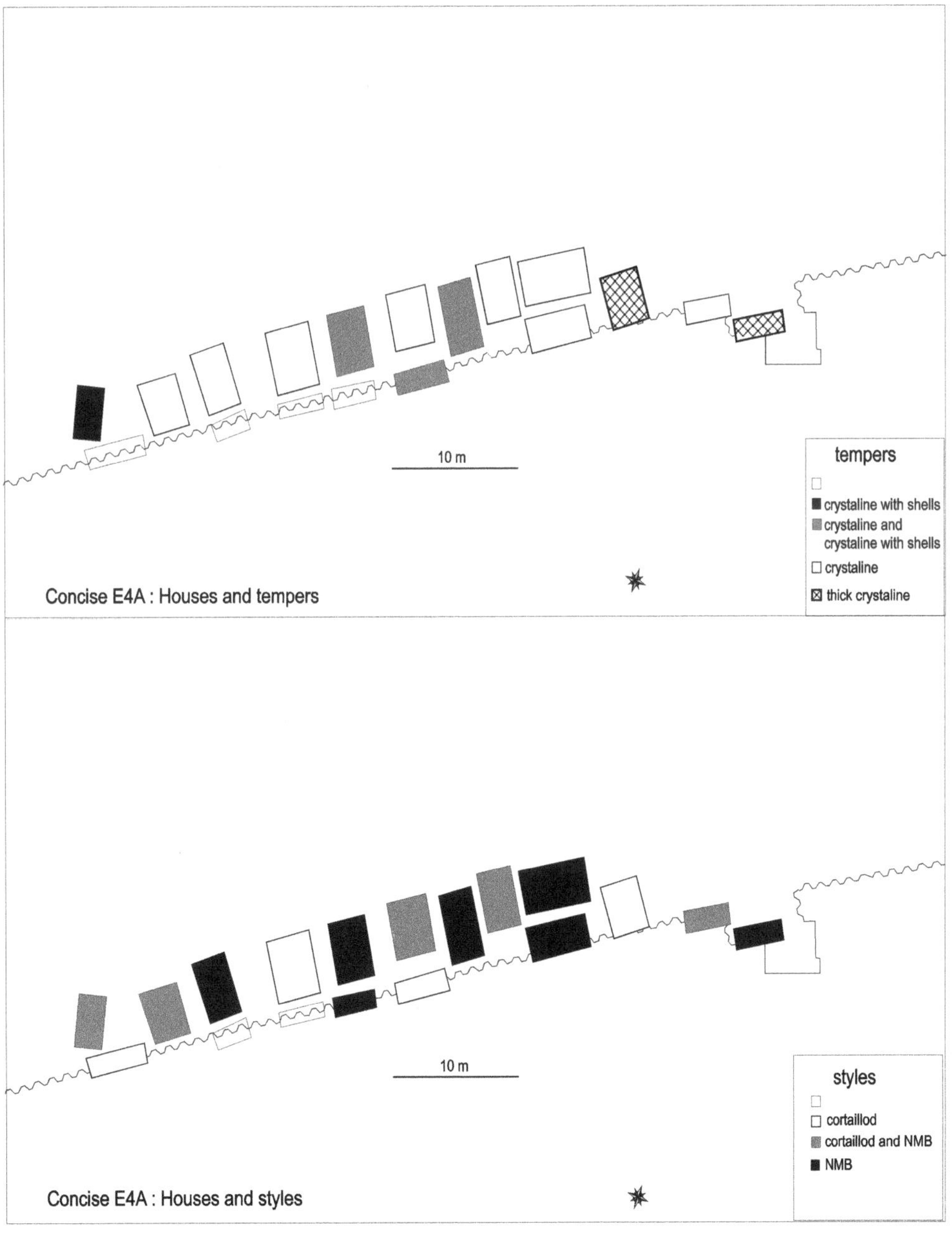

Fig. 5 – The distribution of tempers and styles in the village E4A dated 3645-3635 BC.

7. CONCLUSIONS AND PROSPECTS

As we have seen, the exceptional coexistence of two styles of pottery in lake-shore villages at Concise contributes to the understanding of socio-economic facts. The excellent preservation of these settlements, the extensive excavations and the possibility of precisely dating the strata by dendrochronology allow for planimetric studies. To these planimetric studies we have added an ethnographic interpretation of the material distribution and of the archaeological observations. In these cases, very simple observations, namely the composition of tempers, can be sufficient to reach interesting conclusions, such as the fact that the production of pottery was most likely a domestic activity and that there were quarters of the village that grouped together for the supply of tempers.

The next stage for us will be to try to ascertain who were the manufacturers and the consumers of this pottery and what were the dynamics of the populations who lived at Concise (Burri 2005; Burri to be published). This discussion can only take place once the spatial distribution of technical and stylistic data is understood within the village as well as within a wider regional context and it will also be interesting to have information regarding the manufacturing techniques of the pottery. At a later stage the discussion will be expanded to include ethnoarchaeological data regarding the relationships between pottery styles and population. In the case of Concise and based on this ethnoarchaeological approach, we can assume that at least some of the potters came from elsewhere in the Jura. As for the rest of the population, we need to compare the distribution of the others artefacts with that of the pottery in the individual houses to reach any form of conclusions.

ACKNOWLEDGEMENTS

I thank M. Templer for the correction of the English text, A. Winiger and D. Weidmann for their trust and encouragement, P. Pétrequin, M. Templer and L. Jammet-Reynal for their excellent and useful cooperation, P. Pétrequin and his team at Le Frasnois (Jura, F), those responsible for the SACFR (FR, CH), the SACBE (BE, CH), the Laténium (NE, CH) and the Musée d'Yverdon (VD, CH) for allowing me to look at their collections.

REFERENCES

Burri E. 2005, "La céramique de Concise (VD) au Néolithique moyen et l'influence jurassienne", *Archéologie suisse* 28 (3), p. 24-29.

Burri E. in press, "Concise-sous-Colachoz (VD, CH): des villages du Cortaillod à forte composante NMB au bord du lac de Neuchâtel", *in Actes du colloque de Dijon, 20-21 octobre 2001.*

Burri E. to be published, "Concise (Vaud, Suisse). Les vestiges céramiques d'un village du Néolithique moyen (3645-3636 av. J.-C.): répartitions spatiales et interprétations", *in Actes du 27e colloque interrégional sur le Néolithique, 1er et 2 octobre 2005, université de Neuchâtel.*

Carnes J. 1997, *Die Keramik von Sektor 1 der Grabung Muntelier-Standweg*, Lizentiatarbeit, Universität Bern.

Dufay-Galan A. 1996, "Le Néolithique moyen de la grotte de la Molle-Pierre, à Mavilly-Mandelot (Côte-d'Or): un ensemble NMB", *in* P. Duhamel (dir.), *La Bourgogne entre les Bassins rhénan, rhodanien et parisien: carrefour ou frontière?, Actes du 18e colloque interrégional sur le Néolithique, Dijon 25-27 octobre 1991*, p. 399-405.

Gauthier Y. 1985, *Valeurs attributives des composantes culturelles d'un site cortaillod: Muntelier-Dorf 7*, Mémoire de licence, Université de Berne.

Hafner A., Suter P.J. 2000, *−3400. Die Entwicklung der Bauerngesellschaften im 4. Jahrtausend v. Chr. Am Bielersee aufgrund der Rettungsgrabungen von Nidau und Sutz-Lattrigen,* Ufersiedlungen am Bielersee 6, Berne.

Jaccottey L., Milleville A., Pétrequin P. to be published, "Des camps et des meules: le Néolithique moyen II (4200-3600 av. J.-C.) dans le nord du Jura", *in Actes du 27e colloque interrégional sur le Néolithique, 1er et 2 octobre 2005, université de Neuchâtel.*

Kaenel G. 1976, *La fouille du "Garage Martin 1973": précisions sur le site de Clendy à Yverdon (Néolithique et Âge du Bronze),* Cahiers d'archéologie romande 8, Lausanne.

Lepage L. 1992, *La Vergentière (camp et nécropole) à Cohons (Haute-Marne): du Néolithique moyen au Bronze final,* Musées de la Ville de Langres, Mém. de la Soc. archéol. champenoise 6, Langres.

Liégard S., Urgal A., Fourvel A., Liégard D. 2000, "Étude d'un lot de mobilier du Néolithique moyen II d'affinité nord-orientale découvert à Lapalisse (Allier)", *Revue archéologique du Nord de la France* 39, p. 31-42.

Pétrequin P. 1984, "Les contacts avec le Cortaillod", *in* P. Pétrequin, A. Gallay (dir.), *Le Néolithique Moyen Bourguignon (N.M.B.), Actes du colloque de Beffia (Jura, France), 4 et 5 juin 1983*, Archives suisses d'anthropologie générale 48, 2, p. 57-59.

Pétrequin A.-M., Pétrequin P. 1984, *Habitat lacustre du Bénin: une approche ethno-archéologique*, Recherche sur les civilisations, Mém. 39, Paris.

Pétrequin A.-M., Pétrequin P. 1989, "La céramique du niveau V et le Néolithique moyen bourguignon", *in* P. Pétrequin (dir.), *Les sites littoraux néolithiques de Clairvaux-les-Lacs (Jura). II, Le Néolithique moyen*, Paris, p. 265-284.

Pétrequin A.-M., Pétrequin P. 2005, *Clairvaux-les-Lacs (Jura). Site Néolithique de CL XIV: fouille programmée 2003-2004, rapport de synthèse*, Rapport de fouille, Besançon, université de Franche-Comté, laboratoire de chronoécologie.

Pétrequin P., Gallay A. (dir.) 1984, *Le Néolithique Moyen Bourguignon (N.M.B.), Actes du colloque de Beffia (Jura, France), 4 et 5 juin 1983*, Archives suisses d'anthropologie générale, 48, 2.

Ramseyer D. (dir.) 2000, *Muntelier/Fischergässli. Un habitat néolithique au bord du lac de Morat (3895 à 3820 avant J.-C.),* Archéologie fribourgeoise 15, Fribourg.

Rialland Y. 1991, "L'enceinte néolithique du Champ de la Grange à Bruère-Allichamps (Cher)", *in Actes du 14e colloque international sur le Néolithique, Blois, 16-18 octobre 1987.* p. 243-245.

SCHIFFERDECKER F. 1982, *Auvernier 4. La céramique du Néolithique moyen d'Auvernier dans son cadre régional.* Cahiers d'archéologie romande 24, Lausanne.

SCHWAB H. 1999, *Archéologie de la 2e correction des eaux du Jura, volume 2. Les premiers paysans sur la Broye et la Thielle,* Archéologie fribourgeoise 14, Fribourg.

SEPPEY V. 1991, *La céramique cortaillod de Corsier-Port (Genève),* Travail de diplôme, université de Genève, département d'anthropologie et d'écologie.

STOECKLI W.E. 1981a, *Die Cortaillod-Keramik der Abschnitte 6 und 7.* Die neolithischen Ufersiedlungen von Twann 10, Berne.

STOECKLI W.E. 1981b, *Die Keramik der Cortaillod-Schichten,* Die neolithischen Ufersiedlungen von Twann 20, Berne.

TEMPLER M. 2006, *Analyse typologique, évolution et affinités culturelles de la céramique néolithique de Clairvaux XIV, Jura, France. (1re moitié du IVe millénaire av. J.-C.),* mémoire de licence, université de Neuchâtel.

WERNLI M. 1995, *Le Néolithique moyen II de la Grotte du Gardon,* travail de diplôme, université de Genève, département d'anthropologie et d'écologie.

WINIGER A. 2003, "Concise (Vaud), une stratigraphie complexe en milieu humide", in M. Besse, L.-I. Stahl Gretsch, P. Curdy (dir.), *ConstellaSion. Hommage à Alain Gallay,* Cahiers d'archéologie romande 95, Lausanne, p. 207-228.

WINIGER A. to be published, "Datations et reconstitutions architecturales d'un village néolithique moyen (E4A) entre 3645 et 3636 av. J.-C. à Concise (Vaud, Suisse)", *in Actes du 27e colloque interrégional sur le Néolithique, 1er et 2 octobre 2005, université de Neuchâtel.*

EARLY AND MIDDLE/LATE NEOLITHIC POTTERY PRODUCTION IN NORTHERN CALABRIA (ITALY): RAW MATERIAL PROVENANCE, PASTE PREPARATION AND FIRING TECHNIQUES

ITALO M. MUNTONI[1], PASQUALE ACQUAFREDDA[2], ROCCO LAVIANO[2]

[1] Facoltà di Scienze MM. FF. e NN., Università di Bari, Italy; Facoltà di Scienze Umanistiche, Università "La Sapienza", Roma, Italy; Italo.Muntoni@uniroma1.it

[2] Dipartimento Geomineralogico, Università di Bari, Via E. Orabona 4, 70125 Bari, Italy; p.acquafredda@geomin.uniba.it, rocco.laviano@geomin.uniba.it

THE ARCHAEOLOGICAL CONTEXT

The Neolithic village of Favella della Corte (Corigliano Calabro, CS) lies in the Plain of Sibari, in northern Calabria. It is located on a fluvial terrace along the Crati River, at 20 m above sea level and some 5 km from the present-day Ionian coastline (Guerricchio and Melidoro 1975). Archaeological excavations, regularly carried out from 1990 to 2002 by the University of Genoa and the Soprintendenza al Museo Preistorico Etnografico "L. Pigorini" of Rome (Tinè 2004), identified two main phases of occupation of the site, the earlier dated to the Early Neolithic (5970-5660 cal. BC 2σ) and the later to the Middle/Late Neolithic. Extensive researches brought to light several Early Neolithic pits of different size and shape, and more superficial Middle/Late Neolithic cobble structures. Two different wares are represented in Early Neolithic contexts (Natali 2004): coarse ovoid pots with smoothed and impressed surfaces ("impressa" ware) and fine smaller open vessels with burnished surfaces, sometimes impressed (plain ware). Very fine and smoothed pots, plain or brown painted, are instead typical of Middle/Late Neolithic cobble structures of the village ("figulina" ware). Large quantities of sintered daub fragments were often recorded in many Early Neolithic pits, representing the walls of the collapsed structure. Many fragments show the impressions left by the timber framework.

THE GEOLOGICAL CONTEXT

The Calabrian-Peloritanian Arc represents an exotic belt among the carbonate domains of the Apennines and of the Sicily mountains. Several tectonic units, formed by crystalline rock bodies, which represent segments of Palaeozoic continental crust and Mesozoic oceanic crust, compose it (Fig. 1). Around the site, alluvial deposits and Pleistocene deposits crop out (Fig. 2): they are composed, at the bottom, of conglomerates and sands, blue-grey clays and, at the top, of sands and conglomerates. The geology of the area is also characterised by Calabrian grey marly or silty clays (Dell'Anna and Laviano 1988), with sandy levels interbedded, and by the Palaeozoic dioritic-kinzigitic outcrops of the Sila Massif (Messina *et al.* 1994), consisting mainly of plutonic igneous rocks and medium to high grade micaschists and gneisses.

AIMS OF TECHNOLOGICAL ANALYSES AND SAMPLING

The reconstruction of the working sequence used in pottery manufacturing at Favella, from raw material provenance to preparation of bodies and firing techniques (Muntoni 2004), should help us to obtain insights into: a) the potter's role as an active and controlling agent in the procedure of a specific pottery manufacture; b) the economics of pottery or the socio-economic arrangements involved in the production, use and distribution of ceramics.

Forty-two pottery samples from the Early and Middle/Late Neolithic structures were analysed. Early Neolithic coarse ("impressa" or ware A) and fine (plain or ware B) wares (n=27: FAV01-27) were sampled from two of the many pits (D and E) evidenced on the site. Middle/Late Neolithic fine painted ("figulina" or ware C) ware (n=15: FAV28-42) was sampled from different cobble structures.

Wall daub samples (n=5: FAV43-47) were all sampled from one Early Neolithic pit (E).

ANALYTICAL METHODS

Mineralogical studies were carried out by X-ray powder diffraction analyses (PXRD), using a Philips diffractometer (PW 1710) with Ni-filtered CuKα radiation and employing NaF as internal standard. They were complemented by petrographic observation on thin sections, with a polarized light microscope (OM). Major and trace element determination was performed by X-ray fluorescence (XRF), using a Philips PW 1480/10 spectrometer (Cr anode for major and minor elements, Rh anode for Rb, Sr, Y, Zr, Nb and W anode for Ce, La, Ba, Ni, Cr, V), following analytical techniques outlined by Leoni and Saitta (1976) and Franzini *et al.* (1972, 1975). Two reference standards (AGV-1 from USGS-USA and NIM-G from NIM-South Africa) were used to check the accuracy of the analytical data. Loss on ignition was determined by heating the samples at 1000 °C for 12 hours. This value is affected by H_2O uptake and organic matter absorbed during burial, by the late loss of OH due to mineral dehydroxylation and by CO_2 from carbonate dissociation. PXRD patterns of the same previously heated samples, for the identification of mineralogical changes, were recorded at room temperature.

THIN-SECTION ANALYSES

The temper of Early Neolithic coarse (A) and plain (B) wares has sizes in the range 0.6-1.5 mm (occasionally reaching 4 mm). It is represented by quartz, biotite, muscovite, feldspars, sillimanite and lithic fragments of polycrystalline grains of quartz, two-mica granites, micaschists and sillimanite gneisses. Few samples (n=4), mainly of plain ware, also reveal the systematic presence of grog. The two wares have similar temper-matrix ratio.

The temper of the Middle/Late Neolithic ware (C) has sizes in the range 0.1-0.4 mm and it is represented essentially by quartz, muscovite and plagioclase; some samples (n=4) reveal also the occasional presence of grog.

The temper of Early Neolithic wall daub samples is very similar to that of Early Neolithic coarse (A) and plain (B) wares. It is composed of quartz, feldspars (microcline and plagioclase), sillimanite and lithic fragments of polycrystalline grains of quartz, granites, micaschists, micaschists with static blastesis of biotite and sillimanite gneisses.

POWDER X-RAY DIFFRACTION ANALYSES

PXRD data (Table 1) of Early Neolithic coarse ware (A) confirm the presence of predominant quartz, K-feldspar and micas, with small quantities of plagioclase in all samples. Among clay minerals, weak peaks of smectite and chlorite were also observed. Early Neolithic plain ware (B) shows an almost identical mineralogical composition, with predominant quartz, K-feldspar and micas, with small quantities of plagioclase, and weak peaks of chlorite. The systematic occurrence of new phases, formed during firing, such as diopsidic pyroxenes, gehlenite and hematite, was observed only in Middle/Late Neolithic samples (C). Early Neolithic wall daub samples show a mineralogical composition very similar to the Early Neolithic wares (A and B), with predominant quartz and K-feldspar, with a small quantity of plagioclase and micas.

PXRD analyses of samples heat-treated at 1000 °C confirm the presence of predominant quartz, while feldspars

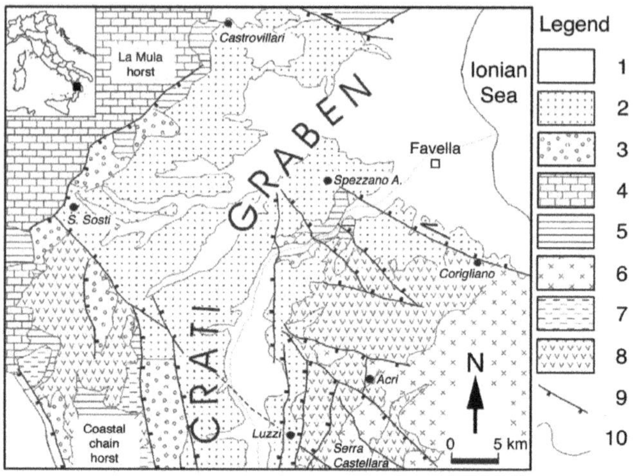

Fig. 1 – Geological sketch map of Northern Calabria (after Tansi et al. *2005, modified). 1. Holocene deposits; 2. Late Pliocene-Pleistocene sediments; 3. Late Miocene-Early Pliocene sediments; 4. Mesozoic carbonate unit; 5. Mesozoic ophiolitic unit; 6. Late Hercynian granitoid rocks; 7. Palaeozoic phyllite unit; 8. Palaeozoic medium-to high-grade gneissic unit; 9. Fault; 10. Stratigraphic boundary.*

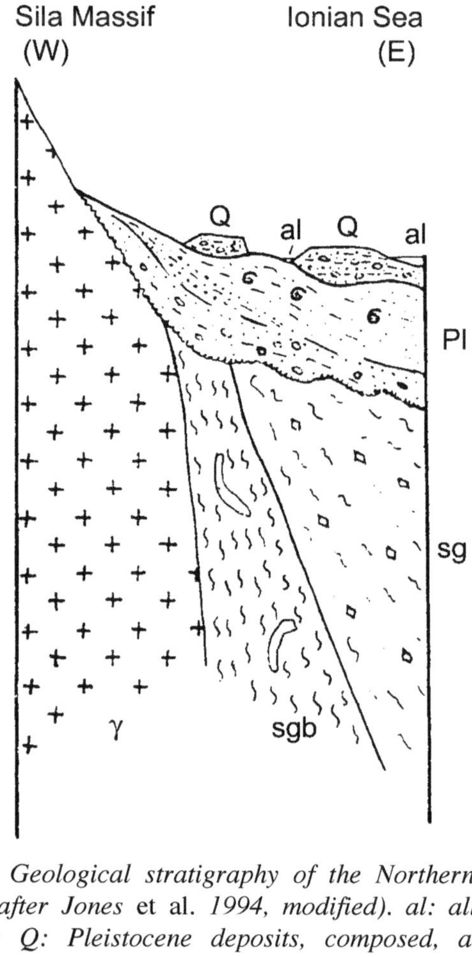

Fig. 2 – Geological stratigraphy of the Northern Sila Massif (after Jones et al. *1994, modified). al: alluvial deposits; Q: Pleistocene deposits, composed, at the bottom, of conglomerates and sands, blue-grey clays and, at the top, of sands and conglomerates; Pl: Calabrian grey marly or silty clays, with sandy levels interbedded; sg: gneisses and micaschists with garnet; sgb: schists and biotite-gneisses; γ: plutonic igneous rocks.*

Sample	Ware	C.M.	Chl	Micas	Qtz	Kfs	Pl	Cal	Px	Amp	Gh	Hem
FAV01	A	tr	tr	XXX	XXXXX	XXXX	x	tr	tr	tr		
FAV02	A				XXXXX	XXXX	tr	tr				X
FAV03	A	tr	tr	XXX	XXXXX	XXXX	x					
FAV04	A	tr	tr	XXX	XXXXX	XXXXX	x	tr		X		
FAV05	A	tr	tr	XXX	XXXXX	XXX	X	tr		tr		
FAV06	A	tr	tr	XXX	XXXXX	XXXX	X	tr		tr		
FAV13	A		tr		XXXXX	XXXX	tr	tr	tr	X		
FAV14	A			tr	XXXXX	XX	tr					X
FAV15	A			tr	XXXXX	XX	tr					tr
FAV17	A	tr		X	XXXXX	XXX	X			tr		tr
FAV18	A	tr	tr	XX	XXXXX	XXXX	X					
FAV21	A		tr	XX	XXXXX	XX	XX	tr				
FAV22	A		tr	XXX	XXXXX	XX	tr					
FAV25	A	tr	tr	X	XXXXX	XX	X					tr
FAV26	A	tr	tr	X	XXXXX	XXX	X					
FAV27	A	tr	tr	X	XXXXX	XXX	tr					tr
FAV07	B		tr	XXXX	XXXXX	XXXX	X		tr	tr		
FAV08	B		X	XXX	XXXXX	XXXX	X			tr		
FAV09	B		tr	XXXX	XXXXX	XXXX	X	tr		tr		tr
FAV10	B	tr	tr	XX	XXXXX	XXXX	X					
FAV11	B		tr	XXXX	XXXXX	XXX	X			X		tr
FAV12	B		tr	XXXX	XXXXX	XXX	X					X
FAV16	B		X	XXX	XXXXX	XXXX	X	tr				
FAV19	B		X	XXX	XXXXX	XX	X					
FAV20	B		tr	XXX	XXXXX	XX	X					
FAV23	B	tr	tr	XX	XXXXX	XX	X					
FAV24	B	tr	tr	XX	XXXXX	XX	tr					tr
FAV28	C		tr	X	XXXXX	X			X		tr	tr
FAV29	C		tr	XX	XXXXX	XXX	X		X		tr	X
FAV30	C		tr	XXX	XXXXX	XXX	X		X		X	tr
FAV31	C		tr	XX	XXXXX	XX	X		tr		tr	tr
FAV32	C		tr	XX	XXXXX	XX	X		X		tr	tr
FAV33	C				XXXXX	XXX	X		XX		tr	X
FAV34	C		tr	tr	XXXXX	XXX	X		X		tr	X
FAV35	C				XXXXX	XXXX	XX		XX		X	X
FAV36	C			X	XXXXX	XXX	X		XX		X	
FAV37	C		tr	X	XXXXX	XXX	X		XXX		XX	tr
FAV38	C		tr	tr	XXXXX	XXX	X		XX		X	X
FAV39	C			X	XXXXX	XX	X		XX		X	X
FAV40	C				XXXX	XXXXX	XX		XX		tr	tr
FAV41	C		X	X	XXXXX	XXX	X		XX		tr	X
FAV42	C		tr	XXX	XXXXX	XXX	X		XX		tr	tr
FAV43	Wall daub	tr	tr	X	XXXXX	XXX	XX			tr		tr
FAV44	Wall daub	tr	tr	XX	XXXXX	XXX	X			tr		tr
FAV45	Wall daub	tr		tr	XXXXX	XXX	X			tr		tr
FAV46	Wall daub	X	tr	XX	XXXXX	XXXX	X			tr		
FAV47	Wall daub	X	tr	XX	XXXXX	XXXX	X			tr		

Table 1 – Mineralogical composition (by PXRD) of pottery and wall daub samples. C.M.: clay minerals; Chl: chlorite; Micas: muscovite + biotite; Qtz: quartz; Kfs: K-feldspar; Pl: plagioclase; Cal: calcite; Px: pyroxene; Amp: amphibole; Gh: gehlenite; Hem: hematite (symbols as in Kretz 1983); number of X is in relationship with mineralogical phase abundance; tr: traces. Archaeological wares: A = coarse; B = plain; C = fine painted.

increase in concentration. The presence of neoformed diopsidic pyroxenes, gehlenite and hematite was detected mainly in Middle/Late Neolithic samples (C).

The four different phase associations detected (Heimann, Maggetti 1981; Maggetti 1981; Maggetti 1982) show that all Early Neolithic pottery samples were fired at a temperature not exceeding 800 °C (Table 2). However, on the basis of the presence in X-ray patterns of very weak peaks of calcite, one could argue that some Early Neolithic samples were fired at temperatures that could have reached 600-700 °C (Table 2, phase association 2). Other Early Neolithic samples, which show a lower amount of mica and the presence of hematite, could have been fired at higher temperatures, but still not over 800 °C (Table 2, phase association 3). In these two groups the coarseness of the grains seems not to have slowed down the reaction processes: thin-section analyses do not show reaction rims with neoformed minerals which might suggest higher firing temperatures. Only Middle/Late Neolithic ware (C), which shows neoformation phases such as diopsidic pyroxenes and hematite only readable in PXRD analyses, were fired at a temperature in a range of 900-1000 °C (Table 2, phase association 4). Phase transformations in this group were obviously fostered by the very fine paste texture. The presence in X-ray patterns of Early Neolithic wall daub of weak peaks of smectite and chlorite and textural features related to a very scarce synterization grade suggests that the maximum temperatures reached during the wattle and daub building firing did not exceeded 600 °C (Table 2, phase association 1).

CHEMICAL ANALYSES

XRF analyses (Table 3) are consistent with mineralogical data and show that SiO_2, Al_2O_3 and Fe_2O_3 are the main oxides, and that CaO concentrations are very low. Different SiO_2 and CaO+MgO concentrations (Fig. 3) allow to divide the samples into three subgroups: Early Neolithic samples (coarse and plain wares), Early Neolithic wall daub and Middle/Late Neolithic ones. Different SiO_2 and Al_2O_3 concentrations between Early Neolithic pottery and wall daub samples could be related to the different temper-matrix ratio. High P_2O_5 concentration was detected in some Middle/Late Neolithic samples from more superficial cobble structures: the increase of phosphorous content could be related to alteration processes during burial.

The concentration of trace elements (Table 4) is also very useful to identify a strong affinity between Early Neolithic wares from the two sampled archaeological pits and Early Neolithic wall daub samples. Only Nb and Zr, linked to clay minerals and/or to resistant phases, are significant to distinguish Early and Middle/Late Neolithic samples (Fig. 4). The lower concentration of Zr in Middle/Late Neolithic pottery is related to the absence of metamorphic and granitoid lithic fragments in the temper. These components are the carriers of minerals such as zircons that are particularly Zr rich.

CONCLUDING REMARKS

The two Early Neolithic wares show similar composition and fabrication technique, while shapes and surface finishing techniques seem to be the major source of variation. The mineralogical and petrological components of the Early Neolithic pottery and wall daub samples fit very well with those of fluvial sands of the western slope of the Sila Massif (Critelli and Le Pera 2003), characterized by bedrock association of plutonic and metamorphic sources (gneisses and schist-phyllites). Mainly Early Neolithic plain ware (B) was occasionally tempered with grog. Early Neolithic pots were fired in open firing, in a semi-controlled oxidising atmosphere. Temperatures in a range of 600-850 °C were obtained.

Middle/Late Neolithic pottery shows the exploitation of a different clay source, probably to be identified with Calabrian grey marly or silty clays (Dell'Anna and Rizzo 1982; Dell'Anna and Laviano 1991). In Middle and Late Neolithic societies, pottery production, mainly of fine brown painted or plain ware, probably evolved from a "domestic mode of production" to an "incipient-specialization stage" (Muntoni 2003). This stage would include an increasing standardization of paste composition, reflecting greater exploitation of particular kinds of clay, and greater skill more consistency evident in manufacturing technology and in an advanced (up to 1000 °C) firing technology.

ACKNOWLEDGEMENTS

We would like to thank Dr. Vincenzo Tinè, from the Soprintendenza al Museo Preistorico Etnografico "L. Pigorini" of Rome, for financial support and providing archaeological materials. One of the authors, Italo M. Muntoni, received a research grant from the Consiglio Nazionale delle Ricerche (CNR) through the Progetto Giovani – Agenzia 2000 (G004AC5). Many thanks to Dr. Elena Natali for Early Neolithic pottery sampling.

	Phase Associations			
	1	2	3	4
C.M.	··········		··········	
Chl	··········	··········	··········	··········
Micas	▪▪▪▪▪▪▪	■■■■■	▪▪▪▪▪▪▪	··········
Qtz	■■■■■	■■■■■	■■■■■	■■■■■
Kfs	■■■■■	■■■■■	■■■■■	▪▪▪▪▪▪▪
Pl	··········	··········	··········	··········
Cal		··········		
Px				▪▪▪▪▪▪▪
Amp	··········	··········		
Gh				··········
Hem			··········	··········
estimated T °C	<600	600-700	700-800	900-1000
Samples	5	14	13	15
Wares	Wall daub (n=5)	A (n=6); B (n=8)	A (n=10); B (n=3)	C (n=15)

Table 2 – The different phase associations, detected by PXRD, with the estimated firing temperatures (abbreviations as in table 1). The dimension of squares is in relationship with mineralogical phase abundance.

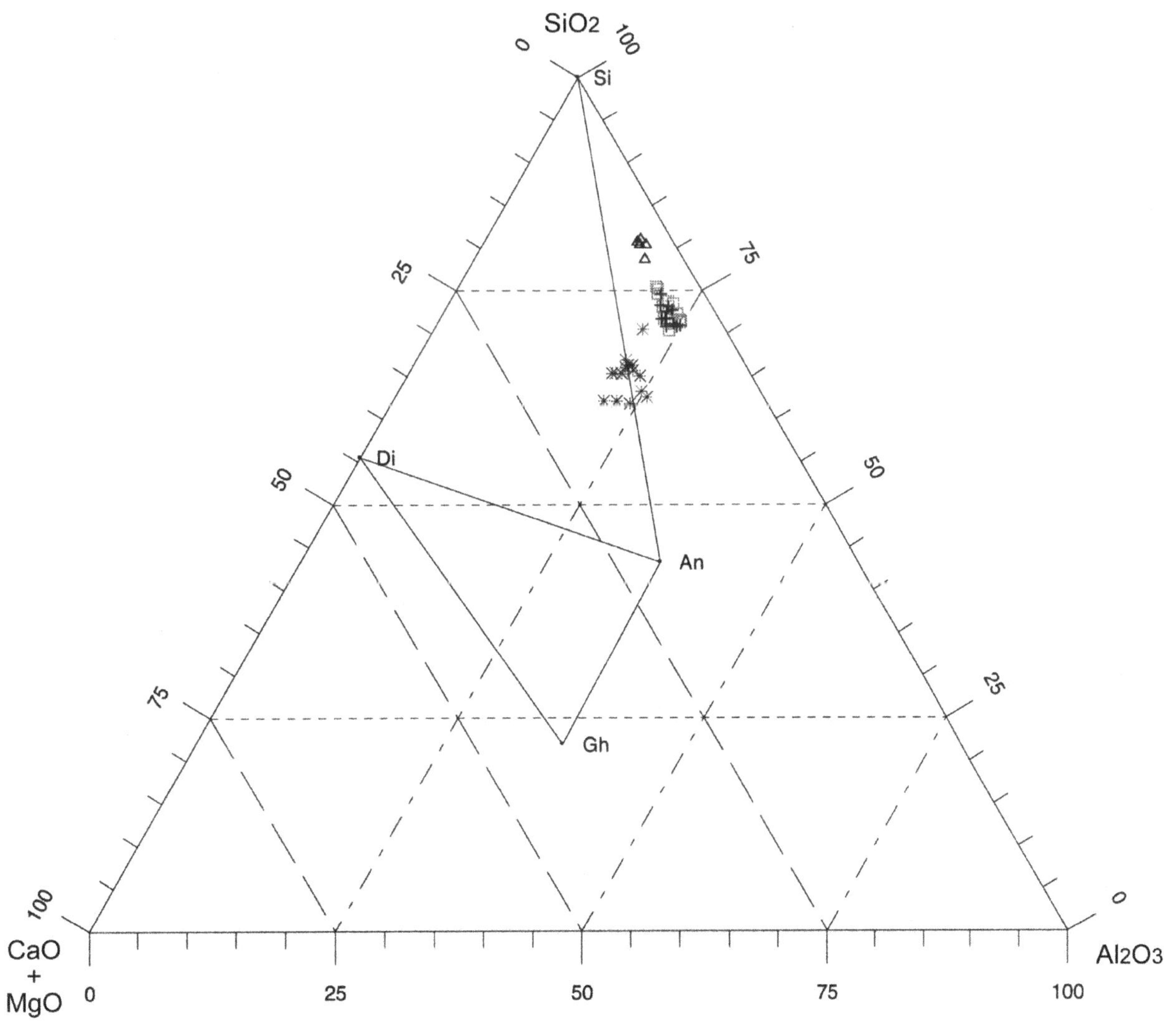

Fig. 3 – (CaO+MgO)/Al_2O_3/SiO_2 wt% diagram (Di: diopside; Gh: gehlenite; An: anorthite). Coarse ware (A): grey squares; plain ware (B): black crosses; fine painted ware (C): black asterisks; wall daub: black triangles.

Sample	Ware	SiO_2	TiO_2	Al_2O_3	Fe_2O_3	MnO	MgO	CaO	K_2O	Na_2O	P_2O_5	LOI
FAV01	A	59.66	0.65	16.25	6.37	0.12	1.63	1.86	2.34	1.31	0.58	9.23
FAV02	A	55.48	0.80	18.30	8.05	0.11	1.85	1.10	2.70	1.26	0.22	10.13
FAV03	A	59.02	0.69	15.77	5.72	0.10	1.51	1.95	2.22	1.31	0.91	10.81
FAV04	A	58.35	0.79	16.18	6.43	0.08	1.96	1.70	2.21	1.37	0.70	10.23
FAV05	A	56.15	0.70	17.34	6.88	0.10	1.44	1.87	2.36	1.10	1.03	11.03
FAV06	A	55.10	0.73	18.06	7.86	0.21	1.94	1.87	2.42	1.16	0.77	9.88
FAV13	A	56.81	0.77	17.08	7.02	0.11	2.02	1.76	2.53	1.08	0.61	10.21
FAV14	A	56.98	0.76	17.73	7.03	0.08	1.70	1.14	2.47	1.31	0.25	10.55
FAV15	A	53.83	0.79	17.83	7.48	0.09	2.15	1.90	2.59	1.02	1.29	11.03
FAV17	A	54.18	0.76	17.06	6.37	0.07	2.13	2.20	1.98	1.02	1.08	13.15
FAV18	A	57.63	0.64	16.28	6.35	0.09	1.43	1.90	1.94	1.12	0.69	11.93
FAV21	A	54.30	0.77	18.86	7.80	0.09	2.04	1.51	2.23	0.85	0.30	11.25
FAV22	A	56.01	0.76	16.55	6.86	0.09	2.21	1.66	2.23	0.88	0.58	12.18
FAV25	A	53.14	0.85	17.99	8.02	0.15	2.02	1.58	2.18	0.78	0.45	12.84
FAV26	A	56.07	0.72	17.28	7.21	0.08	1.45	1.85	1.97	1.05	0.69	11.63
FAV27	A	54.89	0.80	17.12	6.81	0.11	1.70	1.92	2.21	1.12	1.34	11.98
m		56.10	0.75	17.23	7.02	0.11	1.82	1.74	2.29	1.11	0.72	11.13
σ		1.89	0.06	0.87	0.69	0.03	0.27	0.29	0.22	0.18	0.34	1.10
FAV07	B	51.43	0.88	17.54	8.59	0.12	1.50	2.64	2.26	1.00	2.83	11.21
FAV08	B	55.47	0.87	18.32	7.50	0.13	2.14	1.47	2.63	1.06	0.33	10.08
FAV09	B	55.35	0.79	17.59	7.26	0.12	2.11	2.09	2.60	1.07	1.12	9.91
FAV10	B	57.79	0.72	17.44	6.57	0.14	2.03	1.13	2.31	1.38	0.16	10.33
FAV11	B	55.30	0.82	18.91	7.71	0.12	1.95	1.09	2.95	1.15	0.17	9.83
FAV12	B	54.63	0.78	18.92	7.71	0.11	1.98	1.07	2.48	0.93	0.26	11.13
FAV16	B	55.72	0.71	17.75	6.63	0.05	1.37	1.78	1.97	1.02	0.87	12.13
FAV19	B	54.94	0.80	17.55	7.17	0.11	2.10	1.97	2.29	1.05	0.98	11.04
FAV20	B	54.06	0.86	17.93	8.17	0.18	2.41	1.95	2.07	0.94	0.52	10.91
FAV23	B	54.23	0.83	17.39	7.37	0.14	2.44	1.57	2.14	1.04	0.67	12.18
FAV24	B	53.97	0.75	18.93	7.71	0.08	1.77	1.45	2.21	0.93	0.55	11.65
m		54.81	0.80	18.02	7.49	0.12	1.98	1.66	2.36	1.05	0.77	10.95
σ		1.54	0.06	0.63	0.60	0.03	0.33	0.49	0.29	0.13	0.76	0.84
FAV28	C	44.86	0.90	18.36	7.76	0.11	3.09	5.39	2.41	0.53	2.69	13.91
FAV29	C	47.99	0.78	16.55	7.00	0.09	2.72	5.83	2.59	0.77	1.34	14.35
FAV30	C	46.51	0.80	16.91	7.49	0.16	2.76	5.32	2.32	0.92	2.87	13.93
FAV31	C	49.80	0.78	16.34	6.32	0.10	2.83	6.44	2.38	0.87	1.00	13.15
FAV32	C	52.76	0.75	15.89	6.78	0.12	3.18	3.03	2.35	0.69	1.60	12.85
FAV33	C	48.24	0.73	15.49	6.65	0.14	3.20	6.87	2.32	0.80	0.94	14.61
FAV34	C	48.61	0.77	16.32	7.10	0.13	3.21	5.21	2.50	0.89	1.37	13.88
FAV35	C	46.78	0.86	18.26	7.28	0.08	3.18	5.75	2.48	0.82	0.43	14.07
FAV36	C	45.17	0.77	16.54	6.97	0.13	3.76	7.29	2.43	0.86	0.95	15.12
FAV37	C	46.01	0.75	15.88	6.50	0.12	3.98	8.22	2.36	0.72	0.51	14.95
FAV38	C	48.77	0.76	15.43	6.88	0.12	3.42	7.02	2.34	0.84	0.88	13.53
FAV39	C	44.69	0.84	17.54	7.53	0.13	3.52	6.59	2.51	0.56	2.39	13.71
FAV40	C	49.21	0.79	16.34	6.60	0.10	3.11	6.70	2.39	1.09	0.28	13.38
FAV41	C	49.38	0.78	16.29	6.95	0.13	3.46	5.37	2.34	0.92	1.54	12.85
FAV42	C	49.64	0.75	15.83	6.55	0.16	2.64	6.13	2.45	1.13	1.73	12.98
m		47.90	0.79	16.53	6.96	0.12	3.20	6.08	2.41	0.83	1.37	13.82
σ		2.23	0.05	0.90	0.42	0.02	0.38	1.20	0.08	0.16	0.79	0.72
FAV43	Wall daub	69.70	0.70	13.44	3.82	0.09	1.07	1.56	2.66	1.90	0.22	4.84
FAV44	Wall daub	67.76	0.74	13.37	3.92	0.09	1.15	1.77	2.72	1.74	0.53	6.21
FAV45	Wall daub	68.66	0.68	14.12	4.38	0.10	1.04	1.38	2.52	1.76	0.18	5.18
FAV46	Wall daub	64.44	0.79	14.13	4.81	0.10	1.15	2.00	2.56	1.46	1.16	7.40
FAV47	Wall daub	67.43	0.72	12.91	4.06	0.10	1.11	1.87	2.40	1.58	0.92	6.91
m		67.60	0.73	13.59	4.20	0.10	1.10	1.72	2.57	1.69	0.60	6.11
σ		1.97	0.04	0.53	0.40	0.01	0.05	0.25	0.12	0.17	0.43	1.09

Table 3 – Chemical composition of pottery and wall daub samples determined by XRF analysis: major and minor elements and L.O.I. (wt %). m = mean; σ = standard deviation.

Sample	Ware	Ba	Rb	Sr	Y	Zr	Nb	V	Cr	Ni	La	Ce
FAV01	A	671	103	258	24	225	17	101	138	61	34	64
FAV02	A	453	117	141	32	259	21	133	178	64	43	94
FAV03	A	667	104	306	28	249	20	86	106	44	47	91
FAV04	A	685	101	240	28	233	19	99	1175	51	41	75
FAV05	A	729	112	319	26	246	22	105	135	58	42	82
FAV06	A	872	109	257	312	264	19	108	125	56	47	100
FAV13	A	445	111	148	30	231	18	113	139	53	41	84
FAV14	A	418	143	130	34	256	25	130	155	68	51	99
FAV15	A	395	130	123	30	244	24	122	155	63	47	93
FAV17	A	803	105	308	23	223	17	106	135	57	33	68
FAV18	A	530	125	180	27	231	21	126	154	72	43	85
FAV21	A	905	96	269	25	204	19	104	139	58	41	76
FAV22	A	1156	106	313	29	238	19	95	122	51	42	76
FAV25	A	677	108	190	31	181	21	124	144	70	46	84
FAV26	A	634	104	197	27	180	18	111	139	56	38	67
FAV27	A	741	122	222	21	215	21	119	145	81	32	73
m		674	112	225	45	230	20	111	205	60	42	82
σ		203	12	69	71	25	2	13	259	9	5	11
FAV07	B	774	115	228	22	217	18	100	131	51	35	75
FAV08	B	542	108	169	263	265	22	105	133	45	37	86
FAV09	B	962	109	295	30	249	18	102	126	51	39	79
FAV10	B	1373	90	502	32	298	21	122	158	62	44	96
FAV11	B	607	139	176	27	267	22	114	125	58	42	92
FAV12	B	944	118	299	28	236	19	111	122	52	38	79
FAV16	B	927	105	326	22	161	14	108	127	52	35	75
FAV19	B	604	120	214	27	214	18	103	123	62	38	80
FAV20	B	747	136	221	24	251	21	123	146	72	32	85
FAV23	B	1017	96	299	23	233	22	103	134	57	34	71
FAV24	B	1040	109	281	24	210	17	110	131	53	40	76
m		867	113	274	47	236	19	109	132	56	38	81
σ		244	15	93	72	36	2	8	11	7	3	8
FAV28	C	1360	118	343	27	172	19	96	164	83	45	90
FAV29	C	753	132	352	25	154	17	116	148	67	41	81
FAV30	C	1401	94	408	28	193	15	103	145	57	39	81
FAV31	C	717	118	377	25	174	16	108	125	67	38	81
FAV32	C	956	103	234	23	147	15	89	142	73	38	80
FAV33	C	756	123	325	25	157	15	968	149	69	37	71
FAV34	C	938	109	330	27	162	16	112	163	75	41	81
FAV35	C	514	138	208	31	146	17	161	184	84	49	99
FAV36	C	681	120	236	22	136	15	105	161	69	39	74
FAV37	C	515	117	265	20	137	15	112	162	73	44	78
FAV38	C	2771	110	312	24	159	15	93	162	76	41	80
FAV39	C	1413	119	397	25	158	17	94	172	74	41	82
FAV40	C	416	130	225	30	167	16	127	156	66	43	89
FAV41	C	911	117	326	28	192	15	85	1420	611	37	77
FAV42	C	1010	115	362	25	194	14	95	140	58	36	75
m		1007	117	313	26	163	16	164	239	107	41	81
σ		581	11	65	3	19	1	223	327	140	3	7
FAV43	Wall daub	472	91	193	28	210	20	64	76	19	36	64
FAV44	Wall daub	579	91	218	25	195	18	59	84	22	37	61
FAV45	Wall daub	457	99	183	28	206	22	63	72	23	38	68
FAV46	Wall daub	748	75	219	21	167	18	66	84	27	40	72
FAV47	Wall daub	629	96	275	29	226	21	64	78	22	36	64
m		577	90	218	26	201	20	63	79	23	37	66
σ		120	9	36	3	22	2	3	5	3	2	4

Table 4 – Chemical composition of pottery and wall daub samples determined by XRF analysis: trace elements (ppm). m = mean; σ = standard deviation.

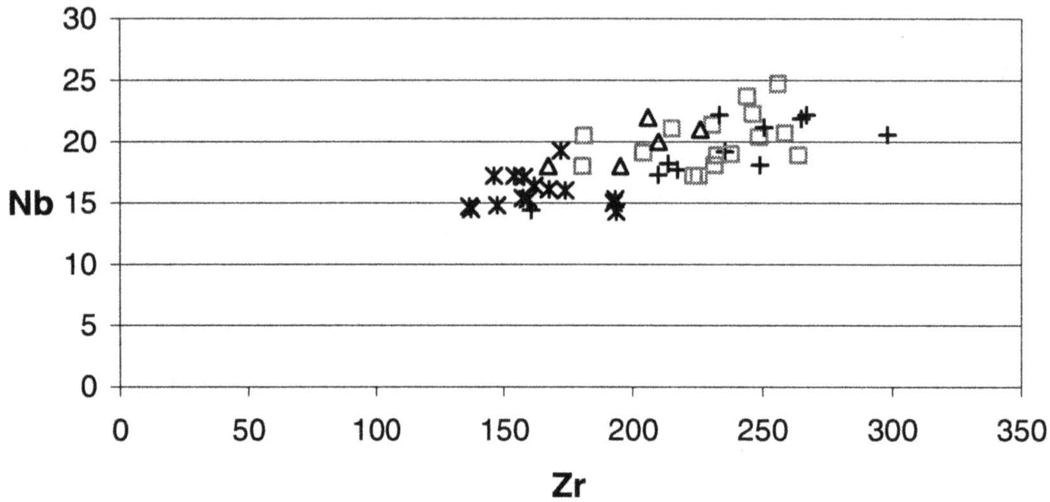

Fig. 4 – Nb vs. Zr plot (ppm). Coarse ware (A): grey squares; plain ware (B): black crosses; fine painted ware (C): black asterisks; wall daub: black triangles.

REFERENCES

Critelli S., Le Pera E. 2003, "Provenance relations and modern sand petrofacies in an uplifted thrust-belt, Northern Calabria, Italy", *in* A. Basu and R. Valloni (eds), *Quantitative provenance in Italy,* Memorie descrittive della Carta Geologica d'Italia, vol. LXI, Roma, p. 25-39.

Dell'Anna L., Laviano R. 1988, "A clay sedimentation picture of the Crati River Basin (Calabria, Southern Italy)", *Geologia applicata e idrogeologia* 23, p. 191-202.

Dell'Anna L., Laviano R. 1991, "Mineralogical and geochemical features of some Miocene clays of Crati River Basin (Calabria, Southern Italy)", *in* M. Störr, K.-H. Henning, P. Adolphi (eds), *Proceedings of the 7th Euroclay conference Dresden '91*, Greifswald, p. 283-288.

Dell'Anna L., Rizzo V. 1982, "Composizione mineralogica e granulometrica e alcune caratteristiche geotecniche delle argille del Pleistocene della media Vale del fiume Crati (Calabria)", *Rendiconti della Società italiana di mineralogia e petrologia* 38 (3), p. 1447-1458.

Franzini M., Leoni L., Saitta M. 1972, "A simple method to evaluate the matrix effects in X-ray fluorescence analysis", *X-ray spectrometry* 1, p. 151-154.

Franzini M., Leoni L., Saitta M. 1975, "Revisione di una metodologia analitica per fluorescenza X, basata sulla correzione completa degli effetti di matrice", *Rendiconti della Società italiana di mineralogia e petrologia* 31, p. 356-378.

Guerricchio A., Melidoro G. 1975, "Ricerche di geologia applicata all'archeologia della città di Sibari sepolta", *Geologia Applicata e Idrogeologia* 10 (1), p. 107-128.

Heimann R.B., Maggetti M. 1981, "Experiments on simulated burial of calcareous terra sigillata (mineralogical changes). Preliminary Results", *in* M.J. Hughes (ed.), *Scientific studies in ancient ceramics*, British Museum occasional paper 19, London, p. 163-177.

Jones R.E., Lazzarini L., Mariottini M., Orvini E. 1994, "Studio minero-petrografico e chimico di ceramiche protostoriche dal Broglio di Trebisacce (Sibari)", *in* R. Peroni and F. Trucco (eds), *Enotri e Micenei nella Sibaritide*, Taranto, p. 413-454.

Kretz R. 1983, "Symbols for rock-forming minerals", *American mineralogist* 68, p. 277-279.

Leoni L., Saitta M. 1976, "Determination of Yttrium and Niobium on standard silicate rocks by X-ray fluorescence analysis", *X-ray spectrometry* 5, p. 29-30.

Maggetti M. 1981, "Composition of Roman pottery from Lousanna (Switzerland)", *in* M.J. Hughes (ed.), *Scientific studies in ancient ceramics*, British Museum occasional paper 19, London, p. 33-49.

Maggetti M. 1982, "Phase analysis and its significance for technology and origin", *in* J.S. Olin and A.D. Franklin (eds), *Archaeological ceramics*, Smithsonian Institution Press, Washington, p. 121-133.

Messina A., Russo S., Borghi A., Colonna V., Compagnoni R., Caggianelli A., Fornelli A., Piccarreta G. 1994, "Il massiccio della Sila – Settore settentrionale dell'Arco Calabro-Peloritano", *Bollettino della Società geologica italiana* 113, p. 539-586.

Muntoni I.M. 2003, *Modellare l'argilla. Vasai del Neolitico antico e medio nelle Murge pugliesi*, Firenze.

Muntoni I.M. 2004, "Analisi archeometriche sulle ceramiche impresse di Favella: caratterizzazione delle materie prime e tecnologia di manifattura", in *Preistoria e protostoria della Calabria, Atti della XXXVII Riunione Scientifica dell'I.I.P.P.*, Firenze, p. 703-709.

Natali E. 2004, "Favella e la facies delle ceramiche impresse arcaiche del sud-est", *in Preistoria e protostoria della Calabria, Atti della XXXVII Riunione Scientifica dell'I.I.P.P.*, Firenze, p. 145-162.

Tansi C., Tallarico A., Iovine G., Folino Gallo M., Falcone G. 2005, "Interpretation of radon anomalies in seismotectonic and tectonic-gravitational settings: the south-eastern Crati graben (northern Calabria, Italy)", *Tectonophysics* 396, p. 181-193.

Tinè V. 2004, "Il Neolitico in Calabria", in *Preistoria e protostoria della Calabria, Atti della XXXVII Riunione Scientifica dell'I.I.P.P.*, Firenze, p. 115-143.

POTTERY PRODUCTION IN THE NEOLITHIC AND COPPER AGE VILLAGE OF MADDALENA DI MUCCIA (MARCHE, CENTRAL ITALY): RAW MATERIAL PROVENANCE AND MANUFACTURING TECHNOLOGY

ROCCO LAVIANO[1], ITALO M. MUNTONI[2]

[1] Dipartimento Geomineralogico, Università di Bari, Via E. Orabona 4, 70125 Bari, Italy; rocco.laviano@geomin.uniba.it

[2] Facoltà di Scienze MM. FF. e NN., Università di Bari, Italy; Facoltà di Scienze Umanistiche, Università "La Sapienza", Roma, Italy; Italo.Muntoni@uniroma1.it

THE ARCHAEOLOGICAL CONTEXT

The village of Maddalena di Muccia is located on a fluvial terrace in the Chienti valley (Marche, East-Central Italy), at 430 m above sea level (Fig. 1). It lies at the edge of the Apennines, some 50 km from the Adriatic coastline. It was densely settled in the Early Neolithic, Middle Neolithic and Late Copper Ages (from VIth to IIIrd millennium cal. BC). Soprintendenza ai Beni Archeologici delle Marche and Rome and Pisa Universities have carried out systematic researches since 2001 (Silvestrini *et al.* 2005; Manfredini *et al.* 2005a; 2005b). They have brought to light several Early (5715-5322 BC cal. 2σ) and Middle (4539-4051 BC cal. 2σ) Neolithic pits of different sizes and shapes and many Copper Age (2881-2196 BC cal. 2σ) hut foundations.

A programme of characterisation analyses was carried out on 49 pottery and 3 wall daub samples from different archaeological contexts (Muntoni 2005). The purpose of the analytical project is to study the technological changes in pottery production in relation to the different pottery wares produced at the same site in the three prehistoric phases. Early Neolithic coarse ("impressa") and plain wares (n=18: MAD01-18) were sampled from pit 114, Middle Neolithic coarse ("impressa"), plain and fine painted wares (n=18: MAD22-39) were sampled from pit 12, Late Copper Age plain burnished and incised wares (n=13: MAD40-52) were sampled from pit 105. Wall daub samples (n=3: MAD19-21) were all sampled from a Middle Neolithic pit (n. 12).

THE GEOLOGICAL CONTEXT

The site is located on Pleistocene terraced alluvial deposits. Around the site (Fig. 2) Miocene arenaceous and clayey marl (*Molasse inferiori*) and Miocene grey and bluish-grey well stratified arenaceous marl, with intercalation of yellowish-grey detrital limestone, marly limestone and light-grey and blue-grey marl with chert nodules (*Formazioni dello Schlier e del Bisciaro*) crop out. The geology of the region (Centamore and Deiana 1986) is characterised also by Upper Eocene-Oligocene marl, marly limestone and clayey marl (*Scaglia cinerea*) and Early-Middle Eocene red micritic limestone with nodules of chert, with intercalation of white subcrystalline limestone and small nummulithic breccias (*Scaglia rossa e bianca*). Early Cretaceous white, compact, well stratified limestone and micritic limestone (*Calcare rupestre*) and Mesozoic marly limestone, marl and limestone with chert nodules (*Scisti ad aptici* and *Rosso ammonitico*) crop out west of the site.

ANALYTICAL METHODS

Mineralogical studies were carried out by X-ray powder diffraction analysis (PXRD) using a Philips diffractometer (PW 1710) with Ni-filtered CuKα radiation and employing NaF as internal standard. They were complemented by petrological observation on thin sections, with a polarizing microscope (OM). Major and trace element determination was performed by X-ray fluorescence (XRF), using a Philips PW 1480/10 spectrometer (Cr anode for major and minor elements, Rh anode for Rb, Sr, Y, Zr, Nb and W anode for Ce, La, Ba, Ni, Cr, V), following analytical techniques outlined by Leoni and Saitta (1976) and Franzini *et al.* (1972, 1975). Two reference standards (AGV-1 from USGS-USA and NIM-G from NIM-South Africa) were used to check the accuracy of the analytical data. Loss on ignition (LOI) was determined by heating the samples at 1000 °C for 12 hours. This value is affected by H_2O uptake and organic matter absorbed during burial, by the late loss of OH due to mineral dehydroxylation and by CO_2 loss from carbonate dissociation. PXRD patterns for the same previously heated samples, to identify mineralogical changes, were recorded at room temperature.

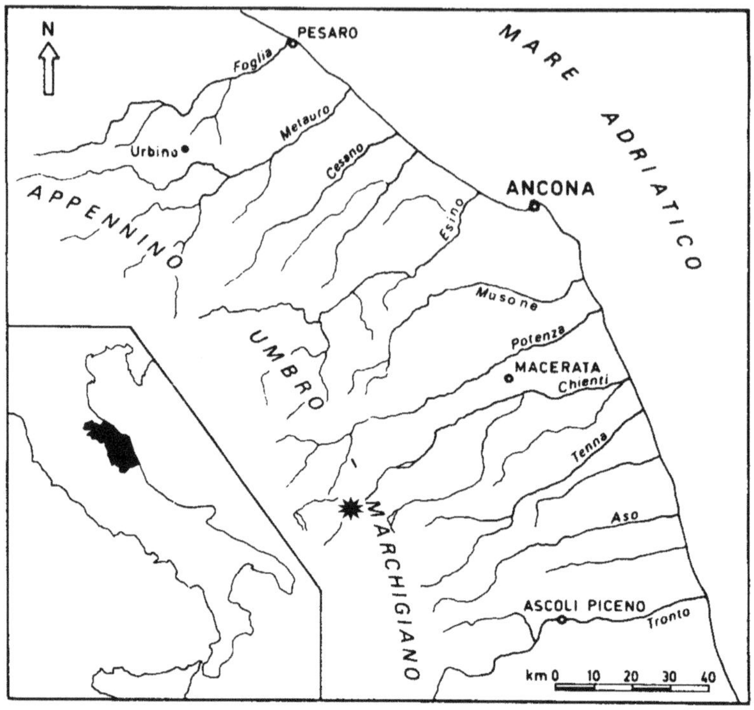

Fig. 1 – Geographical location of the archaeological site of Maddalena di Muccia in the Chienti valley (Marche, East-Central Italy) (after Centamore and Deiana 1986, fig. 45, modified).

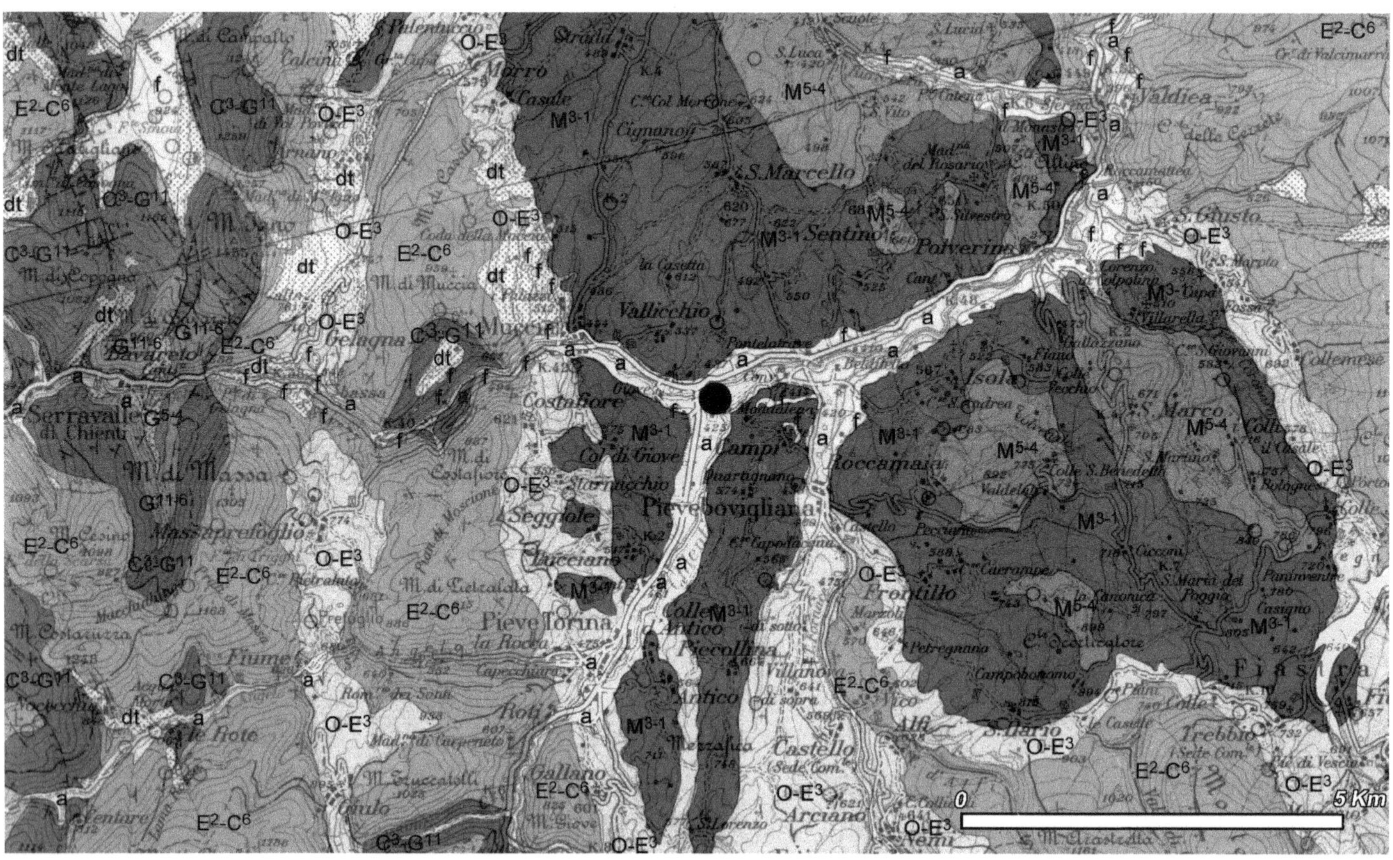

Fig. 2 – Geological location of the archaeological site of Maddalena di Muccia (Marche, East-Central Italy) (after Carta Geologica d'Italia, F. 124, modified). a, dt: Holocene alluvial deposits and talus; f: Pleistocene terraced alluvial deposits; $M^{5\text{-}4}$: Molasse inferiori; $M^{3\text{-}1}$: Formazioni dello Schlier e del Bisciaro; O-E^3: Scaglia cinerea; E^2-C^6: Scaglia rossa e bianca; C^3-G^{11}: Calcare rupestre; $G^{11\text{-}6}$: Scisti ad aptici; $G^{5\text{-}4}$: Rosso ammonitico.

THIN-SECTION ANALYSES

Seven different fabrics were recognised by mineralogical examination on thin sections, with different prevailing clastic constituents (chert, organic and inorganic calcite, quartz and K-feldspar) (Table 1). The groups are very homogeneous, though some degree of variation in grain size and percentage is present mainly in Qtz/Kfs and Grog fabrics. Grog fragments are systematically present only in Copper age samples.

The Middle Neolithic wall daub samples (n=3, MAD 19-21) are characterised by a brown micritic matrix and by fine (100-500 μm) calcite and carbonate rock fragments.

The pottery Cal group (n=6: MAD 12, 23, 29, 34-35 and 37) is characterised by a very high percentage of poorly-sorted and mainly subrounded fine fragments of fossiliferous limestone (50-450 μm).

The Qtz/Cal group (n=4, MAD 2, 11, 31 and 32) is an inhomogeneous group characterised by poorly-sorted subrounded quartz and calcite clasts (up to 1 mm).

The Middle Neolithic fine painted samples (n=3, MAD 36 and 38-39) show a very fine paste texture, with a quite fat clay matrix: of the sheet-silicates only mica crystals are recognisable. Non-plastic inclusions are homogeneous fine-grained (mainly ≤200 μm) quartz; clasts of feldspars and iron oxides/hydroxides are also present.

The Qtz/Kfs fabric (n=15) is very rich in poorly-sorted quartz, feldspar (both K-feldspar and plagioclase), micas, pyroxenes, polycrystalline quartz, micaschists and granitic rock fragments. These pottery samples can be divided in two subgroups on account of some differences in grain size distribution and the temper-matrix ratio. Some Early Neolithic samples (n=4, MAD 6, 14, 17 and 18) show a low incidence of non-plastic materials and a finer granulometry (50-500 μm). The other Early and Middle Neolithic samples (n=11, MAD 8, 10, 13, 15-16, 25-28, 30 and 33) show a high percentage of poorly-sorted clasts (up to 1.5 mm).

The pottery Qtz group (n=8, MAD 1, 3-5, 7, 9, 22 and 24) is characterised by a very iron-rich matrix and by a high percentage of sharp edged and poorly-sorted clasts of microcrystalline chert (from 125 μm to 2.5-3 mm).

All the Copper Age pottery samples (n=13, MAD 40-52) have been heavily tempered with grog (from 150 μm to 1,5–2 mm) and show a very humic dark red matrix with well-sorted and fine subrounded quartz.

POWDER X-RAY DIFFRACTION

The result of X-ray diffraction analysis (Table 1) shows the presence of predominant quartz and calcite, with a small quantity of K-feldspar, plagioclase and muscovite, and confirms the same differences between paste groups as those observed in thin sections. PXRD analysis also revealed the presence of other minerals such as diopsidic pyroxenes and hematite only in Middle Neolithic fine painted pottery (Px fabric). Weak peaks of probable maghemite were detected only in samples belonging to Qtz and Grog fabrics, both characterised by very iron-rich and humic red dark matrix.

Mineralogical data show that all samples were fired at a temperature not exceeding 600-800 °C. Only Middle Neolithic fine painted pottery (Px fabric), which shows diopsidic pyroxenes and hematite neoformations, only readable by PXRD analysis, were fired at a higher temperature, in a range of 850-1050 °C (Heimann and Maggetti 1981; Maggetti 1981; Maggetti 1982). Phase transformations in the Px group were obviously fostered by the very fine paste texture; moreover, for the other groups, thin-section analyses do not show an incipient carbonate dissociation which might suggest a firing temperature higher than 600-800 °C.

CHEMICAL ANALYSES

XRF analyses (Table 2) showed SiO_2 (m=48.49), Al_2O_3 (m=15.11), CaO (m=8.79) and Fe_2O_3 (m=6.15), as the main oxides, with some variations as regards CaO percentage (σ=10.98). LOI percentage appears higher than expected because the specimens were not previously dried at 110 °C in order to eliminate absorbed H_2O and organic matter absorbed during burial. One (MAD 36) of the samples of the Px group, with a 16.55% LOI mean value (n=3), is richer in calcite than the others (Table 1).

Chemical data are consistent with mineralogical data. The $(CaO+MgO)/Al_2O_3/SiO_2$ diagram (*wt%*) show (Fig. 3) a clear separation of samples very rich in CaO (Wall daub) from those characterised by similar amount of the three main oxides (Cal, Qtz/Cal and Px) and a partial overlapping of SiO_2 rich samples (Qtz/Kfs, Qtz and Grog). Some trace element plots (Table 3), like Cr *vs.* V and Ba *vs.* Sr also show the separation of CaO rich samples (Wall daub, Cal, Qtz/Cal and Px) and a partial overlapping of SiO_2 rich samples (Qtz/Kfs, Qtz and Grog) (Fig. 4).

CONCLUDING REMARKS

Six pottery fabrics, with different dominant mineralogical constituents (chert, organic and inorganic calcite, quartz and K-feldspar), were used in the same Neolithic and Copper Age village located in Maddalena di Muccia. Archaeometric data showed some difference in the raw materials used in pottery production. The dominant clastic constituents are all compatible with the geological deposits, cropping out in the hinterland of the site. The mineralogical differences amongst the six pottery fabrics suggest that Early and Middle Neolithic potters used to collect different clays in areas around the settlement itself.

The large amount of sharp edged and very poorly sorted chert fragments, which characterizes the Qtz fabric, might have been intentionally added. It must be pointed out that the limestone formations (*Calcare rupestre*; *Scaglia rossa e bianca*) outcropping around the site are

Samples	Chronology	Fabric	C.M.	Ms	Qtz	Kfs	Pl	Cal	Px	Gh	Hem	Mgh
MAD19	MN	Wall daub	tr.		XX			XXXXX				
MAD20	MN	Wall daub	tr.	X	XXX	tr.		XXXXX				
MAD21	MN	Wall daub	tr.	X	XX	tr.		XXXXX				
MAD12	EN	Cal	X	tr.	XXXX	X		XXXXX				
MAD23	MN	Cal	X	tr.	XXXX	X	tr.	XXXXX				
MAD29	MN	Cal	X	tr.	XXXX	X	tr.	XXXXX				
MAD34	MN	Cal	X	X	XXXX	X		XXXXX				
MAD35	MN	Cal	XX	X	XXXX	X		XXXXX				
MAD37	MN	Cal	X	X	XXXX	tr.		XXXXX				
MAD02	EN	Qtz/Cal	XX	X	XXXXX	X		XXXX				
MAD11	EN	Qtz/Cal	XX	XX	XXXXX	XX	X	XXXX				
MAD31	MN	Qtz/Cal		tr.	XXXXX	X	tr.	XXXX				
MAD32	MN	Qtz/Cal	X	tr.	XXXXX	XXX	tr.	XXXX				
MAD36	MN	Px	tr.	X	XXXXX	X	tr.	XXX	tr.		tr.	
MAD38	MN	Px		X	XXXXX	X		X	XX		tr.	
MAD39	MN	Px	tr.	X	XXXXX	XX	tr.	X	XX		tr.	
MAD06	EN	Qtz/Kfs	XX	X	XXXXX	XX	tr.	X				
MAD08	EN	Qtz/Kfs	XX	X	XXXXX	XXX	tr.	X				
MAD10	EN	Qtz/Kfs	XX	X	XXXXX	XXXX	X	XX		tr.	tr.	
MAD13	EN	Qtz/Kfs	X	X	XXXXX	XXX	X	X			tr.	
MAD14	EN	Qtz/Kfs	XX	X	XXXXX	XXXX	X	X				
MAD15	EN	Qtz/Kfs	X	XX	XXXXX	XX	X	X	tr.	tr.	tr.	
MAD16	EN	Qtz/Kfs	XX	XX	XXXXX	XXXX	X	X				
MAD17	EN	Qtz/Kfs	X	X	XXXXX	XXXX	X	X	tr.	tr.		
MAD18	EN	Qtz/Kfs	XX	XX	XXXXX	XXX	X	tr.				
MAD25	MN	Qtz/Kfs	X	tr.	XXXXX	tr.						
MAD26	MN	Qtz/Kfs	XX	X	XXXXX	XXXXX	X	tr.				
MAD27	MN	Qtz/Kfs	XX	X	XXXXX	XX	X				tr.	
MAD28	MN	Qtz/Kfs	tr.	tr.	XXXXX	XX	X					
MAD30	MN	Qtz/Kfs	XX	X	XXXXX	XXXXX	X		tr.	tr.	tr.	
MAD33	MN	Qtz/Kfs	XX	tr.	XXXXX	XX	X					
MAD01	EN	Qtz	X	tr.	XXXXX	tr.		tr.			tr.	tr.
MAD03	EN	Qtz	X	tr.	XXXXX	tr.		tr.				tr.
MAD04	EN	Qtz	X	tr.	XXXXX	tr.		X				
MAD05	EN	Qtz	XX	X	XXXXX	X		tr.			tr.	tr.
MAD07	EN	Qtz	X	tr.	XXXXX	tr.						tr.
MAD09	EN	Qtz	X	tr.	XXXXX	tr.		X			tr.	tr.
MAD22	MN	Qtz	XX	tr.	XXXXX	tr.		tr.				
MAD24	MN	Qtz	XX	X	XXXXX	X	tr.	XX				tr.
MAD40	C	Grog	X	tr.	XXXXX	tr.	X					tr.
MAD41	C	Grog	X	tr.	XXXXX	tr.						tr.
MAD42	C	Grog	X	tr.	XXXXX	tr.	tr.					tr.
MAD43	C	Grog	tr.	X	XXXXX	tr.					tr.	tr.
MAD44	C	Grog	X	tr.	XXXXX	tr.	tr.					tr.
MAD45	C	Grog	X	tr.	XXXXX	X	tr.				tr.	
MAD46	C	Grog	X	tr.	XXXXX	X	X					tr.
MAD47	C	Grog	X	tr.	XXXXX	tr.						tr.
MAD48	C	Grog	X	tr.	XXXXX							tr.
MAD49	C	Grog	X	tr.	XXXXX	X	tr.					
MAD50	C	Grog	X	tr.	XXXXX	X	tr.					tr.
MAD51	C	Grog	X	tr.	XXXXX	tr.						
MAD52	C	Grog	X	tr.	XXXXX	tr.						tr.

Table 1 – Mineralogical composition of pottery and wall daub samples by PXRD analysis. C.M.: clay minerals; Ms: muscovite; Qtz: quartz; Kfs: K-feldspar; Pl: plagioclase; Cal: calcite; Px: pyroxene; Gh: gehlenite; Hem: hematite; Mgh: maghemite (symbols as in Kretz 1983); number of X is in relationship with mineralogical phase abundance; tr: traces. Archaeological phases: EN = Early Neolithic; MN = Middle Neolithic; C = Copper age.

	SiO_2	TiO_2	Al_2O_3	Fe_2O_3	MnO	MgO	CaO	K_2O	Na_2O	P_2O_5	LOI
					Wall daub fabric (n=3)						
m	16.10	0.30	5.92	2.45	0.20	1.02	39.86	0.77	0.13	0.65	32.60
σ	3.88	0.04	0.88	0.26	0.05	0.08	2.61	0.33	0.02	0.26	1.44
					Cal pottery fabric (n=6)						
m	34.38	0.52	10.14	4.25	0.12	1.68	25.14	1.17	0.28	0.43	21.89
σ	3.98	0.05	0.86	0.67	0.07	0.21	2.94	0.14	0.03	0.22	2.62
					Qtz/Cal pottery fabric (n=4)						
m	44.00	0.70	13.76	5.77	0.14	1.64	13.76	1.76	0.47	0.87	17.12
σ	3.95	0.06	1.37	0.82	0.05	0.47	3.89	0.43	0.22	0.99	1.40
					Px pottery fabric (n=3)						
m	40.03	0.93	16.11	7.83	0.17	2.86	11.08	1.66	0.34	2.44	16.55
σ	3.24	0.08	1.60	0.97	0.04	0.92	5.63	0.15	0.05	0.44	1.28
					Qtz/Kfs pottery fabric (n=15)						
m	55.41	0.76	16.71	5.95	0.09	1.85	2.86	2.20	0.75	0.76	12.67
σ	2.11	0.05	0.94	0.57	0.03	0.46	1.21	0.28	0.21	0.69	1.37
					Qtz pottery fabric (n=8)						
m	57.47	0.69	14.45	6.33	0.15	1.35	3.27	1.72	0.28	1.07	13.22
σ	3.75	0.09	1.30	0.64	0.05	0.34	1.44	0.12	0.17	0.91	0.99
					Grog pottery fabric (n=13)						
m	52.32	0.90	18.26	7.71	0.26	1.54	2.25	1.95	0.31	0.96	13.54
σ	4.87	0.11	2.37	1.28	0.12	0.35	0.50	0.29	0.15	0.38	1.17
					All groups (n=52)						
m	48.49	0.74	15.11	6.15	0.16	1.67	8.79	1.80	0.43	0.92	15.75
σ	11.56	0.18	3.68	1.64	0.09	0.54	10.98	0.47	0.26	0.73	5.32

Table 2 – Chemical composition (wt%) of pottery and wall daub samples by XRF analysis. m = mean; σ = standard deviation.

characterised by abundant nodules of chert. Only in the Late Copper Age did potters heavily temper pots with grog. The very clay-rich matrix of Middle Neolithic Px pottery fabric and the differences in grain composition in the Qtz/Kfs fabric might be explained by a very fine and accurate level of levigation. Micaschist and granitic rock fragments observed in Qtz/Kfs fabric do not occur in the sediments which crop out in the proximity of Maddalena di Muccia. The temper might derive from a different (and non-local) geological context and may have been collected as a fluvial sand from the bed of a river course.

Mineralogical data show that all samples were fired at a temperature not exceeding 600-800 °C. Only Middle Neolithic fine painted pottery was fired at a higher temperature, in a range of 850-1,050 °C.

Thus, raw material variability is evident on both synchronic and diachronic levels (Table 4). The next step of the analytical project will be the systematic correlation of the different fabrics with the pottery types produced, their shapes and possible functions. On a synchronic level such variability might indicate appreciable group choices as a response to local, functional or/and social constraints. On a diachronic level, it may suggest some substantial shift in the production sequence, with significant modification in paste preparation and firing techniques.

ACKNOWLEDGEMENTS

We would like to thank Dr. Mara Silvestrini, from the Soprintendenza ai Beni Archeologici delle Marche, for financial support, and Prof. Alessandra Manfredini, from the University of Rome "La Sapienza", for providing archaeological materials.

	Ba	Rb	Sr	Y	Zr	Nb	V	Cr	Ni	La	Ce
					Wall daub fabric (n=3)						
m	474	40	437	24	58	3	26	64	78	34	41
σ	155	14	39	5	5	0	3	11	11	8	9
					Cal pottery fabric (n=6)						
m	760	73	475	19	101	7	70	114	116	31	52
σ	122	12	98	4	17	2	10	8	15	5	6
					Qtz/Cal pottery fabric (n=4)						
m	934	108	310	29	173	13	97	142	111	45	88
σ	195	20	196	8	30	2	13	22	30	18	36
					Px pottery fabric (n=3)						
m	958	69	182	23	135	13	85	199	102	45	77
σ	69	17	98	6	26	3	29	48	10	4	6
					Qtz/Kfs pottery fabric (n=15)						
m	948	122	111	26	178	14	103	161	115	38	73
σ	176	16	15	12	25	2	10	20	17	16	29
					Qtz pottery fabric (n=8)						
m	1213	108	99	50	159	13	98	127	132	65	93
σ	253	26	26	13	40	2	16	19	13	13	26
					Grog pottery fabric (n=13)						
m	1566	125	102	40	220	19	120	155	133	67	117
σ	307	15	49	16	65	3	13	37	20	45	64
					All groups (n=52)						
m	1093	106	187	32	167	13	97	144	119	49	84
σ	385	30	154	15	59	5	26	38	22	28	44

Table 3 – Trace elements (ppm) of pottery and wall daub samples by XRF analysis.
m = mean; σ = standard deviation.

Pottery fabrics	**Neolithic**		**Copper Age**
	Early	**Middle**	
Qtz/Kfs (finer)	4		
Qtz	6	2	
Qtz/Kfs (coarser)	5	6	
Qtz/Cal	2	2	
Cal	1	5	
Px		3	
Grog			13
	18	**18**	**13**

Table 4 – Chronological variability of pottery fabrics; the significant correlations between pottery fabrics and chronological phases are highlighted.

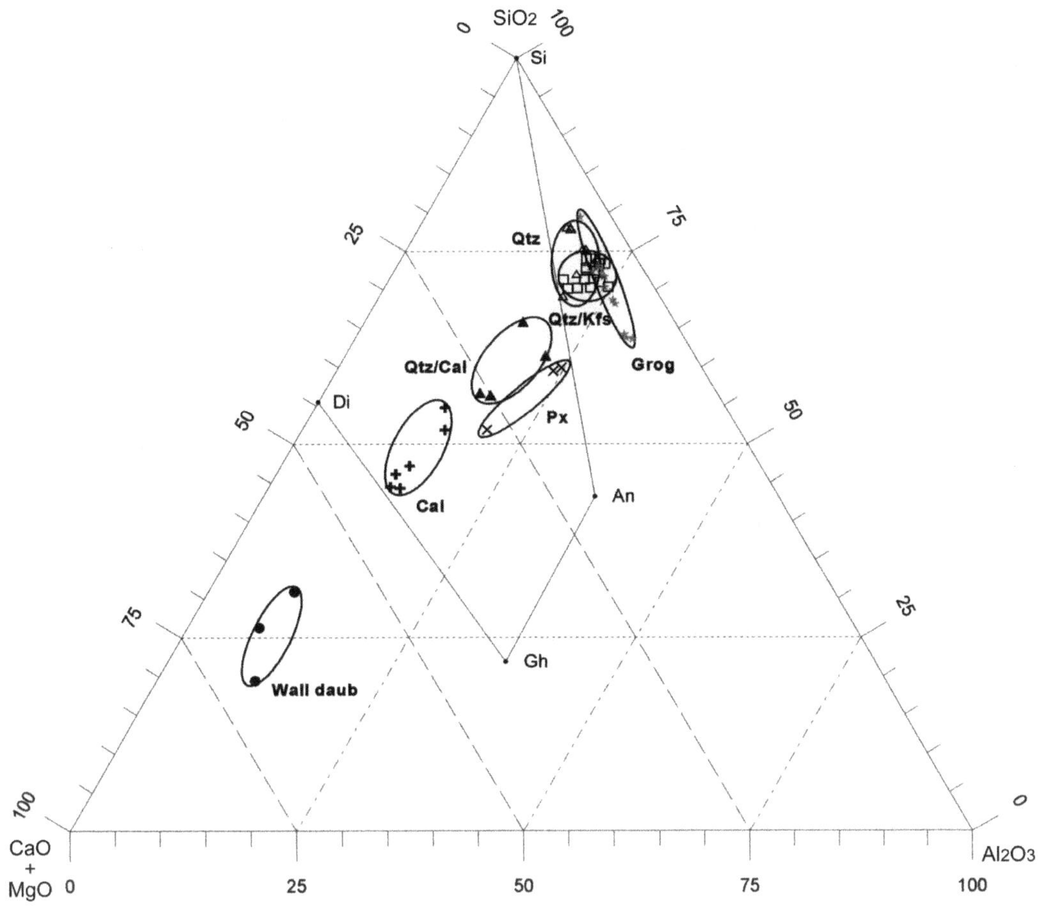

Fig. 3 – (CaO+MgO)/Al$_2$O$_3$/SiO$_2$ wt% diagram (Di: diopside; Gh: gehlenite; An: anorthite). Pottery fabrics: wall daub (dots), Cal (+), Qtz/Cal (solid triangles), Px (×), Qtz/Kfs (squares), Qtz (empty triangles), Grog (grey asterisks).

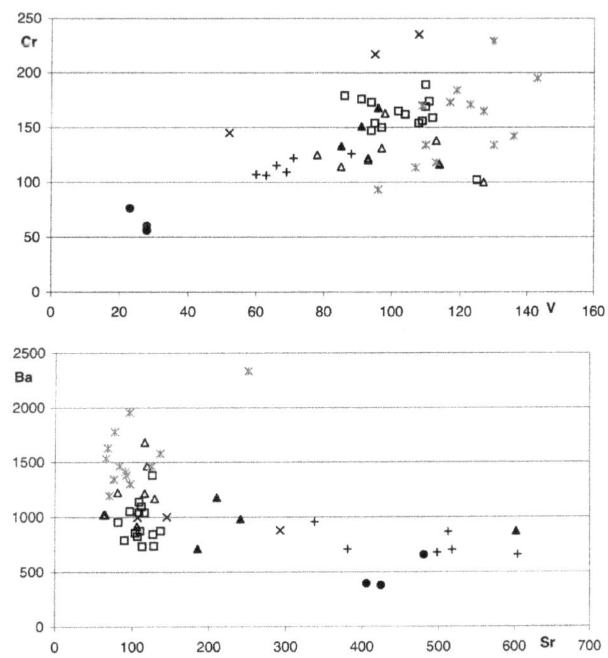

Fig. 4 – Cr vs. V and Ba vs. Sr plots (ppm). Pottery fabrics: wall daub (dots), Cal (+), Qtz/Cal (solid triangles), Px (×), Qtz/Kfs (squares), Qtz (empty triangles), Grog (grey asterisks).

REFERENCES

CENTAMORE E., DEIANA G. (ed.) 1986, *La Geologia delle Marche*, Studi Geologici Camerti (num. spec.), Camerino.

FRANZINI M., LEONI L., SAITTA M. 1972, "A simple method to evaluate the matrix effects in X-ray fluorescence analysis", *X-ray spectrometry* 1, p. 151-154.

FRANZINI M., LEONI L., SAITTA M. 1975, "Revisione di una metodologia analitica per fluorescenza X, basata sulla correzione completa degli effetti di matrice", *Rendiconti della Società italiana di mineralogia e petrologia* 31, p. 356-378.

KRETZ R. 1983, "Symbols for rock-forming minerals", *American mineralogist* 68, p. 277-279.

LEONI L., SAITTA M. 1976, "Determination of Yttrium and Niobium on standard silicate rocks by X-ray fluorescence analysis", *X-ray spectrometry* 5, p. 29-30.

HEIMANN R.B., MAGGETTI M. 1981, "Experiments on simulated burial of calcareous terra sigillata (mineralogical changes). Preliminary results", *in* M.J. Hughes (ed.), *Scientific studies in ancient ceramics*, British Museum occasional paper 19, London, p. 163-177.

MAGGETTI M. 1981, "Composition of Roman pottery from Lousanna (Switzerland)", *in* M.J. Hughes (ed.), *Scientific studies in ancient ceramics*, British Museum occasional paper 19, London, p. 33-49.

MAGGETTI M. 1982, "Phase analysis and its significance for technology and origin", *in* J.S. Olin and A.D. Franklin (eds.), *Archaeological ceramics*, Smithsonian Institution Press, Washington, p. 121-133.

MANFREDINI A., CARBONI G., CONATI BARBARO C. 2005a, "La struttura 12 di Maddalena di Muccia: un'ipotesi interpretativa", in *Preistoria e protostoria delle Marche, Atti della XXXVIII Riunione Scientifica dell'I.I.P.P.*, Firenze, p. 335-346.

MANFREDINI A., CARBONI G., CONATI BARBARO C., SILVESTRINI M., FIORENTINO G., CORRIDI C. 2005b, "La frequentazione eneolitica di Maddalena di Muccia (Macerata)", in *Preistoria e protostoria delle Marche, Atti della XXXVIII Riunione Scientifica dell'I.I.P.P.*, Firenze, p. 433-444.

MUNTONI I.M. 2005, "La produzione ceramica dalle strutture d'abitato neolitiche ed eneolitiche di Maddalena di Muccia (Macerata): primi dati sulla provenienza delle materie prime e le tecnologie di manifattura", in *Preistoria e protostoria delle Marche, Atti della XXXVIII Riunione Scientifica dell'I.I.P.P., Firenze*, p. 347-361.

SILVESTRINI M., MANFREDINI A., RADI G. 2005, "L'abitato di Maddalena di Muccia (Macerata). Problemi e prospettive di ricerca", in *Preistoria e protostoria delle Marche, Atti della XXXVIII Riunione Scientifica dell'I.I.P.P.*, Firenze, p. 221-229.

BLACK-ON-RED PAINTED POTTERY PRODUCTION AND DISTRIBUTION IN LATE NEOLITHIC MACEDONIA

Z. TSIRTSONI[1], D. MALAMIDOU[2], V. KILIKOGLOU[3], I. KARATASIOS[3], L. LESPEZ[4]

[1] UMR 5138 Archéométrie et Archéologie, CNRS-Universités Lyon 1 et Lyon 2, 7 rue Raulin, 69007, Lyon, France
[2] 18th Ephorate of Prehistorical and Classical Antiquities, 17 Erithrou Stavrou str., 65110, Kavala, Greece
[3] Institute of Materials Science, NCSR Demokritos, 15310, Agia Paraskevi, Attiki, Greece
[4] Geophen, CNRS-Université de Caen, Esplanade de la Paix, BP 5186, 14032, Caen Cedex, France

Black-on-red painted pottery is one of the most distinctive ceramic categories of the Neolithic Aegean. Particularly attractive, thanks to its vivid dark-on-light contrast frequently combined with an elaborate motif design (Fig. 1), this pottery is also remarkable for its high technical quality, witnessed by the fineness of the fabric, the uniformity of the surface colours, the regularity of shapes, the density and hardness of the vessels' walls. Pottery of this kind is found at settlements in Northern Greece (Thessaly, Macedonia and Thrace), dated to an advanced stage of the Late Neolithic period, roughly between 4800/4700-3900/3800 BC. Although it was early recognised as a diagnostic ware for that period in the areas of its occurrence (Welch 1918-19, p. 44-46; Heurtley 1939, p. 74-75; French 1961, p. 114-116; Deshayes and Garašanin 1964), no previous attempt was made to construct a systematic classification or to study its production technology and distribution patterns.

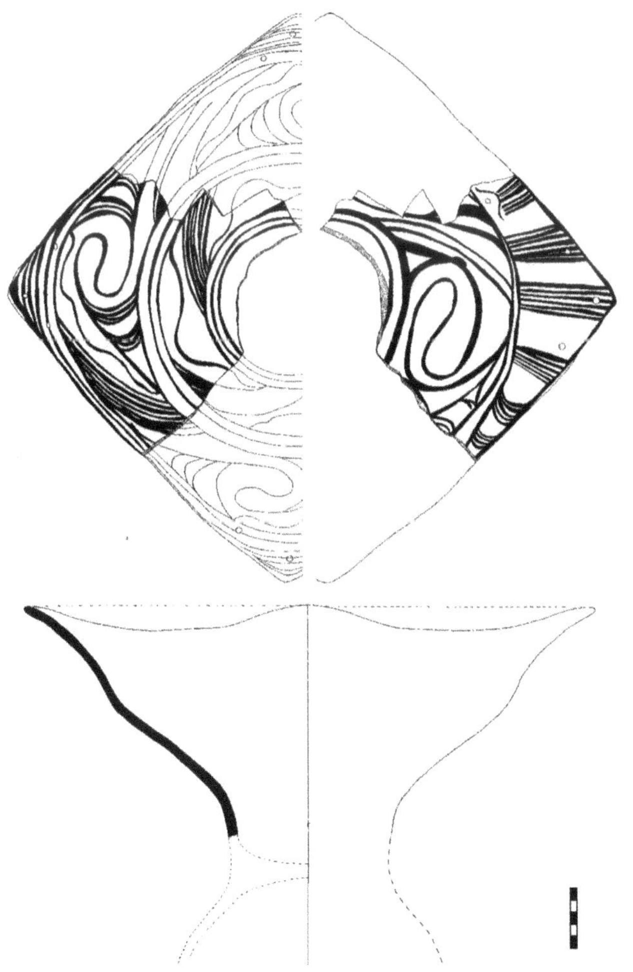

Fig. 1 – Black-on-red painted open vessel from Kryoneri (drawing D. Malamidou).

In this paper, we summarise the main results of a series of studies carried out between 2000 and 2004, within the framework of a multidisciplinary research program supported by the French School at Athens (ÉFA) and the French National Centre for Scientific Research (CNRS).[1] The main body of the analytical results, concerning the production areas and distribution of the pottery, has been presented in detail elsewhere (Malamidou *et al.* 2006; for the NAA analyses, see also Kilikoglou *et al.* 2007). Here emphasis is given to the scanning electron microscopy examination for the technological characterisation of this pottery class.

ARCHAEOLOGICAL AND CHEMICAL GROUPS: REGIONAL DISTRIBUTION AND POTTERY TRADITIONS

Preliminary macroscopical analysis of archaeological ceramics from more than one hundred Neolithic settlements, according to technological (fabric, surface treatment, paint consistency) and morphological criteria (vessel shapes, decoration), resulted in their classification into four main varieties which also correspond to different

1 – We express our thanks to the two successive directors of the ÉFA, Roland Étienne and Dominique Mulliez, for financing this program within the frame of the ÉFA-CNRS conventions. Besides the authors, P. Yiouni and A. Tsolakidou have contributed to some parts of the analytical program.

geographical distributions (Malamidou 2005). The distribution areas of varieties 1, 3 and 4 coincide roughly with the physical subdivisions of Greek Macedonia into eastern, central and western regions, whereas variety 2 is limited to the middle and upper Struma valley, across the Greek-Bulgarian border.

The overall presence of black-on-red decorated pottery in Late Neolithic Northern Greece reveals the existence of some common technical standards beyond common aesthetic preferences, such as the use of iron-rich clays for the vessels, firing to high temperatures under oxidising conditions, and the use of manganese-based paints. Most of these features have a long history, at least in Eastern Macedonia, where sophisticated dark-on-light painted pottery was produced as early as 5200/5000 BC. (Tsirtsoni 2000, sp. p. 13-18; Yiouni 2000; Yiouni 2001, sp. p. 5-11). On the other hand, the existence of at least four regional varieties suggests an equal number of local pottery traditions, that is, different "recipes" resulting in similar, but not identical, products. These "recipes" differ from one another in several aspects (clay choice and preparation, surface treatments, paints, etc.), reflecting variations in the production processes from one centre to another. Certainly, this variability is related to environmental factors, namely local resources, but the role of tradition should not be underestimated.

Further research focused on the Eastern Macedonian variety 1, which is by far the most abundant and best documented. It is also the most widely distributed, since it is also found (in small quantities) in many other parts of Northern Greece, including Aegean Thrace to the east, the Axios valley to the west and Thessaly to the south, as well as in SW Bulgaria to the north.

Neutron activation analysis (NAA) of some 200 ceramic samples from 14 settlements allowed us to distinguish four main chemical groups (A-D) for the black-on-red ceramics. A fifth group (E) corresponds to samples of the so-called brown-on-cream painted pottery, which is contemporaneous with and technically similar to the black-on-red, but features a light grey-brown decoration on a yellowish background. The distribution pattern of the four black-on-red ceramic groups (Fig. 2) suggests that pottery production took place in at least four separate production centres in Eastern Macedonia. Each production centre could include more than one workshop, exploiting the same, or mineralogically very similar, clay sources.

Extensive geological survey revealed a substantial number (more than 50) possible clay sources, all in strata dated to the Plio-Pleistocene period. However, none of the analysed samples matched the composition of the ceramics found in the same area. Although a straightforward chemical match between clays and ceramics is very rare, the great differences observed lead to the conclusion that the Neolithic clay sources were either exhausted or covered by younger formations. Future survey should take into consideration not only the general geological features, on a regional scale, but the particular characteristics of each valley.

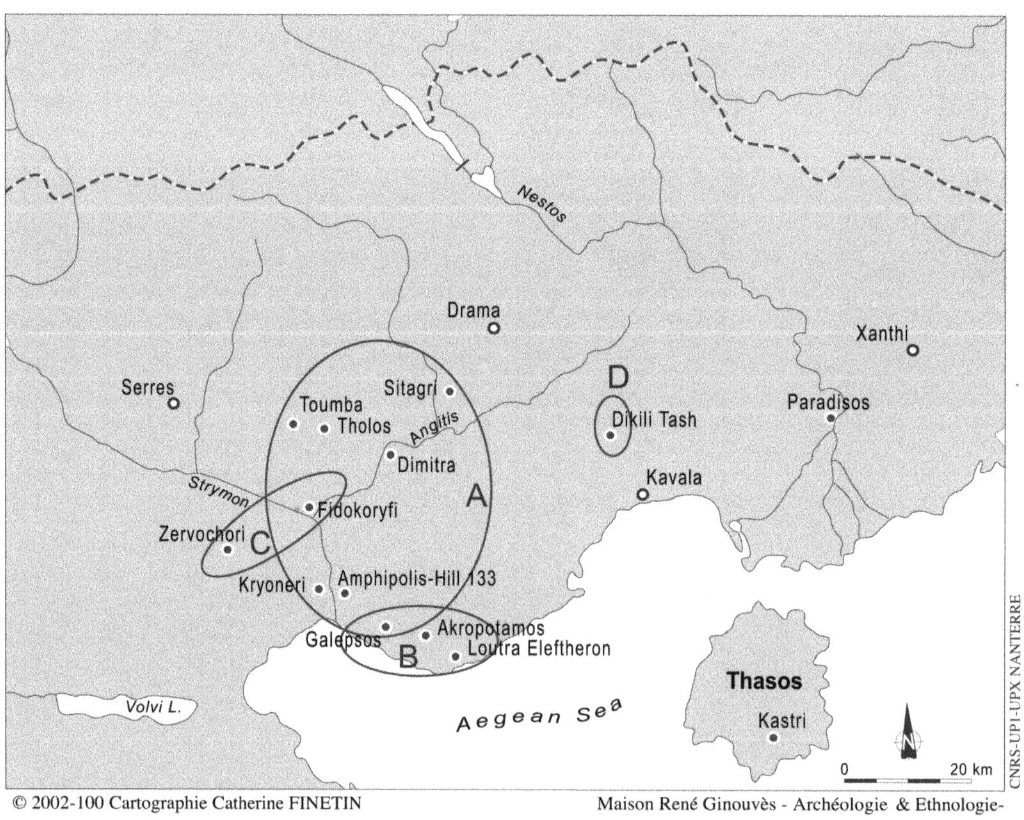

© 2002-100 Cartographie Catherine FINETIN — Maison René Ginouvès - Archéologie & Ethnologie-

Fig. 2 – Geographical distribution of the chemical groups A-D in Eastern Macedonia, suggesting black-on-red pottery production in at least four separate centres.

MICROMORPHOLOGICAL ANALYSIS AND TECHNOLOGICAL EVALUATION

A representative number of 14 samples was examined under the scanning electron microscope (SEM). The examination of the micromorphology of the ceramic body and surface along with their chemical compositions provided information about the selection and manipulation of the raw materials, as well as the pyrotechnology (Maniatis and Tite 1981). This information provided the necessary evidence to reconstruct the production process of the ceramic ware under study. The main question generated by the results of the previous chemical analysis was whether the selection of raw materials and firing parameters were the same in all the production centres. The answer should provide evidence about the level of standardisation in black-on-red pottery production during the Late Neolithic period in Eastern Macedonia.

As far as the clay is concerned, analysis shows a definite preference for non-calcareous clays of a fine nature containing moderate amounts of quartz. Twelve of the fourteen samples studied were found to have been manufactured with clays containing low amounts of CaO (2-6%) (Table 1). Actually, the two samples which are clearly calcareous (CaO up to 12%) do not belong to black-on-red, but to brown-on-cream ceramics (*supra*). The low calcareous (that is, the true black-on-red) group can be further subdivided into two: a virtually non-calcareous sub-group (CaO around 2%), fitting almost perfectly with the chemical group A, which is mainly found in settlements along the Angitis and in the lower part of the Strymon valley, and a second group with CaO between 5-6%, which agrees quite well with the group B, typical of coastal sites.

Sample	Chemical group	CaO content(%)	Vitrification stage	Paint
KRY 01/06	A	1.6	V	med+Mn
KRY 01/16	A	2.6	V	med+Mn
DIM 01/142	A	1.9	V	med+Mn
LUE 01/91	B	5.4	V	fine+Mn
GAL 01/80	B	5.6	V	fine+Mn
AKR 01/72	B	5.6	V	fine+Mn
DT 01/160	D	3.6	V	fine+Mn
KRY 01/64	E	11.5	V	
KRY 01/63	E	11.9	V	
FID 01/101	C	2.0	IV	med+Mn
KRY 01/08	A	2.0	IV	med+Mn
DT 01/161	D	5.9	V+	
KRY 01/09	A	2.7	TV	
KRY 01/02	A	3.9	V+	

Table 1 – Main results of the SEM examination of black-on-red and brown-on-cream ceramic samples; the second column shows correspondence with the chemical groups identified by NAA analysis.

The majority of the samples examined showed vitrification of the clay minerals in their microstructure, in most cases well developed, as evidenced by the extensive networks of glass filaments present. This applies to ceramics coming from both A and B chemical groups, in other words, to samples produced in different production areas. In most cases the glass network is composed of fine filaments (Fig. 3), as a result of the small grain size of the original clay particles, while the firing temperature suggested, for this type of microstructure, is between 900 and 1000 °C. Extensively vitrified microstructure was also developed in the ceramic bodies of the calcareous group E, but the microstructure here is cellular due to the decomposition of calcium carbonate during firing (Fig. 4).

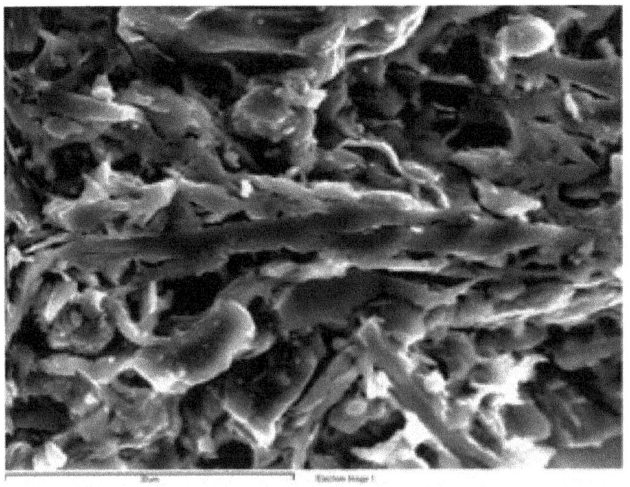

Fig. 3 – SEM microphotograph of a black-on-red pottery sample (DIM 01/142) with a well-developed microstructure, containing an extensive network of fine glass filaments; the estimated firing temperature lies between 900 and 1000 °C. Photo width = 60 μm.

Fig. 4 – SEM microphotograph of a brown-on-cream pottery sample (KRY 01/63): the extensively vitrified cellular microstructure is due to the decomposition of calcium carbonate during firing at high temperatures (900-1000 °C). Photo width = 60 μm.

This microstructure is consistent with firing temperatures of more-or-less the same range (900-1000 °C). Slightly lower firing temperatures are assumed for two samples whose microstructures did not exhibit such an extensive state of vitrification as the others.

In all cases the black decoration was obtained by the addition of manganese oxide to a clay suspension and its application prior to firing. This resulted in a paint layer less than 10 μm thick, which does not appear vitrified in the SEM examination, although, as mentioned above, the firing temperatures were relatively high (Fig. 5a, b). A closer look at the semi-quantitative analytical data obtained by SEM (especially the Al to Si ratio) reveals two patterns: one with the clay suspension of the paint layer (Fig. 5c) being very fine compared to the clay of the body (Fig. 5e), and one with the clay chemistry being the same in both the paint layer (Fig. 5d) and the body (Fig. 5f). In the first case, a fine suspension of the clay used for the body was obviously prepared, then mixed with Mn-oxide

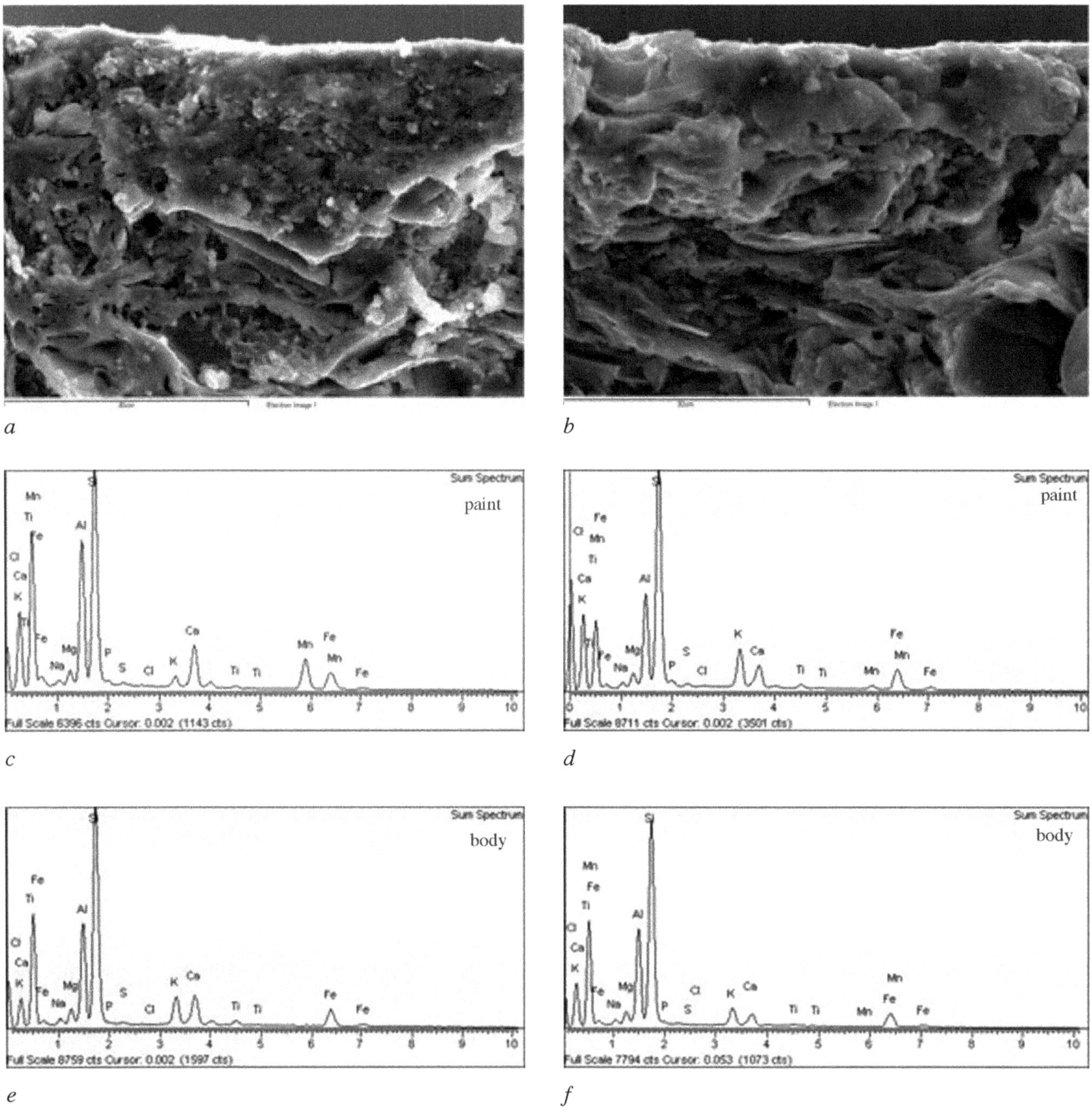

Fig. 5 – SEM microphotographs and comparative spectra of two black-on-red pottery samples (LUE 01/91, KRY 01/16). a-b: paint layers on top of the vitrified body. Photo width = 60 μm; c-f: comparison of the chemical compositions of the paint and body (especially the Al/Si ratio) suggests differences in the preparation of the paint mixtures for the two samples.

to produce the paint, whereas in the second, clay was mixed with water and Mn-oxide, then applied to the body, without prior refinement. It is important to note that the fine suspension pattern is only found in samples coming from the coastal group B, while the untreated one is typical for group A. These differences also had a visual effect, with the black decoration of the "fine" paints being brighter and more durable than the others.

One of the questions that SEM analysis tried to resolve was that of "black" paints that appear grey or white, a feature observed in many specimens of the variety 1 (Fig. 6). Actually, in these cases, the paint does not appear white everywhere, but only in patches—most commonly on the outer surface of the vessel. White motifs are usually surrounded by a black outline, forming a sort of contour. Close examination of samples with such decoration revealed that they had been originally exposed to very high temperatures, well over 1000 °C, as attested by their microstructures (Fig. 7). Another characteristic of these paint layers is that they are low in alkalis and Mg, but rich in Ca, although in the surrounding black lines the situation is exactly the opposite. One possible explanation is that the whitish colour is a result of the effect of the burial environment on the high temperature phases developed in the paint layers: the latter being unstable, they could undergo leaching of K, Na and enrichment of Ca. This is very similar to what happens in glass where a whitish corrosion layer is also formed during burial (Douglas and El-Shamy 1967; Grambow 1992). Whether these high temperatures were reached during the original firing or during the destruction of the settlement, is not known for certain. It seems however more probable that it happened during the original firing, as suggested by the extent of the white zones on whole vessels.

To summarise, it is obvious that there is essentially one technological tradition all over Eastern Macedonia. However, two variations exist in paint preparation : the first, observed mainly in settlements close to the coast, uses a very fine suspension taken from the same clay as the body, whereas the second one uses this same clay, but without prior refinement. Everywhere there is very good control of firing conditions and high temperatures are applied (usually between 900-1000 °C), in order to obtain hard-fired fabrics with bright, uniform colours. Accidents do happen, however, since we find vessels that have been exposed locally to temperatures above 1000 °C. This should certainly not lead to a depreciation of the Neolithic potters' skill, but rather to a re-evaluation of the constraints with which they had to deal, and to extra appreciation of the overall quality of their products.

Black-on-red vases feature, indeed, a remarkably standardised quality in all stages of the *chaîne opératoire*. Cases like this invite us to reconsider our way of thinking about Neolithic pottery production and Neolithic society in general, which may have been less "unspecialised" and less "self-sufficient" than usually thought.

Fig. 6 – Black-on-red vessel from Dikili Tash (h. ca 11 cm): the black paint appears white in some parts of the vessel's exterior (photo: Ph. Collet/ÉFA).

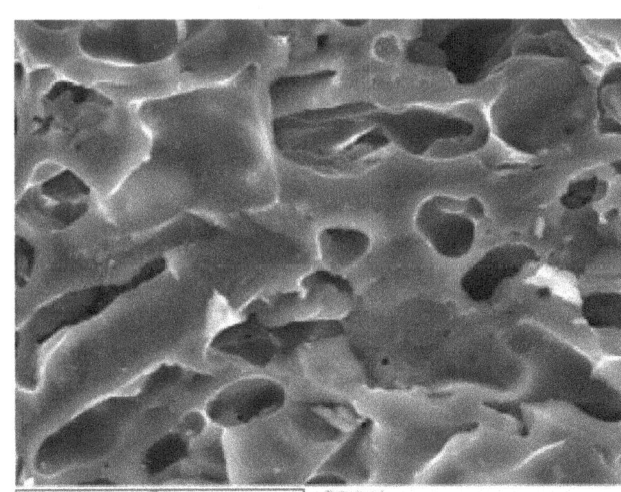

Fig. 7 – SEM microphotograph of a pottery sample with "white-painted" decoration (KRY 01/09): its micro-structure suggests firing well above 1000 °C. Photo width = 60 μm.

REFERENCES

Deshayes J., Garašanin M. 1964, "Note sur la céramique de Galepsos", *BCH* 88, p. 51-66.

Douglas R.W., El-Shamy T.M.M. 1967, "Reactions of glasses with aqueous solutions", *Journal of the American ceramic society* 50, p. 1-8.

French D.H. 1961, "Late Chalcolithic Pottery in Northwest Turkey and the Aegean", *AnatSt* 11, p. 99-142.

Grambow B. 1992, "Geochemical approach to glass dissolution", *in* D.E. Clark and B. Zoitos (ed.), *Corrosion of glass ceramics and ceramic semiconductors. Principles, testing, characterisation and applications*, New Jersey, p. 124-148.

Heurtley W.A. 1939, *Prehistoric Macedonia,* Cambridge.

Kilikoglou V., Grimanis A.P., Tsolakidou A., Hein A., Malamidou D., Tsirtsoni Z. 2007, "Neutron activation patterning of archaeological materials at the National Centre for Scientific Research 'Demokritos': the case of black-on-red Neolithic pottery from Macedonia, Greece", *Archaeometry* 49 (2), p. 301-317.

Malamidou D. 2005, *La céramique à décor peint "noir sur rouge" du Néolithique Récent II en Grèce du Nord : production, distribution et utilisation*, Ph. D., Université Paris I Panthéon-Sorbonne.

Malamidou D., Tsirtsoni Z., Yiouni P., Lespez L., Kilikoglou V., Tsolakidou A. 2006, "Les poteries néolithiques à décor peint "noir sur rouge" en Grèce du Nord: matières premières et production", *BCH* 130 (in press).

Maniatis Y., Tite M.S. 1981, "Technological examination of Neolithic-Bronze Age pottery from central and southeast Europe and from the Near East", *Journal of archaelogical science* 8, p. 59-76.

Tsirtsoni Z. 2000, "Les poteries du début du Néolithique Récent en Macédoine. 1, Les types de récipients", *BCH* 124, p. 1-55.

Welch F.B. 1918-19, "Macedonia: Prehistoric Pottery", *BSA* 23, p. 44-50.

Yiouni P. 2000, "Painted pottery from East Macedonia, in North Greece: technological analysis of decorative techniques", *Documenta Praehistorica* XXVII, p. 199-214.

Yiouni P. 2001, "Surface treatment of Neolithic vessels from Macedonia and Thrace", *BSA* 96, p. 1-25.

BELL BEAKERS BONE BASED DECORATIONS FROM GUADIANA RIVER MIDDLE BASIN (BADAJOZ, SPAIN)

CARLOS ODRIOZOLA[1], ANGEL JUSTO ERBEZ[1], VÍCTOR HURTADO PÉREZ[2]

[1] Instituto de Ciencia de Materiales de Sevilla (ICMSE-CSIC)

[2] Dpto. de Prehistoria y Arqueología, Universidad de Sevilla

ABSTRACT

The aim of this work is to give prominence to the analytical procedure developed for the incised Bell Beaker "incrusted pottery" coming from Guadiana River Middle Basin (Badajoz, Spain), and to determine the nature of those white inlays. The analytical procedure developed in this paper will reveal its nature to be burnt bone and in fewer cases to be calcium carbonate, reversing the usual trend in the Iberian Peninsula where calcium carbonate has been regarded as the only kind of inlay. The analytical facilities used are: XRF microprobe, XRD and FTIR. It has been difficult to establish the bone nature of the inlay, as it was necessary to develop a procedure using several analytical techniques and reference materials to establish its true nature.

KEYWORDS: BELL BEAKER POTTERY, INCRUSTED POTTERY, BONE, XRF MICROPROBE, XRD, FTIR

1. INTRODUCTION

The aim of this work is to characterize and determine the nature of the very typical white inlaid pastes that are present on Bell Beaker pottery in a large number of Late Prehistoric decorated pottery from all over Europe. We have studied particularly pottery coming from the Spanish Guadiana River Middle Basin (Badajoz, Spain), focusing on two of the largest 3rd millennium BC settlements throughout all Iberia: La Pijotilla (80 ha) (Hurtado 1984) and San Blas (30 ha) (Hurtado 2004) (see Fig. 1). These two large settlements, together with some other smaller fortified settlements, were politically organized around La Pijotilla, and form the Tierra de Barros area into a Copper Age organized territory. San Blas is one of the most outstanding settlements on the periphery.

Generally many of the Bell Beaker pots are decorated with a white inlaid paste, which has been characterized as calcium carbonate in France (Salanova 2000), NE Spain

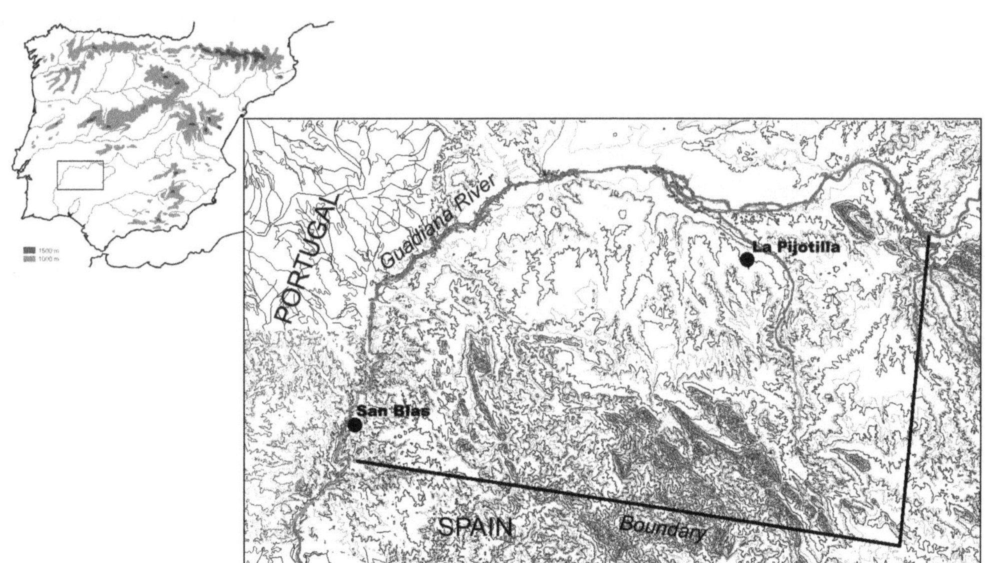

Fig. 1 – The Middle Guadiana River basin landscape within Iberia.

(Martín and Delibes 1989) and the Spanish Meseta (Blasco ed. 1994), whereas the analysis developed in the Iberian Peninsula, on inlaid Bell Beaker pottery coming from Pajares de Adaja and Fuente Olmedo (Valladolid, Spain) (Martín and Delibes 1989), and the Ciempozuelos necropolis (Madrid, Spain) (Blasco ed. 1994) has given prominence to calcium carbonate inlays.

Other kind of inlay have been characterized in other chronological contexts, *e.g.* French Bronze-Iron Ages, or Hungarian Neolithic-Bronze Ages where the use of bone as an inlay have been pointed out (Constantin 2003; Gherdán *et al.* 2005; Sziki *et al.* 2003), although it has not been certainly proved.

Determining the nature of the continental style Bell Beaker inlays from Guadiana River Middle Basin is a complex issue due to the multiple interactions and transformations suffered from production to deposition by the original material in its behavioural chain (Schiffer 1976). These interactions result in structural transformations in the original material, making extremely difficult to assess the true nature of the primary material.

The difficulty is due to several possible origins and interactions of the inlaid material, calcium carbonate, biogenic apatite—bone—, burned biogenic apatite—burnt bone (Stiner and Kuhn 1995; Surovell and Stiner 2001), or non-biogenic apatite—a rock-forming mineral. This uncertainty about its origin makes it essential to use several analytical techniques to assess the raw material used in the manufacture of these white inlaid Bell Beaker pots.

2. METHODOLOGY

This work was carried out on 16 sherds of continental style Beaker pottery (incised Bell Beakers) coming from La Pijotilla and San Blas archaeological sites[1] (see Fig. 2).

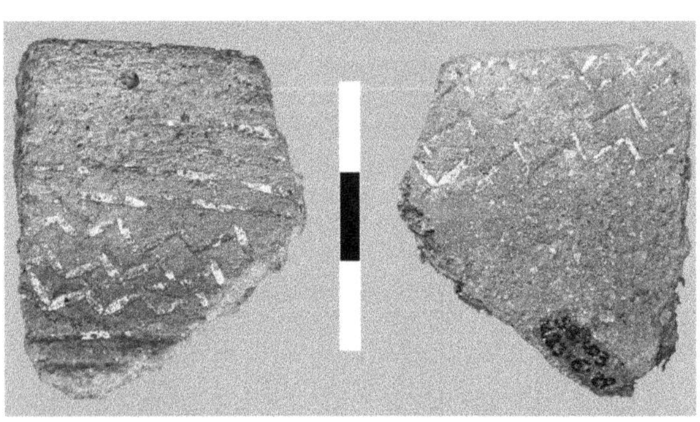

Fig. 2 – White Bell Beaker "incrusted pottery" from San Blas.

Previous analyses of some decorated sherds have rejected apatite for the inlaid paste. It was necessary to sample archaeologically contextualised bones, as well as non-biogenic apatites, in order to have reference materials to differentiate between biogenic and non-biogenic apatites. The 16 samples and the reference materials were analysed as follows to determine the nature of the inlaid paste.

2.1. XRF microprobe

Compositional data of the inlaid material was obtained with an EDAX Eagle III XRF microprobe. After focusing directly on the sample with an optical microscope, an area of 100 μ squared was analyzed using the Kα radiation produced by a Rh tube operated at 40 kV. The elemental composition of the samples is expressed as weight percent of the corresponding oxides. The reported data are the averages of five measurements in order to avoid inhomogeneities.

Compositional data revealed a CaO/P_2O_5 concentration ratio range between 1.79 and 2.88 for 14 of the 16 studied samples, which corresponds to calcium phosphate apatite compositional ratios. The other 2 samples correspond to calcium carbonate, based on their CaO concentrations above 81% in weight (Table 1). This verifies the presence of two categories of inlaid paste, one calcium phosphate, the other calcium carbonate.

Indeed these compositional CaO/P_2O_5 ratios correspond to calcium phosphate apatites, they may also correspond to three species of apatite, on the basis of their compositional CaO/P_2O_5 ratios: biogenic apatite [*ca.* 2.51], non-biogenic apatite [1.53-1.98] and burnt biogenic apatite [1.83-1.95] (Shiegl *et al.* 2003). The average value of the CaO/P_2O_5 ratio of the samples is 1.82 (σ=0.51), in agreement with Shiegl *et al.* (2003), rejecting the biogenic apatite this leaves burnt biogenic apatite and non-biogenic apatite as the only two possible species of apatites.

In Fig. 3 a compositional map of the sample P-SUP-31 is shown, in which it is possible to observe how the calcium phosphate is restricted to the space left by the decoration on the pottery, and how it fills the gap.

2.2. XRD

Powder diagrams were obtained with a Siemens D5000 diffractometer, using a copper anode (γ= 1.5406Å). The diffractogram was registered in the scanning range °2θ [25, 55], with a step size of 0.02° and a counting time of 10 s per step.

1 – Samples labelled H22, H23, IIIF022, J27, R7 and U7 derive from San Blas, the terms separated by hyphens correspond to the stratigraphic unit and the inventory number. Samples labelled P-SUP are from the La Pijotilla survey, and the term separated by hyphens is the inventory number.

Label	CaO/P_2O_5	P_2O_5	CaO	Label	CaO/P_2O_5	P_2O_5	CaO
Apatite [Sr]	2.02	32.41	65.59	**J27-45-932**	1.79	19.99	35.36
Apatite [Cl]	2.09	31.92	66.68	**IIIF022-2-64**	1.93	32.21	62.22
Fossil bone (1)	2.62	27.62	72.41	**P-SUP-22**	2.06	19.16	39.43
Fossil bone (2)	2.35	28.72	67.46	**P-SUP-23**	2.88	20.46	36.93
Fossil bone (3)	2.64	27.45	72.55	**P-SUP-24**	2.09	25.42	53.28
H22-11-457	-----	0.00	81.23	**P-SUP-26**	2.66	15.29	54.95
H22-8-607	1.95	30.92	60.21	**P-SUP-31**	2.04	23.19	47.76
H23-10-154	1.89	26.85	50.79	**P-SUP-32**	----	0.00	87.11
H23-6-28	2.12	31.73	67.44	**P-SUP-34**	2.49	19.57	38.43
J27-28-717	1.79	19.99	35.36	**P-SUP-35**	1.95	22.61	44.70
J27-45-931	1.93	25.10	48.47				

Table 1 – XRF microprobe compositional data for CaO, P_2O_5 and CaO/P_2O_5 ratios for the 16 analyzed samples and the analyzed reference materials.

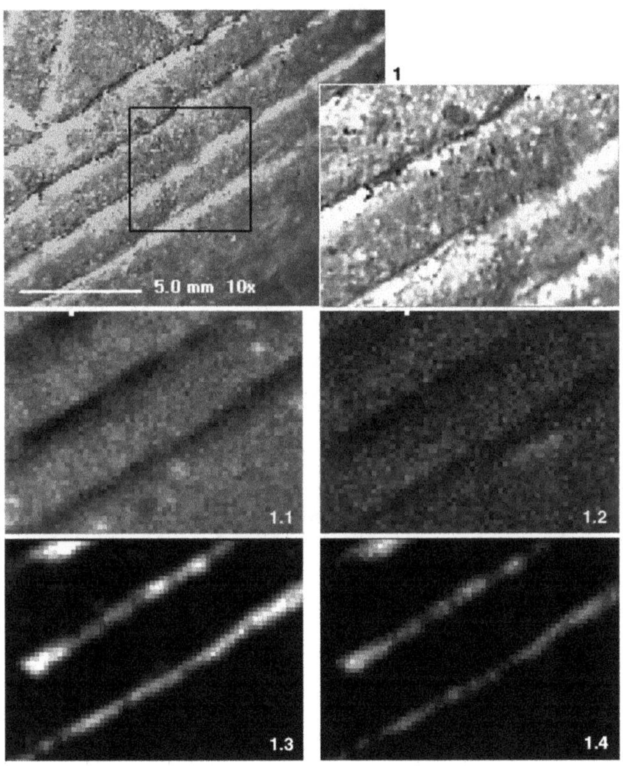

Fig. 3 – XRF microprobe compositional map of sample P-SUP-31. (1.1) Si, (1.2) Al, (1.3) Ca and (1.4) P.

XRD analysis was carried out on powdered samples of the incrustations and determined hydroxyapatite as the crystalline phase, which is the mineral phase of both, biogenic[2] and geological apatites. The XRD diagrams show sharp peaks for the samples as well as for the geological reference material, while it is possible to observe how the biogenic reference materials have broader and less defined peaks if compared, pointing a possible geological origin.

An experimental calcination was performed on a reference bone material in order to investigate the recrystallization process described for bone under high temperatures (Rogers and Daniels 2002). It showed peaks as sharp as the ones for the geological reference material, making either origin possible, geological or calcined bone.

Fig. 4 shows the powder diagrams recorded for sample P-SUP-22, and biogenic and non-biogenic reference materials, revealing clear differences between the major planes of hydroxyapatite [211], [112] and [300].

2.3. FTIR

The infrared spectra were collected using a Nicolet 510P FTIR spectrometer equipped with a DTGS detector. All spectra were recorded after 32 scans with 4 cm^{-1} resolutions.

FTIR constitutes the last step in the analytical procedure, throwing out differential FTIR spectra for geological, fossil bone and experimentally calcined bone, on the basis of the experimentally recorded 1458, 1418, 1388 & 870 cm^{-1} absorption bands of the biogenic material, that are completely absent in the non-biogenic apatites (Farmer 1974; Fowler 1974; Dauphin 1993). While all the biogenic reference materials show bands at *c*. 603 & 570 cm^{-1} (with a combination of two overlapping bands at 562 & 574 cm^{-1}), the geological apatites show a more intense band at 574 cm^{-1}, with the band at 560 cm^{-1} appearing as a shoulder. Contrary to the situation for biogenic apatites where the main band peaks at 560 cm^{-1}, the non-biogenic apatite shows the band at 574 cm^{-1} as a shoulder. The real difference between samples is the presence of a band at 630 cm^{-1} in the calcined fossil bone. This band is assigned to the translational mode of the hydroxyl group in hydroxyapatite and appears upon air calcination at high temperature (Fig. 5).

2 – Hydroxyapatite is the major inorganic constituent of bone of 40% by weight; the remaining 60% are water and organic constituents.

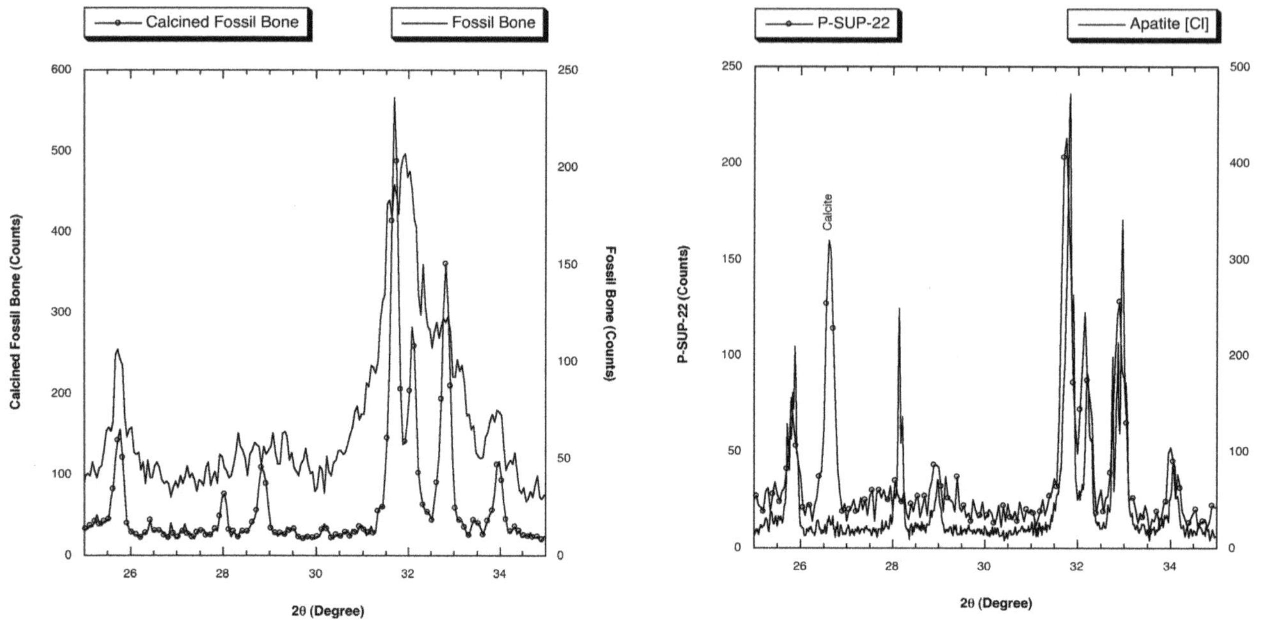

Fig. 4 – XRD powder diagrams of the studied sample P-SUP-22 and the reference materials.

The analytical procedure developed in this work demonstrates that the nature of the white inlaid paste is burnt bone, on the basis of the combined use of three analytical techniques.

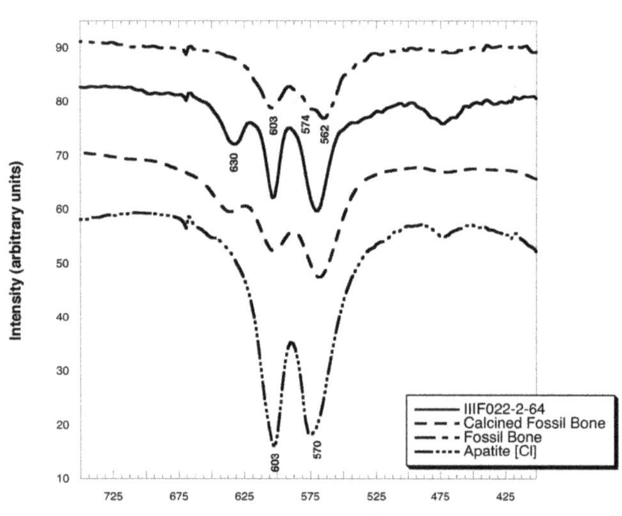

Fig. 5 – FTIR spectra of the sample IIIF022-2-64 and the reference materials.

3. CONCLUSIONS

The difficulty of assessing the nature of the paste was one of the issues highlighted in this research, due to the difficulty of facing up the nature of the paste as burnt bone, due to the interactions and structural transformations suffered during its behavioural chain.

From our point of view a crushed bone paste has been inlaid before firing the pot; and as the exclusive result of the interactions that took place upon the firing, the bone suffered a transformation process to β-tricalcium phosphate that happens around 700-900 °C. This transformation is responsible for the appearance of the FTIR band at 630 cm^{-1} as well as the different intensities that depend on the transformation stage. This stage is also responsible for the sharpening of the peaks in the XRD diagram [hkl 211, 112, 300]. In this sense we are able to see different intensities in the band at 630 cm^{-1} as well as in the XRD peaks [hkl 211, 112, 300] due to the different conditions that prevailed in the manufacturing process of these pots.

The bone was reduced to a powder and was probably mixed either with water or some kind of organic agent such as bone marrow, in order to increase its workability.

This decoration technique has been observed, till today, exclusively in the Guadiana River Middle Basin, and it is seen as a sign of the technological identity of Tierra de Barros Landscape for the Copper Age, contrasting with other neighbouring territorial identities such as the Spanish Meseta ones, defined by calcium carbonate pastes.

REFERENCES

BLASCO C. (ed.) 1994, *El horizonte campaniforme de la región de Madrid en el centenario de Ciempozuelos*, Universidad Autónoma de Madrid, Madrid.

CONSTANTIN C. 2003, "*À propos des décors des céramiques protohistoriques incrustées de pâtes colorées", Bulletin de la Société préhistorique française* 100 (1), p. 135-139.

DAUPHIN Y. 1993, "Potential of the Diffuse Infrared Fourier Transform (DRIFT) method in paleonthological studies of bone", *Applied spectroscopy* 47 (1), p. 52-55.

FARMER V.C. 1974, The infrared spectra of minerals, Mineralogical Society, Monograph 4, Londres.

FOWLER B.O. 1974, "Infrared studies of apatites", *Inorganic chemistry* 13 (1), p. 194-207.

GHERDÁN K., BIRÓ K.T., SZAKMÁNY, G, TÓTH, M., SÓLYMOS K.G. 2005, "Analysis of incrusted pottery from Vörs, Southwest Hungary", *in* M.I. Prudencio, M.I. Dias and J.C. Warenbourgh (eds.), *Understanding people through their pottery. Proceedings of the 7th European meeting on ancient ceramics (EMAC'03)*, Trabalhos de arqueologia 42, p. 103-110.

HURTADO V. 1984, *El yacimiento calcolítico de La Pijotilla. Estudio de relaciones culturales*, PhD., University of Sevilla.

HURTADO V. 2004, "El asentamiento fortificado de San Blas (Cheles, Badajoz). III milenio AC", *Trabajos de prehistoria* 61 (1), p. 141-155.

MARTIN R., DELIBES G. 1989, *La cultura del vaso campaniforme en las campiñas meridionales del Duero: el enterramiento de Fuente-Olmedo*, Junta de Castilla y León, Valladolid.

ROGERS K.D., DANIELS P. 2002, "An X-ray diffraction study of the effects of heat treatment on bone mineral microstructure", *Biomaterials* 23, p. 2577-2585.

SALANOVA L. 2000, *La question du campaniforme en France et dans les Îles Anglo-Normandes. Productions, chronologie et rôles d'un standard céramique*, CHTS, Société préhistorique française, Paris.

SCHIFFER M.B. 1976, *Studies in Archaeology, Behavioral Archaeology*, Academic Press, New York.

SHIEGL S., GOLDBERG P., PFRETZSCHNER H.A., CONARD N.J. 2003, "Palaeolithic burnt bone horizons from the Swabian Jura : Distinguishing between in situ fireplaces and dumping areas", *Geoarchaeology* 18 (5), p. 541-565.

STINER M.C., KUHN S.L. 1995, "Differential burning, recrystallization, and fragmentation of archaeological bone", *Journal of archaeological science* 22, p. 223-237.

SUROVELL T.A., STINER M.C. 2001, "Standardizing infra-red measures of bone mineral crystallinity: an experimental approach", *Journal of archaeological science* 28, p. 633-642.

SZIKI G.A., BIRÓ K.T., UZONYI I., DOBOS E., KISS A.Z. 2003, "Investigation of incrusted pottery found in the territory of Hungary by micro-PIXE method", *Nuclear instruments and methods in physics research B* 210, p. 478-482.

ARCHAEOMETRICAL INVESTIGATIONS OF *IMPASTO* POTTERY FROM *TERRAMARA* OF GORZANO (MODENA, ITALY)

A. CARDARELLI, G. CARPENITO, S.T. LEVI, S. LUGLI, S. MARCHETTI DORI, G. VEZZALINI

Dipartimento di Scienze della Terra, Università degli Studi di Modena e Reggio Emilia, Largo S. Eufemia 19, 41100 Modena, Italy

ABSTRACT

Bronze Age *Impasto* pottery from Gorzano *Terramara* was analysed to identify raw materials and manufacturing techniques. *Impasto* ceramic paste is a heterogeneous mixture of fine sediments and tempers. Ceramics and sediments surrounding the site were characterised through petrographic analysis, chemical analysis of major, minor and trace elements and X-ray powder diffraction. The results show that the fine fraction of the pottery is low in carbonates, while most sediments are calcareous and show different clay mineral/carbonate ratios. Only three sediment samples of the Niviano unit are chemically and mineralogically similar to the fine fraction of the pottery. However the higher concentration of Al_2O_3 in the pottery suggests the use of a raw material richer in clay minerals and hence the possibility that sediment from the Niviano unit were partially levigated before use.

INTRODUCTION

Terramare are fortified villages that occupied the central area of the Po valley extensively during the Middle and Late Bronze Ages. The *Terramare* economic and socio-cultural system was successful for five centuries (17th-12th century BC). Results can be seen in the population growth and in the development of crafts: pottery production was relatively standardized and specialized (Bernabò Brea *et al.* 1997).

This study is based on the site of Gorzano and other *Terramare* in the same area. Gorzano settlement was located on a little terrace close to Tiepido river; the extent of the village was about 7000 m^2. The site was investigated in the second half of the 19th century and again, with a small trench, in 1954: a ditch and bank and four occupation layers dated Middle-Late Bronze Age were discovered. A large amount of artefacts were evidence of a flourishing pottery and metallurgical industry.

Impasto is the coarse handmade pottery typical of Italian protohistory (Levi 1999, 2004). The *Terramare* ceramic paste is a heterogeneous mixture of fine sediments and tempers, mainly grog and rock fragments (Amadori *et al.* 1996; Levi and Loschi Ghittoni 1997; Jenkins *et al.* 1999; Jones *et al.* 2002). The definition of "paste compositional reference units" (Rice 1987) is complicated by the high variability in the mixtures.

We present some preliminary results based on the comparison between the pottery of Gorzano, the local sediments, and other sites in the same area. This work is part of a multidisciplinary project to define a specific methodology for the study of inhomogeneous ceramic products like *Impasto* (Di Pierro 2003).

MATERIALS AND ANALYTICAL METHODS

Pottery samples

A great variety of shapes and functions (transport, storage, cooking, pouring and drinking) are represented in Gorzano vessels (Fig. 1). The open shapes are mainly

Fig. 1 – Examples of vessels from Gorzano: the original illustrations (from Bernabò Brea et al. *1997, p. 100).*

tablewares whilst the closed ones are suitable for cooking, storage and transport. 30 open (5 bowls, 9 cups) and closed (9 jugs, 7 jars) vessels have been analysed.

The samples, dating from the Middle to Late Bronze Age, are all handmade (coil, molds or paddle and anvil), the surfaces are smoothed or burnished and sometimes decorated with impressed geometric patterns, surface colours are often variegated with black and brown hues.

84 samples from 6 *Terramare* (Castiglione di Marano, Pontenuovo, Cà de Monesi, Montebarello, Montale and Casinalbo) in the same area (at a distance < 8 km from Gorzano), that have already been analysed (Levi and Loschi Ghittoni 1997), were reconsidered for a direct comparison and to see if similar raw material was used in different settlements.

Sediment samples

With the objective of identifying the source of raw materials, argillaceous sediments were sampled from the sedimentary succession exposed in the vicinity of the Gorzano Terramare site (Fig. 2). A total of 20 samples were taken from 6 different geological units containing fine-grained sediments that were possibly available for the ancient potters (Fig. 3): modern fluvial sediments (sample n. 1), Niviano unit (upper Pleistocene) (samples n. 2, 3, 4, 5, 6), Lower Emiliano-Romagnola Alloformation (middle Pleistocene) (sample n. 7), Sabbie gialle Formation (lower-middle Pleistocene) (sample n. 8), Argille azzurre Formation (Zanclean–lower Pleistocene) (samples n. 9, 10, 11) and Val Tiepido-Canossa argillaceous breccias (Oligocene-Miocene). With the exception of the last two units, which crop out extensively in badlands for a few hundred meters upstream of the site, all other units are exposed exclusively downstream along the Tiepido stream incision (Fig. 3). The chronology and stratigraphy of these units are taken from Gasperi and Pizziolo (in press). We sampled all the fine-grained facies consisting of clay sediments, silty clays, clayey silts and sandy silts. Some samples were taken also from other non-argillaceous facies to test the possibility that sand was used as temper.

Analytical techniques

— Standard thin sections were prepared from the ceramics and from the clays for mineralogical and petrographical characterization.
— Bulk sample X-ray fluorescence analyses (XRF) for 10 major elements as well as 16 minor and trace elements were performed using a Philips PW 1480 instrument, sited at Dipartimento di Scienze della Terra of Modena and Reggio Emilia University.
— X-ray powder diffraction patterns (XRPD) were collected on pottery sherds and on raw and physico-chemically treated clays (air dried, glycolated, heated at 375 °C and 550 °C oriented mounts) using a Philips Panalytical X'PERT Pro diffractometer with detector RTMS (Real Time Multiple Strip), sited at Centro Interdipartimentale Grandi Strumenti (CIGS) of Modena and Reggio Emilia University.

RESULTS

1. Petrography

Four main groups were identified in the pottery from Gorzano (Fig. 4): three are characterized by the abundance of clasts added as temper (mean diameter of the clasts 0.2 mm-2 mm), the last one is finer (mean diameter of the clasts < 0.1 mm).

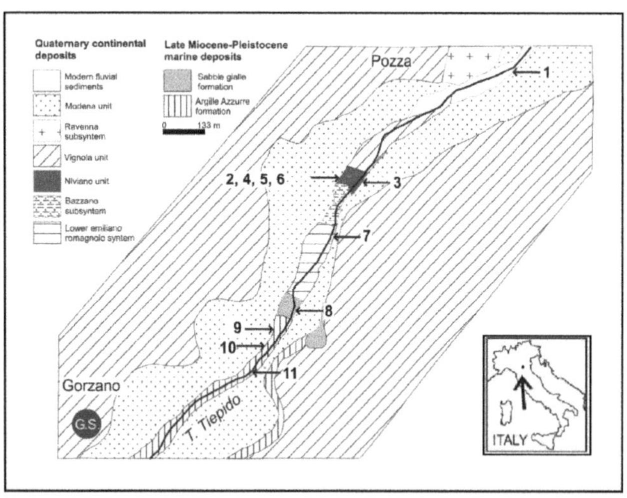

Fig. 2 – Simplified geological map showing the sampling sites of the argillaceous sediments in the vicinity of the Gorzano Terramara (G.S.) (Carta Geologica della Regione Emilia Romagna).

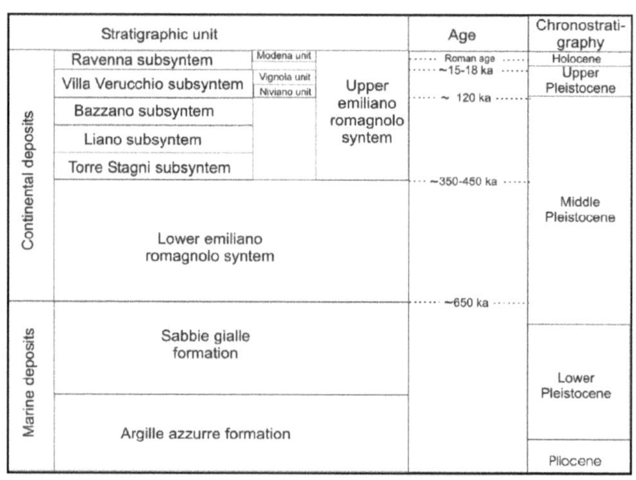

Fig. 3 – Chronostratigraphic sketch of the sampled units (modified from Gasperi and Pizziolo, in press).

A) clasts of grog and calcite (angular): 3 samples (1 jar, 2 jugs);
B) clasts of grog: 12 samples (3 bowls, 5 cups, 1 jar, 3 jugs);
C) clasts of calcite (angular): 3 samples (1 bowl, 2 jars);
D) only fine fraction (micromass): 12 samples (1 bowl, 4 cups, 3 jars, 4 jugs).

Spathic calcite fragments have angular, sharp outlines and their dimensions can reach 2 mm: we can assume that ancient potters added crushed calcite as temper. This practice was very common in pottery manufacture, as demonstrated by the presence of analogous fragments inside the grog clasts.

The matrix is dark, with its colour ranging from brown to black, rarely reddish. Cross sections of some sherds show a colour gradient in the matrix ranging from an outer reddish to an inner grey portion. The common natural aplastic mineral components, with fine and unimodal granulometry, are: quartz, plagioclase, K-feldspar, muscovite, and opaque minerals. Some ceramics also contain clay pellets. Secondary calcite is observed along the fractures.

The fine and grog tempered groups (D and B) are represented in open and closed vessels, the calcite tempered groups (A and C) are more frequent in closed vessel (3 jars, 2 jugs and only 1 bowl).

Pottery from the Terramare in the same area (Levi and Loschi Ghittoni 1997) has similar characteristics of the ceramic paste, similar compositions (grog and calcite: 17 samples; grog: 40 samples; calcite: 5 samples; only fine fraction: 22 samples) and the same trend in the use of calcareous temper mainly for closed vessels.

The four groups identified can be preliminarily proposed as "paste compositional reference units" for Bronze Age pottery production in the Modena area, although these groups show a certain variability in terms of dimension and abundance of the coarser components.

The petrographical examination of the sediments shows different granulometries in the various geological units. Only the Niviano unit, Alloformazione Emiliano Romagnola inferiore and Argille azzurre display a granulometry compatible with the matrix of the pottery. Those sediments are characterized by a fine fraction, mainly related to clay minerals, and a coarser fraction (always < 0.1 mm) consisting of quartz, feldspars, calcite, opaque minerals and amphiboles. The yellow sand sediments, where the clay fraction is very rare, are mainly composed of grains > 0.1 mm, while those of the Val Tiepido-Canossa argillaceous breccias show the presence of larger carbonate (calcite and dolomite) fragments.

2. Chemistry

Two main chemical compositions can be recognized in Gorzano pottery; the first one includes A and C petrographic groups, tempered with calcite, the second one B and D groups. The main discriminant chemical element is Ca, which is higher and more variable in the groups A and C (Fig. 5). B and D groups usually show CaO concentrations < 2% and a large variability in the Al_2O_3/SiO_2 ratio. No obvious chemical difference is present, neither in the major nor in the trace elements, between B and D groups, thus indicating that the grog is chemically very similar to the fine matrix of the pottery.

No relationship was observed between chemical composition and ceramic shape or function.

The P_2O_5 value ranges from 0.12 to 2.63%. Generally it is higher than 0.5%, the value reported in literature as the maximum content derived from natural soils and clays (Lemoine and Picon 1982; Collomb and Maggetti 1996).

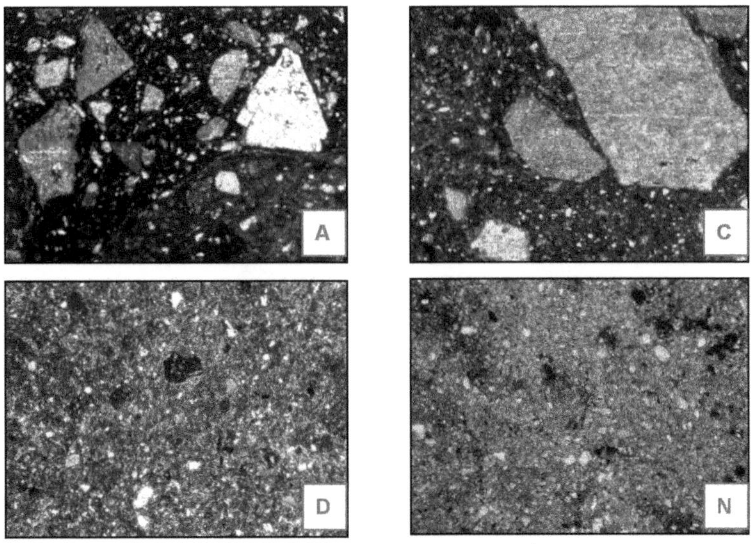

Fig. 4 – Thin section photographs (frame size 1.2 x 1.8 mm) of pottery groups from Gorzano site and of a sediment. A) pottery tempered with grog and calcite, C) pottery tempered with calcite, D) pottery with only fine fraction, N) Niviano unit sample n. 2.

This result is consistent with contamination events during burial in soils characterized by a high level of organic waste (Maritan and Mazzoli 2004). Until the 19th century Terramare soils were extracted and utilized as fertilizers.

The chemical variability of the sediments is very high, even in the same geological unit, in particular as regards CaO and MgO contents relative to the amount of carbonate, and the Al_2O_3/SiO_2 ratio, relative to the clay component. The sediments are more calcareous than the pottery except for three samples of the Niviano unit, which however show lower concentrations of Al_2O_3, due to a lower amount of the clay component (Fig. 5).

In order to compare Gorzano pottery with other Terramare in the same area, only fine pottery, not influenced by different amounts of calcite added as temper, were used in the statistical elaborations. In Fig. 6 the pottery studied by Levi and Loschi Ghittoni (1997) shows a low concentration of CaO and again a high variability in the Al_2O_3/SiO_2 ratio (the higher range of variability is probably due to the larger sample set studied by these authors). This variability is evident at every site and is not linked to ceramic shapes and functions, but it may be related to the compositional variability of the clays used as raw materials.

3. Mineralogy

XRPD analyses on Gorzano pottery give the following mineralogical composition for every sample: quartz, albite, K-feldpar, illite (muscovite) (Fig. 7). Amphiboles, calcite, rutile (or anatase) and hematite are present only in few samples. The four petrographic groups show qualitatively very similar patterns, the differences being related to the calcite content, the quartz/illite ratio and the different levels of amorphous phases. No newly formed calcium-silicates (gehlenite, diopside/wollastonite, anorthite) were found in the samples studied. This can be explained by the use of non-calcareous clay and by the low firing temperature, which can be assumed to be lower than 850 °C because of the presence of fresh calcite and illite (Maggetti 1982, Capel *et al.* 1985, Duminuco et al. 1996, Riccardi et al. 1999).

The mineralogical composition of all the sediments is rather similar, the principal components being: kaolinite, chlorite, smectite, quartz, albite, K-feldpar, illite (muscovite) and carbonates. The main differences concern the clay mineral/carbonate ratio. As a result only three samples of the Niviano unit are compatible with the pottery mineralogy—in particular the fine (group D) and the grog-tempered (group B)—because they contain low levels of chlorite and carbonates are absent (Fig. 7).

CONCLUSIONS

— Gorzano pottery technology, based on the use of fine sediments often tempered with grog and calcite, is consistent with the traditional Bronze Age pottery production of the Po Valley. Calcareous inclusions are used mainly for closed vessels.

— The pottery production of several *Terramare* of Modena area is chemically and mineralogically homogeneous.

— The four main groups identified may represent different "paste compositional reference units" from a petrographic perspective (and probably from the craftsmen's one point of view), but mineralogically and chemically the main division is between the calcite tempered pots and the others.

— The fine fraction of the pottery is low in carbonates.

— The firing temperature is estimated to be lower than 800-850 °C.

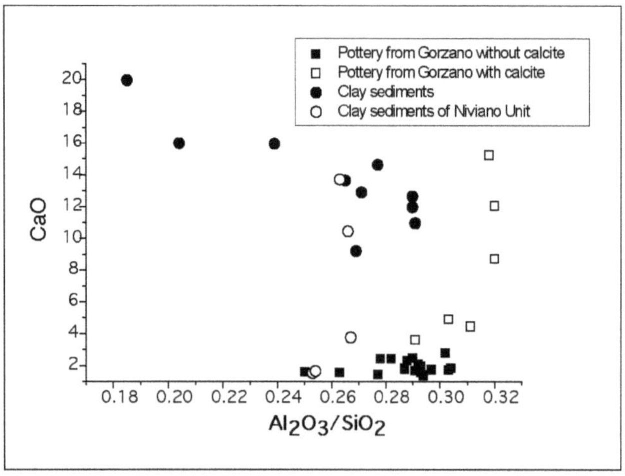

Fig. 5 – Al_2O_3/SiO_2 ratio - CaO wt% correlation diagram of the four ceramic groups from Gorzano site and of the sampled sediments.

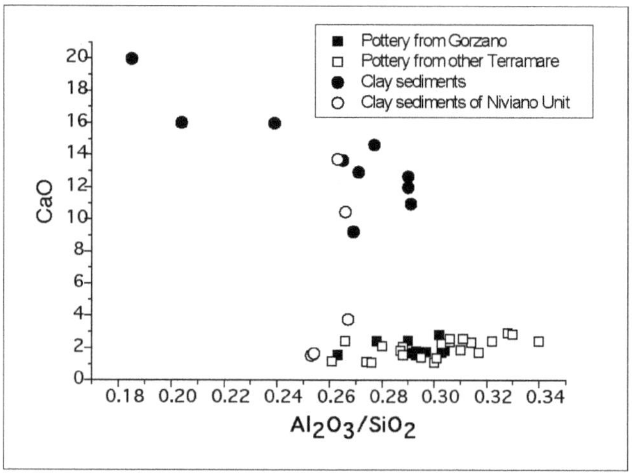

Fig. 6 – Al_2O_3/SiO_2 ratio - CaO wt% correlation diagram of the fine pottery from Gorzano and of other Terramare settlements and of the sampled sediments.

— Only some sediments of Niviano unit are chemically and mineralogically similar to the fine pottery. The higher level of Al_2O_3 in the pottery, however, suggests the use of a raw material particularly abundant in clay minerals. Maybe sediments of the Niviano unit were partially levigated, or the locality of the raw material source needs to be further investigated.

— All the other sediments considered are calcareous, with different clay mineral/carbonate ratios, and also different from the final products.

— The characterization of fine *Impasto* pottery is confirmed as the most appropriate way to investigate the provenance of the raw materials.

— As the grog has a chemical composition similar to the fine matrix, the pottery tempered with grog can be used in provenance studies.

— Work is in progress to perform the quantitative mineralogical analysis of pottery and sediments, in particular on the clay fraction and the amount of amorphous phase.

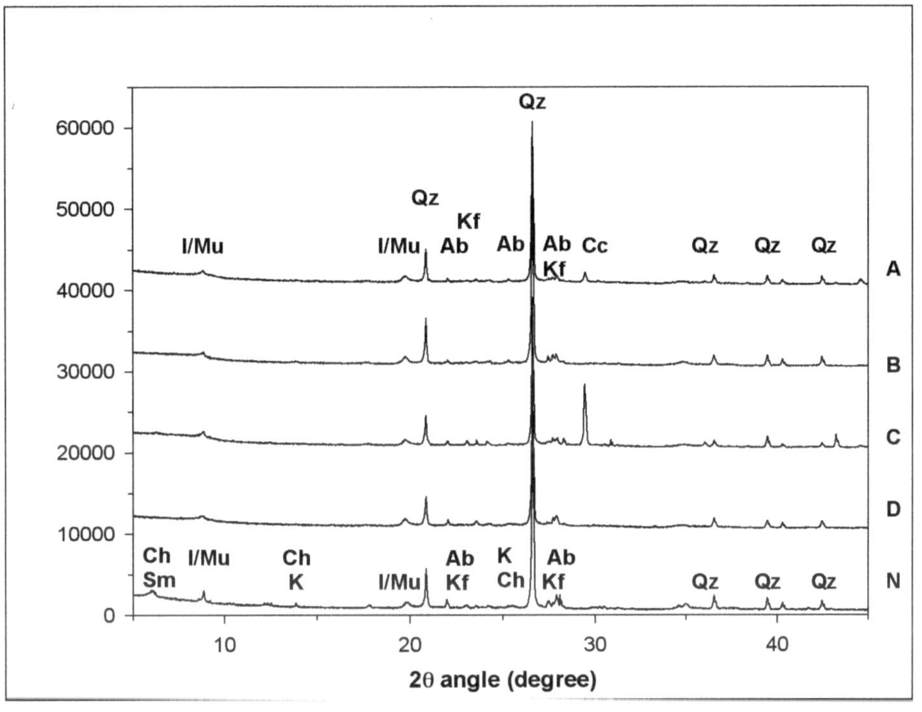

Fig. 7 – XRPD patterns of: A) pottery tempered with grog and calcite, B) pottery tempered with grog, C) pottery tempered with calcite, D) pottery with only fine fraction, E) one of the sediments of Niviano unit. Qz: quartz; Cc: calcite; Ab: albite; Kf: Kfeldspar; I/Mu: illite/muscovite; Sm: smectite; K: kaolinite; Ch: chlorite.

REFERENCES

AMADORI M.L., DESOGUS P., FABBRI B., LEVI S.T., PACCIARELLI M. 1996, "La tecnologia di produzione delle ceramiche da Monte Castellaccio e S.Giuliano indagata attraverso analisi archeometriche", *in* Musei civici di Imola (ed.), *La collezione Scartabelli-2 preistoria*, Bologna, p. 355-415.

BERNABO' BREA M., CARDARELLI A., CREMASCHI M. 1997, "L'insediamento collinare e montano", *in Le Terramare. La più antica civiltà padana*, Milano, p. 275-295.

BETTELLI G., GASPERI G., PANINI F., PIZZIOLO M., PRETI D. (2005). Tavola 1:25000 RER Nuova Carta Geologica Italia, fogli: 219 NE Formigine e 219 SE Marano sul Panaro.

CAPEL J., HUERTAS F., LINARES J. 1985, "High temperature reactions and use of Bronze Age pottery from La Mancha, central Spain", *Mineralogica Petrographica Acta* 29A, p. 563-575.

COLLOMB Ph., MAGGETTI 1996, "Dissolution des phosphates présents dans des céramiques contaminées", *Revue d'archéométrie* 20, p. 69-75.

DI PIERRO S. 2003, "Matrix-temper separation of Neolithic ceramics: an experimental approach to characterize the original raw materials an determine their provenance", *in* S. Di Pierro, V. Serneels and M. Maggetti (ed.), *Ceramic in the society. Proceedings of the 6th European meeting on ancient ceramics, 3-6 October 2001, Fribourg, Switzerland*, p. 109-131.

Duminico P., Riccardi M.P., Messiga B., Setti M. 1996, "Modificazioni tessiturali e mineralogiche come indicatori della dinamica del processo di cottura di manufatti ceramici", *Ceramurgia* 5, p. 281-288.

Gasperi G., Pizziolo M. (in press), Note illustrative della Carta Geologica d'Italia a scala 1:50.000. Foglio 201 "Modena": Regione Emilia-Romagna, Servizio Geologico d'Italia.

Jenkins D., Williams J.L., Levi S.T. 1999, "Petrographic analysis in provenancing and classification of sherds: the lower Po valley: a case study", *in* M. Maggetti and M. Vendrell-Saz (ed), *Atti 5e Curs d'Arquelogia d'Andorra: 4th European meeting on ancient ceramics, Estudis arqueològics i arqueomètrics, 12-15 November 1997, Andorra*, p. 175-181.

Jones R.E., Vagnetti L., Levi S.T., Williams J., Jenkins D., De Guio A. 2002, "Mycenaean pottery from Northern Italy. Archaeological and archaeometric studies", *Studi Micenei ed Egeo-Anatolici* XLIV/2, Istituto di Studi sulla Civiltà dell'Egeo e del Vicino Oriente, Roma, p. 221-261.

Lemoine C., Picon M. 1982, "La fixation du phosphore par les céramiques lors de leur enfouissement et ses incidences analytiques", *Revue d'archéométrie* 6, p. 101-112.

Levi S.T. 1999, *Produzione e circolazione della ceramica nella Sibaritide protostorica. Vol. I. Impasto e dolii*, Firenze.

Levi S.T. 2004, "La ceramica. Circolazione dei prodotti e organizzazione della manifattura", *in* D. Cocchi Genick (ed.), *Atti di L'età del Bronzo recente in Italia, 26-29 October 2000, Viareggio*, p. 233-242.

Levi S.T., Loschi Ghittoni A.G. 1997, "Gli impasti ceramici di siti terramaricoli del territorio modenese", *in Le Terramare. La più antica civiltà padana*, Milano, p. 497-506.

Maggetti M. 1982, "Phase analysis and its significance for technology and origin", *in* J.S. Olin and A.D. Franklin (ed), *Archaeological ceramics*, Smithsonian Institution, Washington, p. 121-133.

Maritan L., Mazzoli C. 2004, "Phosphates in archeological finds: implications for environmental conditions of burial", *Archaeometry* 46, p. 673-683.

Riccardi M.P., Messiga B., Duminico P. 1999, "An approach to the dynamics of clay firing", *Applied clay science* 15, p. 393-409.

Rice P.M. 1987, *Pottery analysis: a sourcebook*, Chicago and London.

EXPLORING PATTERNS OF INTRA REGIONAL POTTERY DISTRIBUTION IN LATE MINOAN IIIA-B EAST CRETE: THE EVIDENCE FROM THE PETROGRAPHIC ANALYSIS OF THREE CERAMIC ASSEMBLAGES

ELENI NODAROU

INSTAP Study Center for East Crete, Pacheia Ammos, Ierapetra 72200, Crete, Greece; enodarou@yahoo.gr

INTRODUCTION

Studies of the Late Minoan IIIA-B period (hereafter LM IIIA-B, *c.* 1400-1200 BC) have traditionally used the possible presence of the Mycenaeans in Crete as an important frame of reference. The starting point for many scholars is the generalized wave of destructions that hit the palaces and major sites at the end of the LM IB period. Opinions about the reason for this collapse vary from natural disasters and internal conflicts to external invasions by groups from the Mycenaean Mainland (Driessen and MacDonald 1997; Rehak and Younger 1998; Warren 2001). The subsequent period (LM II-early LM IIIA) is marked by the reoccupation of many sites and a series of transformations in the material culture (particularly in north central Crete), from mortuary (Preston 2004) and administrative practices (*e.g.* the introduction of Linear B script) to pottery shapes and decorative styles. The most distinctive style for the period appears to have developed at Knossos (Popham *et al.* 1984) and the presence of this Knossian or Knossian style pottery at sites outside north central Crete (*e.g.*, Malia, Kommos, Chania) has been interpreted as an attempt by a new Knossian elite (whether Minoan or Mycenaean) to influence or even exercise control over parts of the island (Brogan *et al.* 2002; Driessen and Farnoux 1997). That this Knossian administration eventually succeeded in extending its authority over large parts of western and central Crete is suggested by the existence of place names of cities from both parts of the island in Linear B tablets found at Knossos (Bennett 1990).

Within this framework the study of the LM IIIA-B period has been traditionally dominated by the preeminence of Knossos, impeding the study of secondary centers and the relations that developed between them (Preston 2004). This was particularly the case of East Crete where no place-names are mentioned in the Linear B tablets, suggesting that this part of the island constituted a separate polity or incorporated a group of polities that were not subjected to the Knossian administrative system (Bennet 1987; 1990, p. 209). In the last decade, however, the growing body of evidence from areas in East Crete opened the way to more regional approaches. Stylistic studies of pottery from well dated contexts, such as the tombs of Mochlos and Myrsini (Smith 2002; Brogan *et al.* 2002) and the settlement at Palaikastro (MacGillivray 1997a; b), demonstrated that major centers had developed in this part of the island as early as the beginning of LM IIIA, if not earlier.

These approaches have contributed greatly to the establishment of local chronological and stylistic sequences through the study of the individual ceramic assemblages. They opened the way to comparative regional studies exploring differences and similarities in the material culture, intra regional influence and interaction in a medium and small scale (*e.g.* Brogan *et al.* 2002; Smith 2002; Banou 2005), on an island like Crete where centers develop only a few kilometers from each other. However, this shift of stylistic studies towards intra-regional connections has not been followed by a similar shift of interest in the analytical studies. The issue that has been dealt with analytically is that of the transport stirrup jars (especially those bearing Linear B inscriptions) and their distribution, as an indication of the relationships between Mycenaean Greece and Crete (Day and Haskell 1995; Haskell 2005).

This paper resulted from three petrographic projects, carried out at the INSTAP Study Center for East Crete, and it attempts to investigate the nature and the individual characteristics of three ceramic assemblages: Chrysokamino, Mochlos and Petras (Fig. 1). From these separate studies there emerged a converging pattern related to the consumption of LM IIIA-B ceramics in East Crete and led to a synthetic study whose preliminary results are presented here. By combining analytical and stylistic information from three settlement sites of different size, history and importance, we explore how pottery moves on small or medium scale in East Crete and how this movement is interwoven with economic and social relationships.

THE SITES

Chrysokamino

The site of Chrysokamino is situated on a hill overlooking the coast, at the eastern side of the bay of Mirabello. The architectural remains suggest the presence of a small farmhouse whose main phase of occupation is LM IIIA and early LM IIIB (Betancourt *et al.* 1999, p. 343, 350). The pottery that characterizes this phase is primarily of domestic character, intended for storage, cooking and serving purposes, along with a limited amount of fine wares. The farmhouse at Chrysokamino represents a rather provincial site in East Crete during the LM III period but its geographical position makes it interesting. Located close to the narrow and relatively flat Isthmus of Ierapetra, Chrysokamino overlooks the northern end of an important overland crossroad for traffic between the north and the south coasts of Crete.

Mochlos

The small island of Mochlos lies on the north coast of Crete at the eastern limit of the Bay of Mirabello, *ca.* 10 km east of Chrysokamino. Geological studies suggest that in antiquity the island was connected to the mainland of Crete by a low isthmus, forming one of the best harbors on the north coast and providing easy access to nearby sources of water and the fertile coastal plain (Soles 2003, p. 1). Excavation on the island has revealed an extensive and multi-phased town occupied from the Early Bronze Age to the LM IB period (Soles and Davaras 1992; 1994; 1996). Destruction and abandonment seem to have resulted in an occupational gap until the late LM II/early LM IIIA period when the site was reoccupied (Brogan *et al.* 2002). The LM III architectural remains consist of at least thirteen houses constructed on top of earlier (LM IB) houses within the streets of the earlier settlement. They date from early LM IIIA to LM IIIB (Brogan *et al.* 2002, p. 98). The pottery assemblage of this re-occupation phase is substantial, comprising domestic vessels as well as considerable amounts of fine wares (Smith 2002). It is marked by both continuity in certain pottery shapes known since the LM IB period and changes in others, most notably with the presence of new Knossian elements. The typological study of the material revealed potential connections with north-central and east Crete and allowed the formulation of hypotheses regarding the formation of regional identity during this period (Smith 2005). Moreover, the emergence of east Cretan regional traditions implies that Mochlos takes active part in the economy of the period and the re-emerging maritime trade, taking advantage of the site's position as a natural port and a gateway (Brogan *et al.* 2002, p. 117-118).

Petras

Petras is a low coastal hill in the bay of Siteia. The site was occupied from the Early Bronze Age and prospered particularly in the Old and New Palace periods as indicated by the presence of a large palatial building on top of the hill and a relatively large town on the side hills, of which two houses have been fully excavated. Unlike the earlier periods, the LM IIIA-B period is rather problematic since there are very few architectural remains preserved, namely two fragmentary buildings on top of the earlier palatial building (Tsipopoulou 2005, p. 314). However, the recent discovery of a double fortification wall in the foothills (Tsipopoulou 2005, p. 312) may suggest that in the LM IIIA-B period Petras was still an important settlement, acting as a gateway to the sea for the agricultural products from the Siteia plain and a controlling point for coastal maritime traffic. While the architectural remains are meager, the ceramic evidence of the LM IIIA-B period is significant in both quantity and quality. The shapes comprise a variety of domestic coarse and semi-coarse vessels intended for storage, cooking and transport purposes, as well as a multitude of fine wares some of which are thought to be imports (Tsipopoulou 1997). The material analyzed derives from House I (Tsipopoulou 1990).

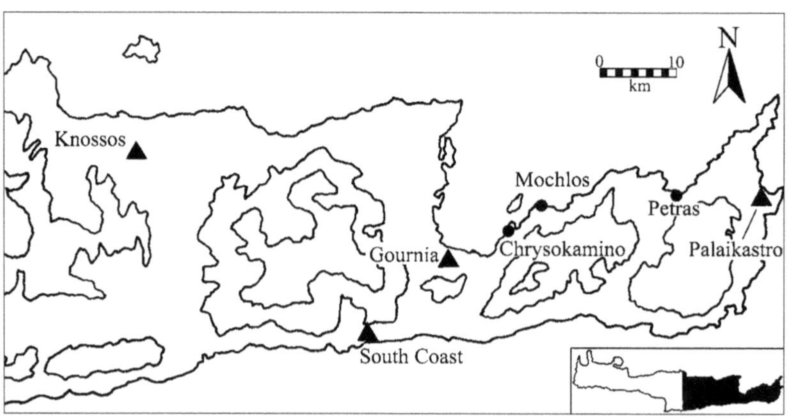

Fig. 1 – Map of East Crete and sites mentioned in the text.

THE LOCAL MATERIAL

In order to discuss the presence and distribution of pottery fabrics in East Crete it is essential to present a brief account of what is considered to be the local component of each assemblage, *i.e.* the fabrics that according to stylistic and petrographic criteria represent local production. The concept of local pottery production is in itself rather ambiguous since in most cases there is no production workshop within the site under study. It is only indirectly, through petrography and macroscopic study of pottery fabrics, that the presence of pottery workshops in the broad vicinity is deduced.

Chrysokamino

There is only one local fabric at Chrysokamino. It is coarse and is characterized by a fine grained, orange-red, non calcareous matrix. The non plastics consist primarily of low-grade metamorphic rock fragments, namely phyllite, along with small fragments of monocrystalline quartz, sandstone and very rare igneous rock fragments (Fig. 2). This fabric forms a very "tight" and homogeneous group in terms of composition and texture. All the samples of the group are very similar to each other, which, as we will see, is not always the case in the other sites. This fabric is related to the Phyllite-Quartzite series (IGSR 1959a) outcropping in the vicinity of the site, in the area of Kavousi.

The non-calcareous, phyllitic raw material has been used for the manufacture of coarse and semi-coarse vessels intended for domestic purposes, mainly cooking and storage, but there are also vessels for other kinds of domestic activities such as beehives and bowls.

Mochlos

In contrast to the provincial farmhouse of Chrysokamino, the town of Mochlos is characterized by both coarse and fine local fabrics.

The most abundant and hence the main coarse fabric is characterized by a non calcareous, fine grained matrix and the presence of coarse and semi coarse inclusions consisting primarily of metamorphic rock fragments, namely phyllite and quartzite (Fig. 3). The other non-plastic components consist of chert, sandstone, rare plagioclase feldspar and altered igneous rock fragments. This rock and mineral suite is characteristic of the Phyllite-Quartzite series outcropping in the area of Mochlos in close proximity to the coast (IGSR 1959a). This fabric occurs at Mochlos from at least the Late Minoan IB period (Day *et al.* 2003); in the LM III assemblage it is used primarily for storage and cooking vessels, including basins, jars, pithoi, cooking dishes and cooking trays.

However, this is not the only phyllitic fabric present at Mochlos. Unlike the homogeneity of the Chrysokamino metamorphic fabric, at Mochlos there is a multitude of phyllitic fabrics that differ from each other primarily in texture. Through petrographic analysis an array of metamorphic fabrics has been established indicating that semi-coarse decorated vessels intended primarily for transport were manufactured using different clay recipes than the plain coarse wares (Fig. 4). The same differentiation between decorated and plain coarse wares is also attested at Petras.

Fig. 3 – Mochlos, phyllite fabric (x25).

Fig. 2 – Chrysokamino, phyllite fabric (x25).

Fig. 4 – Mochlos, phyllite fabric for decorated vessels (x25).

The second major local fabric is used for fine vases. It is characterized by a fine calcareous matrix and semi-fine to fine inclusions, namely micrite, fossils and a few shell fragments (Fig. 5). There are also rare metamorphic rock fragments and textural features that lead us to suggest that it constitutes the product of mixing of the local red phyllitic clay with a fossiliferous marl. Clay mixing is widely practiced in Minoan Crete and it is attested in pottery manufacture up to present (Day 2004; Blitzer 1984). In this fabric are included fine and semi-fine vessels such as kylikes, bowls, cups, jugs, and dippers, *i.e.* shapes connected with serving and drinking liquids.

Petras

Like Mochlos, Petras includes both coarse and fine local fabrics. The main coarse fabric is characterized by the presence of low-grade metamorphic rocks, namely phyllite and quartzite, set in a fine groundmass composed primarily of monocrystalline quartz (Fig. 6). There is also some sandstone, chert and biotite mica. A similar fabric has already been identified in the Neopalatial pottery (Day 1995, p. 157) and taking into account the geology of the area, it is very likely that it is local (I.G.S.R. 1959b). The vessels made in this fabric are cooking pots and storage jars.

The analysis of the ceramic assemblage reflects the variability of the metamorphic geology of the area. As was the case at Mochlos, although the raw materials used for the various types of pottery are all of metamorphic origin, the clay recipes change between the plain and the decorated coarse and semi-coarse vessels (Fig. 7).

As at Mochlos the second most common fabric at Petras is the product of a totally different recipe that is based on calcareous materials and used for fine and semi-fine vessels. It is characterized by a fine grained matrix and contains a few metamorphic rock fragments, mainly phyllite, along with some quartzite, sandstone, micrite and fossils (Fig. 8). This composition is similar to the coarse metamorphic fabric but the presence of the micrite, the fossils and the textural features indicates clay mixing of a red non-calcareous phyllitic clay and a Miocene calcareous marl. The vessels represented are medium and small sized, almost all decorated, used primarily for serving, drinking and storage purposes, such as jars, jugs, amphorae, small

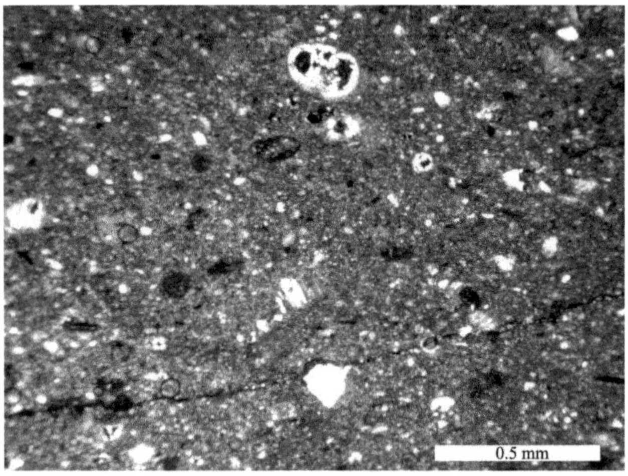

Fig. 5 – Mochlos, fine calcareous fabric (x50).

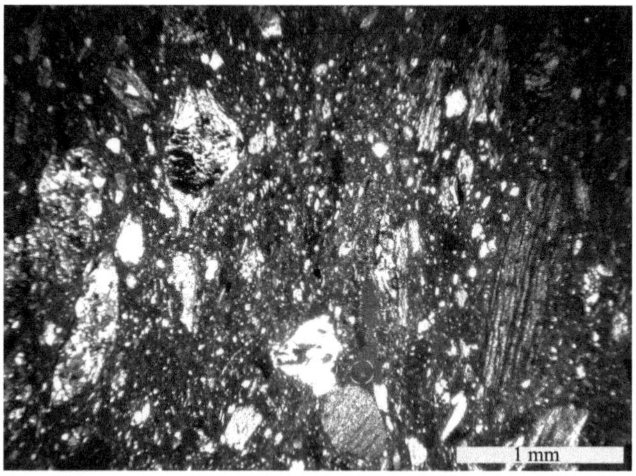

Fig. 6 – Petras, phyllite fabric (x25).

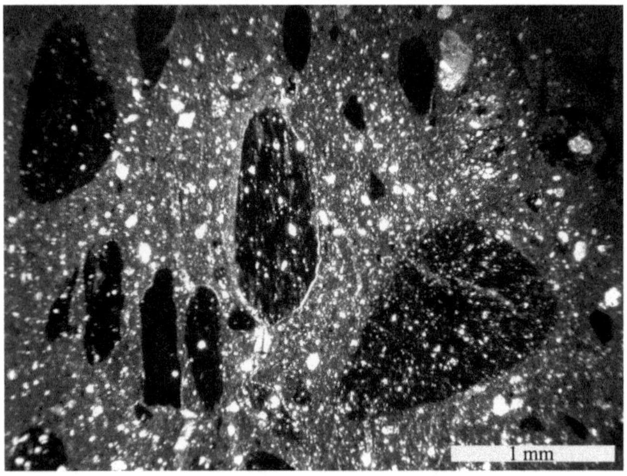

Fig. 7 – Petras, phyllite fabric for decorated wares (x25).

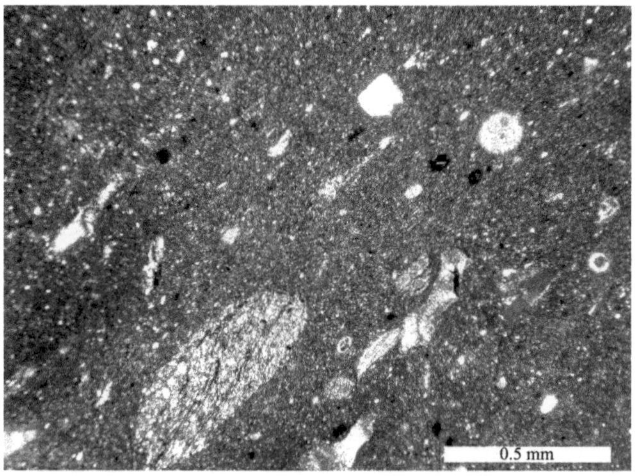

Fig. 8 – Petras, fine calcareous fabric (x50).

pithoi, fruitstands, and cups. It is not possible to be certain whether this fabric represents a different workshop but it is quite clear that there are different recipes for the coarse domestic wares and the fine decorated vessels.

THE IMPORTED MATERIAL

In addition to the material identified as local, all three sites contain imported pottery, *i.e.* a part of the assemblage that stylistically and/or fabric wise does not conform to the local component. In what follows, the imported material is presented by fabric (and not by site) in order to discuss provenance and distribution.

Serpentinite fabric

This is a coarse fabric characterized by the presence of angular and poorly-sorted serpentinite fragments in a reddish brown calcareous matrix (Fig. 9). The bimodal distribution and angularity of the non plastics indicate that serpentinite was deliberately added in a levigated clay paste as temper. Serpentinite occurs in the ophiolite series and the flysch mélange of the south coast of Crete, just west of Myrtos, which is possibly the place of origin of these vessels (IGSR 1993). This fabric has been used for storage and transport jars and is encountered at Chrysokamino and Petras, but is totally absent from Mochlos.

Sedimentary and igneous rock fabric

A second fabric can also be associated with the ophiolite series and the flysch mélange of the south coast containing sedimentary and igneous rock fragments (Fig. 10). A similar fabric has been identified at the Early Minoan settlement of Myrtos Fournou Korifi (Whitelaw *et al.* 1997, p. 268) and in the Early Minoan material from the Kavousi survey (Day *et al.* 2005, p. 180). Unlike the serpentinite fabric, this fabric occurs only at Chrysokamino and is used also for coarse amphorae and jars. Its presence at the small site of Chrysokamino is not surprising given the position of the site on the north coast of the Isthmus that provided one of the easiest and shortest points of travel (*ca.* 15 km) between the north and south coasts.

Granodiorite fabric

The granodiorite is one of the most common fabrics in East Crete. It has been given the name "Mirabello fabric" because it reflects the rock and mineral suite encountered in the area of Gournia-Kalo Chorio, around the bay of Mirabello (Day 1991). It is characterized by the presence of acid igneous rock fragments, namely granodiorite, set in a very fine, red firing matrix, along with some plagioclase, biotite and significant amounts of clay pellets (Fig. 11). The vessels represented in this fabric are mainly jugs, amphorae, and pithoid jars. The granodiorite fabric is encountered in a number of sites in East Crete since the Early Minoan II period, throughout the Middle Minoan and to a lesser extent during the Neopalatial. In the LM IIIA-B period it occurs in significant quantities at Chrysokamino and Petras but it is totally absent from Mochlos.

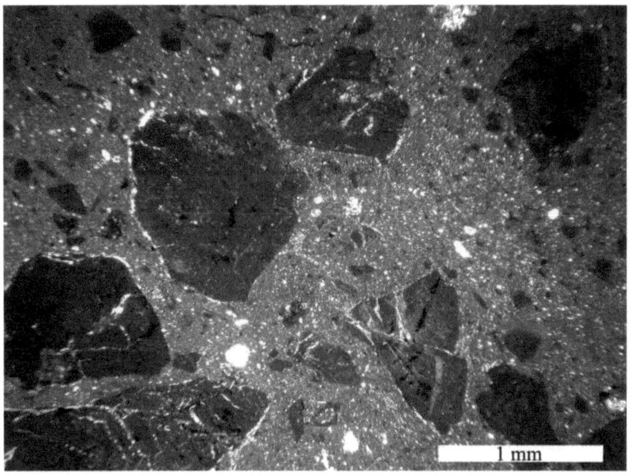

Fig. 9 – Serpentinite fabric (x25).

Fig. 10 – Sedimentary and igneous rock fabric (x25).

Fig. 11 – Granodiorite fabric (x25).

Fine red, quartz-rich fabric

This is a fine fabric characterized by a red matrix with frequent inclusions of monocrystalline quartz and a few fragments of phyllite and micrite, indicating raw materials of metamorphic and sedimentary origin respectively (Fig. 12). This fabric stands out macroscopically because of the pink color of the clay and the high firing. Its existence is known since the Neopalatial period and its provenance has been assigned to Palaikastro at the far east of Crete (Day 1995, p. 161). In the LM IIIA-B period it was found at all three sites examined. The shapes represented are mainly pulled-rim bowls and transport vessels, such as amphorae and jars.

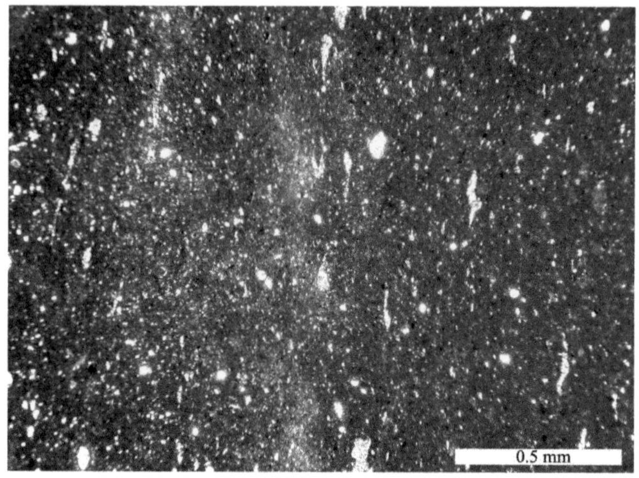

Fig. 12 – Fine red quartz-rich fabric (x50).

Fine fabric with clay pellets

This is a very fine fabric characterized primarily by clay pellets. The rare non-plastics comprise a few small fragments of monocrystalline quartz and very rare fragments of fine-grained sandstone (Fig. 13). This fabric occurs at all three sites and on typological grounds it is considered to be an import from north central Crete. Until more data become available one has to assume the presence of one or more workshops at Knossos or its broader area near the north coast manufacturing wares intended for the transportation and consumption of liquids. The vessels represented consist of fine wares, mainly kylikes and small decorated transport jars.

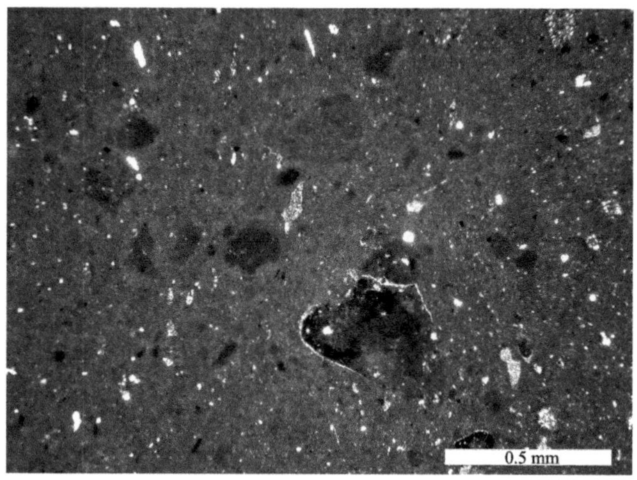

Fig. 13 – Fine fabric with clay pellets (x50).

Type of Fabric	Sites		
	Chrysokamino	Mochlos	Petras
Metamorphic for plain vessels	15	30	15
Metamorphic for decorated vessels		14	12
Metamorphic with fossils		10	12
Sedimentary& igneous rock	6		
Serpentinite	2		2
Granodiorite	18		12
Fine, red, quartz-rich	4	13	5
Fine with clay pellets	2	4	4
Other	11	29	8
TOTAL SAMPLES EXAMINED	58	100	70

Table 1 – Account of the LM III A-B samples examined per site and fabric.

DISCUSSION

This paper has dealt with the results of the petrographic analysis of three ceramic assemblages from East Crete dating to the LM IIIA-B period (for a detailed account of the samples examined per site and fabric, see Table 1). Despite the preliminary character of our study and the small number of sites included, it is possible to discern clear patterns in the production and distribution of the local and imported pottery.

With regard to the local pottery, all three sites share common traditions of pottery production both in the selection of raw materials and the clay recipes used for the domestic coarse and fine vessels. More specifically, the main coarse wares (mainly cooking and storage vessels) are manufactured with metamorphic raw materials, which reflect the Phyllite-Quartzite series of the north east coast. Although manufactured in similar fabrics, the decorated coarse and semi-coarse vessels, intended mainly for transporting liquids, seem to follow a different recipe from their undecorated counterparts. Finally, the local fine wares which consist primarily of small sized vessels intended for serving and drinking, exhibit common recipes based on clay mixing of a phyllitic non calcareous clay and a calcareous Neogene marl.

In contrast to the common trends observed in the local production, the three sites present clear differences in the consumption of imported pottery (Table 2). *Chrysokamino* has the closest connection to the South Coast as indicated by the presence of coarse and semi-coarse vessels in the serpentinite fabric and the fabric with sedimentary and igneous rock fragments. The former occurs also at Petras but not at Mochlos, whereas the latter is not seen in any other site except Chrysokamino. Despite its small scale and remote location, Chrysokamino imports also pottery from other sources. There is a substantial amount of jars and pithoid jars in the granodiorite fabric originating from the neighboring Gournia/Kalo Chorio area. Finally, a small part of the fine wares consists of imports from further afield, such as Palaikastro and north central Crete.

Unlike Chrysokamino, the imported pottery encountered at *Mochlos* does not derive from any of the neighboring production areas, *i.e.* the South Coast and Gournia/Kalo Chorio region to the west. The absence of the latter (*i.e.* the vessels in the granodiorite fabric) continues the local consumption pattern of the later Neopalatial period when imports of granodiorite fabrics decreased dramatically from numbers seen in the EM III-MM III phases at Mochlos (Barnard 2003, p. 8; T. Brogan personal communication). Instead, the imported pottery of Mochlos is limited to fine wares deriving from north central Crete and Palaikastro. This suggests a highly selective strategy of pottery importation and consumption, involving medium and small sized fine wares, either for their visual appearance or their valuable content.

Finally, *Petras*, despite its apparent isolation in the far east of Crete, receives ceramic products from all production centers mentioned so far: transport jars from the South Coast and Gournia/Kalo Chorio, as well as an array of fine wares from north central Crete and Palaikastro. Mochlos and Petras are both long-lived and important settlements, occupying a strategic geographical position and controlling the adjacent plains and the maritime routes of the north coast. They differ markedly, however, in their preferences in ceramic imports. Products from Gournia/Kalo Chorio and the South Coast were imported to Petras, but not to Mochlos, although the latter is geographically closer to these production areas.

Moving from consumption sites to the centers of production and the distribution of their products across East Crete, it appears from the analysis that the South Coast products are limited to transport amphorae and jars, which were presumably exchanged for their content. The Gournia/Kalo Chorio products, which in earlier phases were exported in a wide range of cooking as well as transport shapes, change in the LM IIIA B period to a more limited repertoire of transport jars. The latter maintain their popularity at Chrysokamino and Petras but are totally absent at Mochlos. The imports from north central Crete (*i.e.* the Knossos area) consist of shapes that exist also in local fabrics and seem to represent valued items. They include drinking vessels such as kylikes and decorated small transport jars, which may have been used for transporting perfumes or other prestigious liquids (Brogan *et al.* 2002, Fig. 7, nos. P3390, P3393, P373, P3273). Finally, Palaikastro appears as an important and dynamic center in East Crete for the production of fine table wares. The vessels are also connected with the transport and consumption of liquids, such as pulled rim bowls, amphorae and jars (Smith 2005, fig. 7).

CONCLUSIONS

One of the primary aims of this study was to examine the distribution of various types of pottery and explore potential economic and social relationships between producers and consumers. The study of the local fabrics revealed the existence of a common language across these regions in terms of potting technologies, raw materials, and clay recipes, which cannot be simply attributed to the common geology of the areas. Possible explanations for this continuity include common traditions and practices, frequent contacts and the exchange of technologies and ideas.

Various archaeological and ethnographic studies have discussed the symbolic role of pottery as factor defining regional and personal identities and how regional variation in the material culture reflects not only preferential interaction between regions but also an attempt to establish and communicate one's idiosyncratic identity (Wiessner 1983; 1984; Shennan 1989; Gosselain 1999; 2000). Within this framework, the importation of specific ceramic products at the three settlements not only reflects trade routes but delineates potential spheres of interaction on a west-east and south-north axis where the presence of certain ceramic vessels is as significant as their absence (*e.g.* the absence of Gournia/Kalo Chorio pottery at Mochlos). It can be suggested that the preferential attitude observed in the consumption of imported ceramics at sites of similar status such as Mochlos and Petras reflects a complex network of sociopolitical relationships developed between the various East Cretan centers.

This is reinforced by the fact that, despite their geographical proximity, none of the three sites imports from the other. For the coarse wares this might imply that there was little or no exchange of agricultural products between these sites, whereas for the fines wares it might mean that the local table wares produced at Mochlos and Petras were not considered as prestigious as those of

		Sites		
Fabric	**Origin**	Chrysokamino	Mochlos	Petras
Serpentinite	South Coast	x		x
Sedimentary and igneous	South Coast	x		
Granodiorite	Gournia/Kalo Chorio	x		x
Fine, red, quartz-rich	Palaikastro	x	x	x
Fine with clay pellets	North central Crete	x	x	x

Table 2 – Distribution of imported fabrics.

north central Crete or Palaikastro. Whatever the case, the above evidence reinforces the idea that the sociopolitical organization of East Crete was far more fragmented and that this part of the island, having perhaps a preference for smaller polities, followed a different trajectory from Central and West Crete (Bennet 1987, p. 87; MacGillivray 1997b, p. 279).

The three assemblages presented here offer a preliminary view of the patterning that characterizes the distribution of pottery in East Crete during the LM IIIA-B period. In order to establish a more secure explanatory framework, a broader study is needed encompassing a larger number of sites. More specifically, it should include more sites lying close to or within the production zones mentioned in the paper, *e.g.* Gournia and Palaikastro, as well as sites that are situated on major routes on the Isthmus of Ierapetra and the south coast. The study of the pottery should also be sustained by contextual information and the study of other classes of material culture. Through the combination of stylistic study and petrographic analysis a more refined approach to the archaeological material can be achieved exploring patterns of inter- and intra-regional interaction and sociopolitical organization in this part of the island.

ACKNOWLEDGEMENTS

I should like to express my warm thanks to the excavators of the three sites, namely Prof. J. S. Soles (Mochlos), Dr M. Tsipopoulou (Petras) and Dr C. R. Floyd (Chrysokamino) for allowing me to deliver this paper in advance of the primary publications of the sites. Dr T. M. Brogan and Dr Y. Papadatos are gratefully acknowledged for their comments on earlier drafts of the paper. I also wish to thank Mrs S. Apostolakou, Director of the 24th Ephoreia of Prehistoric and Classical Antiquities, for providing sampling permits.

REFERENCES

Banou E. 2005, "LM III Mokhlos (East Crete) versus LM III Viannos (Central Eastern Crete): differences and similarities", *in* A.L. D'Agata and J. Moody (eds.), *Ariadne's threads: connections between Crete and the Greek mainland in Late Minoan III (LM IIA2 to LM IIIC)*, Tripodes 3, Scuola Archaeologica Italiana di Atene. Atene, p. 145-178.

Barnard K.A. 2003, "A macroscopic analysis of the Neopalatial fabrics", *in* K.A. Barnard and T.M. Brogan (eds.), *Mochlos IB, Period III. Neopalatial settlement on the coast. The Artisans' Quarter and the Farmhouse at Chalinomouri. The Neopalatial Pottery*, Philadelphia, INSTAP Academic Press, p. 3-11.

Bennet J. 1987, "The Wild Country East of Dikte: the problem of East Crete in the LM III Period", *in* J.T. Killen, J.L. Melena and J.-P. Olivier (eds.), *Studies in Mycenaean and Classical Greek presented to John Chadwick*, Minos 20-22, p. 77-88.

Bennet J. 1990, "Knossos in context: comparative perspectives on the Linear B administration of LM II-III Crete", *AJA* 94, p. 193-211.

Betancourt P.P., Muhly J.D., Farrand W.R., Stearns C., Onyshkevych L., Hafford W.B., Evely D. 1999, "Research and excavation at Chrysokamino, Crete, 1995-1998", *Hesperia* 68, p. 343-370.

Blitzer H. 1984, "Traditional pottery production in Kentri, Crete: workshops, materials, techniques and trade", *in* P.P. Betancourt (ed.), *East Cretan White-on-Dark Ware: studies of a Handmade Pottery of the Early to Middle Minoan periods*, University Museum Monograph 51, Pennsylvania, p. 143-157.

Brogan T.M., Smith R.A.K., Soles J.S. 2002, "Mycenaeans at Mochlos? Exploring culture and identity in the Late Minoan IB to IIIA1 transition", *Aegean Archaeology* 6, p. 89-118.

Day P.M. 1991, *A petrographic approach to the study of pottery in Neopalatial East Crete*, Ph.D., University of Cambridge.

Day P.M. 1995, "Pottery production and consumption in the Siteia Bay area during the New Palace period", *in* M. Tsipopoulou and L. Vagnetti, *Achladia: scavi e ricerche della Missione Greco-Italiana in Creta Orientale (1991-1993)*, Rome, p. 149-173.

Day P.M. 2004, "Marriage and mobility: traditions and the dynamics of the pottery system in Twentieth century East Crete", *in* P.P. Betancourt, C. Davaras and R.H. Simpson (eds.), *Pseira VIII: The archaeological survey of Pseira Island, Part 1*. INSTAP Academic Press, p. 105-142.

Day P.M., Haskell H.W. 1995, "Transport stirrup jars from Thebes as evidence of trade in Late Bronze Age Greece", *in* C. Gillis, C. Risberg and B. Sjöberg (eds.), *Trade and production in premonetary Greece: aspects of trade,* Studies in Mediterranean Archaeology and Literature Pocket-book 134, p. 87-109.

Day P.M., Joyner L., Relaki M. 2003, "A petrographic analysis of the Neopalatial pottery", *in* K.A. Barnard and T.M. Brogan (eds.), *Mochlos IB, Period III. Neopalatial settlement on the coast. The Artisans' Quarter and the Farmhouse at Chalinomouri. The Neopalatial Pottery*, INSTAP Academic Press, p. 13-32.

DAY P.M., JOYNER L., KIRIATZI E., RELAKI M. 2005, "Petrographic analysis of some Final Neolithic-Early Minoan II pottery from the Kavousi area", *in* G.C. Gesell and L.P. Day (eds.), *Kavousi I, The Archaeological Survey of the Kavousi Region*, INSTAP Academic Press, Appendix 3, p. 177-195.

DRIESSEN J., MACDONALD C.F. 1997, *The troubled Island: Minoan Crete before and after the Santorini eruption*, Aegaeum 17, Liège, Université de Liège.

DRIESSEN J., FARNOUX A. (eds.) 1997, *La Crète mycénienne, Actes de la table ronde internationale organisée par l'École française d'Athènes, 26-28 Mars 1991*, BCH, Suppl. 30, Paris.

GOSSELAIN O.P. 1999, "In Pots We Trust: the processing of clay and symbols in Sub-Saharan Africa", *Journal of material culture* 4 (2), p. 205-230.

GOSSELAIN O.P. 2000, "Materialising identities: an African perspective", *Journal of archaeological method and theory* 7 (3), p. 187-217.

HASKELL H.W. 2005, "Region to region export of transport stirrup jars from LM IIIA2/B Crete", *in* A.L. D'Agata and J. Moody (eds.), *Ariadne's threads: connections between Crete and the Greek mainland in Late Minoan III (LM IIA2 to LM IIIC)*, Tripodes 3, Scuola Archeologica Italiana di Atene, Atene, p. 205-221.

IGSR 1959a, *Geological map of Greece*: 1:50 000, Kato Chorion (Ierapetra) sheet, Institute for Geology and Subsurface Research.

IGSR 1959b, *Geological map of Greece*: 1:50 000, Siteia sheet, Institute for Geology and Subsurface Research.

IGSR 1993, *Geological map of Greece*: 1:50 000, Ierapetra sheet, Institute for Geology and Subsurface Research.

MACGILLIVRAY J.A. 1997a, "Late Minoan II and III pottery and chronology at Palaikastro: an introduction", *in* E. Hallager and B.P. Hallager (eds.), *Late Minoan III pottery chronology and terminology, Acts of a meeting held at the Danish Institute at Athens, August 12-14, 1994*, Monographs of the Danish Institute at Athens 1, Aarhus, p. 193-202.

MACGILLIVRAY J.A. 1997b, "The re-occupation of eastern Crete in the Late Minoan II-IIA1/2 periods", *in* J. Driessen and A. Farnoux (eds.), *La Crète mycénienne, Actes de la table ronde internationale organisée par l'École française d'Athènes, 26-28 Mars 1991*, BCH, Suppl. 30, Paris, p. 275-279.

POPHAM M.R., BETTS J.H., CAMERON M., CATLING H.W. and E.A., EVELY D., HIGGINS R.A., SMYTH D. 1984, *The Minoan unexplored mansion at Knossos*, BSA Suppl. 17, London.

PRESTON L. 2004, "A mortuary perspective on political changes in Late Minoan II-IIIB Crete", *AJA* 108, p. 321-348.

REHAK P., YOUNGER J.G. 1998, "Neopalatial, Final Palatial, and Postpalatial Crete", *AJA* 102, p. 149-166.

SHENNAN S.J. 1989, "Introduction: archaeological approaches to cultural identity", *in* J.S. Shennan (ed.), *Archaeological approaches to cultural identity*, London, Unwin Hyman, p. 1-32.

SMITH R.A.K. 2002, "The tombs of Mochlos and Myrsini: pottery and cultural regionalism in Late Minoan III Crete", Ph.D., Bryn Mawr College, Bryn Mawr.

SMITH R.A.K. 2005, "Minoans, Mycenaeans and Mokhlos: the formation of regional identity in Late Minoan III Crete", *in* A.L. D'Agata and J. Moody (eds.), *Ariadne's threads: connections between Crete and the Greek mainland in Late Minoan III (LM IIA2 to LM IIIC)*, Tripodes 3, Scuola Archeologica Italiana di Atene, Atene, p. 185-204.

SOLES J.S. 2003, "Introduction", *in* J.S. Soles (ed.), *Mochlos IA, Period III. Neopalatial settlement on the Coast. The Artisans' Quarter and the Farmhouse at Chalinomouri, The Sites*, Philadelphia, INSTAP Academic Press, p. 1-3.

SOLES J.S., DAVARAS C. 1992, "Excavations at Mochlos, 1989", *Hesperia* 61, p. 413-445.

SOLES J.S., DAVARAS C. 1994, "Excavations at Mochlos, 1990-1991", *Hesperia* 63, p. 391-336.

SOLES J.S., DAVARAS C. 1996, "Excavations at Mochlos, 1992-1993", *Hesperia* 65, p. 175-230.

TSIPOPOULOU M. 1990, "Νέα στοιχεία για τη Μινωική κατοίκηση στην περιοχή της πόλης της Σητείας", *in Proceedings of the 6th Cretological Congress*, A2, Chania, p. 305-321.

TSIPOPOULOU M. 1997, "Late Minoan III reoccupation in the area of the palatial building at Petras, Siteia", *in* E. Hallager and B.P. Hallager (eds.), *Late Minoan III pottery chronology and terminology, Acts of a meeting held at the Danish Institute at Athens, August 12-14, 1994*, Monographs of the Danish Institute at Athens 1, Aarhus, p. 209-252.

TSIPOPOULOU M. 2005, "'Mycenoans' at the Isthmus of Ierapetra: some (preliminary) thoughts on the foundation of the (Eteo)Cretan cultural identity", *in* A.L. D'Agata and J. Moody (eds.), *Ariadne's threads: connections between Crete and the Greek mainland in Late Minoan III (LM IIA2 to LM IIIC)*, Tripodes 3, Scuola Archeologica Italiana di Atene, Atene, p. 303-333.

WARREN P. 2001, "Review of The Troubled Island, by J. Driessen and C.F. MacDonald", *AJA* 105, p. 117-118.

WHITELAW T.M., DAY P.M., KIRIATZI E., KILIKOGLOU V., WILSON D.E. 1997, "Ceramic traditions at EM IIB Myrtos Fournou Korifi", *in* R. Laffineur and P.P. Betancourt (eds.), *TEXNH: craftsmen, craftswomen and craftsmanship in the Aegean Bronze Age*, Aegaeum 16, Liège, Université de Liège, p. 265-275.

WIESSNER P. 1983, "Style and social information in Kalahari San projectile points", *American antiquity* 48, p. 253-276.

WIESSNER P. 1984, "Reconsidering the behavioural basis for style: a case study among the Kalahari San", *Journal of anthropological archaeology* 3, p. 190-234.

PRELIMINARY RESULTS OF ARCHAEOMETRIC ANALYSIS OF AMPHORAE AND GNATHIA-TYPE POTTERY FROM RISAN, MONTENEGRO

MAŁGORZATA DASZKIEWICZ[1], PIOTR DYCZEK[2], GERWULF SCHNEIDER[3], EWA BOBRYK[4]

[1] ARCHEA, Warsaw, Poland
[2] Institute of Archaeology, Warsaw University, Poland
[3] Arbeitsgruppe Archäometrie, Freie Universität Berlin, Germany
[4] Faculty of Chemistry, Warsaw University of Technology, Poland

INTRODUCTION

In 2001, the Centre for Archaeological Research – Novae – of Warsaw University initiated explorations at a new site in Montenegro. It lies in the territory of the modern town of Risan, on the northern coast of the Kotor Gulf. The finds from excavations consisted of several artefacts: coins mainly of King Ballaios' issue (c. 167-135 BC),[1] huge quantities of Gnathia-type ware, amphorae and other ceramic products. Archaeometric analysis of selected pottery fragments began in 2004. These laboratory analyses were intended to present a basis for the archaeological interpretation of pottery finds by distinguishing local and imported pottery products. For comparison, besides chemical analyses in the Berlin databank of analyses of ancient pottery, data published by other groups were used bearing in mind the fact that the latter cannot always be compared when important details are missing.[2] This is true both for chemical and for petrographic analysis. In this preliminary paper, not all the possibilities of a detailed comparison for each of the analysed samples were dealt with completely, partly due to missing hypotheses on their provenances but also because of a lack of comparison data for the most probable region of provenance along the Adriatic coasts of Italy, Croatia, Montenegro and Albania.

ARCHAEOLOGICAL BACKGROUND

The variants of the town name appearing in the ancient written sources are *Rhizon, Rhizinium,* or *Risinium* (CIL VIII 2581). Archaeological investigations of Risan were initiated by A. Evans in the 1870s (Evans 1884, p. 42 ff). Regular excavations started in 1988 (Kovačević 1997a, p. 17-21; Kovačević 1997b, p. 18-25) yielded fragments of Gnathia-type skyphoi, amphorae with stoppers and coins of Ballaios. Excavations in 2001–2004 detected an extensive building complex covering over 600 m^2 (Dyczek 2005, p. 115-118). One wing took the form of a long building consisting of a series of small narrow rooms. Gnathia-type ware prevailed here, together with coins of Ballaios. The building, however, predates the reign of the king, C^{14} dates indicating a range from the 4th to the 2nd c. BC. The pottery is typologically uniform: mostly skyphoi, oinochoe, plates of differing size, small bowls and two kinds of gutti. Of the macroscopically distinguishable two groups of Gnathia-type vessels, the one characterized by finer quality, was probably imported from centres in Italy and/or Greece, and the other, chronologically later, may have been produced in Croatia, assuming that the archaeological and historical evidence has been read correctly. The imported group represents Late Gnathia, that is, from 325-270 BC (Green 1976, p. 10-13; Curti 1998, p. 14). The other group could correspond to the reign of Ballaios, that is, after 168 BC. It could even be referred to as post-Gnathia, considering the liberties that the potters making these vessels took with the original style (Vikić and Damaevski 1998, p. 97-100).

Wine amphorae constitute another important group of vessels. Indeed, wine seems to have been the main product imported from (southern) Italy and Greece. Sherds of amphorae were discovered in various places inside the building and the stores. In terms of amphora typology, they belong to the Greco-Italic group, representing different variants dated to different periods. The Risan amphorae clearly cover two chronological periods: from the 4th to the late 3rd c. BC and from the 3rd to the late 2nd c. BC.

1 – Dating of the reign of king Baillos is still in discussion.

2 – This is especially true when in an article on results of chemical analysis the original data are not presented and are only used in the text or in diagrams after multivariate statistics.

SAMPLES

MD3008 is the rim of a Dressel 1A amphora, a type dated generally to the late 2nd c. BC (Sciallano and Sibella 1991, p. 32; Peacock and Williams 1986, p. 86-87; Panella 1998, p. 531 ff). Amphorae of this type were produced on the Tyrrhenian coast of Italy from Etruria to Campania as containers for wine. MD3007 represents Lamboglia 2, which is a derivative of Dressel 1. Amphorae of this type were produced on the Adriatic coast of Italy at the end of the 2nd and in the early 1st c. BC (Beltrán Lloris 1970, fig. 129). MD3009, 3010, 3012, 3013, 3014, and 3016 are too fragmentarily preserved for any more substantial attribution. They represent a Greco-Italic type produced in the 3rd-2nd centuries BC (Peacock and Williams 1986, p. 84-85; Sciallano and Sibella 1991 p. 30; Py 1993, p. 47-48). The clay fabrics suggests several centres of production. MD3017 is a Will b type of vessel produced in 250-200 BC (Will 1982, p. 338-340). From a typological point of view MD3011 is a product of Magna Graecia workshops (MGS II). MD3015 was also manufactured in Magna Graecia; it can be attributed to the MGS V type. It was used to transport both wine and olive oil. MD3018 is identified as *Solocha I/l'épave de la Tour d'Angello sim.* Amphorae of this type are recorded in the 3rd and 2nd centuries BC (Zeest 1960, p. 151, pl. 15, 32). Typologically, MD3004 and MD3006 belong to the Rhodos 1 type which was produced between 325 and 275 BC (Py 1993, p. 42).

The Gnathia-type vessels, mostly bowls, two skyphoi and an oinochoe, are generally dated to the 3rd-2nd c. BC. Without delving into the intricacies of typological differentiation, we can say that bowls MD3020-3026 have parallels among products from Hellenistic workshops of the late 3rd through the end of the 2nd c. BC (Bernardini 1961, p. 28, pl. 66; Forti 1965, p. 49 f, pl. 15; Ujes 1999, p. 207-208, pl. 6). Skyphoi MD3027 and MD3029 were manufactured presumably in Italic workshops sometime in the 3rd c. BC (Vikić and Damaevski 1998, p. 106-107, 125, table 16). Oinochoe MD3028 is likely to be of the same origin (Bernardini 1961, p. 19 f, pl. 39, 40). On the other hand, the small plate MD3030 could be from the late 4th or early 3rd c. BC (Bernardini 1961, p. 29 f, pl. 59; Forti 1965, p. 101, pl. 28).

Provenance and type for the Gnathia-type ware is impossible to detect, mostly due to the fragmentary state of the decoration. Possibly some vessels are imports from Greek or Italic workshops, while others were manufactured in pottery centres situated in the territory of present-day Croatia, in Stari Grad on Hvar Island and at Vis and Lumbarda on Korali. These workshops were known to produce Gnathia-type ware in 360-270 BC (Vikić and Damaevski 1998, p. 106-107).

ARCHAEOMETRIC ANALYSIS

Through archaeometric studies we want to find a distinction between local and imported production and to get information about the main centres of production of amphorae and Gnathia-type pottery relating to Risan. The first procedure undertaken was chemical analysis,[3] which reveals the geochemical characteristics of both the plastic and non-plastic components of a pottery fabric. The primary purpose of additional thin-section studies is to specify the mineralogical and petrographic content, as well as the texture and grain size distribution of the non-plastic components of the body. However, it can only provide very general information about the matrix due to the limited resolution of the microscope, the small size of the clay minerals making up the plastic part of the body and the fact that they undergo transformation when fired. In view of these facts a third method was employed, which we call MGR-analysis[4] (Matrix Group by Refiring). This was used to identify matrix types, given that the thermal behaviour of plastic components during firing is governed by their chemical composition and, moreover, by their phase composition (*e.g.* Daszkiewicz and Schneider 2001). It is essential to carry out this type of procedure as chemical analysis provides only a general summary of the content of individual elements regardless not only of the phases in which they occur, but also regardless of the "position" of specific phases in a particular sherd, *e.g.* calcium in the matrix vs. calcium in added temper. The case where calcium is both in the matrix and in the temper can only be seen by MGR-analysis.

3 – Wavelength-dispersive X-ray fluorescence analysis (Philips PW1400) was used to determine the content of major elements (given as oxides in per cent by weight), including phosphorus and a rough estimation of sulphur and chlorine (the latter are not included in the Tables). It was also used to determine a series of fifteen trace elements (given in ppm), six of which, however, could only be ascertained with poor precision (Cu, Nb, La, Ce, Pb, Th). Total iron was calculated as Fe_2O_3. Samples were prepared by pulverising fragments weighing 1.5-2 g having first removed their surfaces and cleaned the remaining fragments with distilled water in an ultrasonic device. The resulting powders were ignited at 880 °C (heating rate 200 °C/h, soaking time 1 h), melted in Pt-Au crucibles with a lithium-borate mixture (Merck Spectromelt A12) and cast on Pt-Au support into small discs for measurement. This data is, therefore, valid for ignited samples but, with the ignition losses (l.o.i.) given in the table, may be recalculated to a dry basis. For easier comparison the major elements are normalised to a constant sum of 100% the original totals being between 97 and 101%. Calibration is based on some 50 international geochemical reference samples.

4 – From amphorae fragments four thin slices were cut from each sample in a plane at right angles to the vessel's main axis. One of these sections was left as an indicator of the sample's original appearance, whilst the remaining three were fired in a laboratory electric chamber furnace, each one at a different temperature. Firing was carried out at three different temperatures: 1100, 1150 and 1200 °C in air, static, with a heating rate of 200 °C/h and a soaking time of 1 h at the peak temperature. From Gnathia-type wares nine thin slices were cut from each sample, firing was carried out at 400, 600, 700, 800, 900, 1100, 1150 and 1200 °C to estimate matrix type as well as the equivalent original firing temperature.

ALTERATION OF SHERDS

Some high contents of P together with high losses of ignition indicate secondary influences changing the concentrations of Ba, Sr and sometimes also of Mn and Zn (*e.g.* sample MD3023). For the other elements, *e.g.* Mg, Cr and Ni, this effect, based on our experience with thousands of analyses of ancient pottery, can be ignored.

One fragment of a Greco-Italic type amphora (MD3014), also showing an elevated P-content, provides a typical example of alteration. Examined macroscopically, a paler colour (brownish-yellow) is visible in fresh fracture at the outer edges of the sherd, whilst the interior portion of the sherd is beige-red, which might be indicative of a higher concentration of carbonates at the edges. In reality the situation is quite the reverse: the outer edges contain fewer carbonates. Fig. 1 shows the appearance of the sample in its original state, and after refiring at 1150 °C and 1200 °C. After refiring at 1150 °C the outer portion of the sample becomes over-fired and its central part slightly over-melted. After refiring at 1200 °C the whole sample becomes melted (with a semi-melted matrix type), but still does not have a uniform colour. At both temperatures the central part fires yellowish-green. This thermal behaviour is typical of a ceramic body with a high content of calcium carbonates. The outer edges of the sample turn brown, indicating a lower concentration of carbonates.

In thin-section the outer and inner parts of the fragment look like two different samples. A highly calcareous matrix with large quantities of cryptocrystalline aggregates, some grains of polycrystalline quartz and very fine plagioclases is visible in the central part. An almost carbonate-free matrix occurs in the outer sections; only isolated very fine aggregates of cryptocrystalline carbonates and coarse iron-rich clay aggregates are visible. Seen under a polarising microscope this sample, and in particular its appearance after refiring, illustrates the problems which can derive from the results of chemical analysis if only small samples are taken for analysis, the powder being extracted by chance from only a single layer. As we removed about 1 mm of the outer layers before powdering our analysis still represents a mixture of the original core and of the altered surface layers where, for example, additional phosporus is absorbed from the surrounding earth and carbonates with calcium, strontium, and, maybe, magnesium are leached.

RESULTS OF ANALYSIS

The results of chemical analysis for amphorae are given in Table 1 and for Gnathia-type wares in Table 2. The first step when dealing with the chemical data consisted of grouping all of the samples using multivariate clustering procedures. Amphorae and Gnathia-type wares were regarded together because it was hypothesized that they may have the same origin somewhere on the eastern Adriatic coast or islands. The groups resulting from this first clustering mainly depend on Mg, Cr and Ni concentrations (Fig. 2) whereas Ti, K and Zr, in many other cases being the most characteristic elements, vary only within a very limited range. Based on Mg, Cr and Ni three major groups were distinguished: samples with low concentrations of these elements (*e.g.* Cr < 200 ppm), samples with medium concentrations (Mg about 3.5 to 5% MgO, Cr > 200 and Ni > 100 ppm), and samples with high concentrations (Mg > 6% MgO, Cr and Ni > 400 ppm).

In a next step, the results of cluster analysis were compared with the results of MGR-analysis and thin-section studies. The grouping resulting from chemical analysis fully corresponds with matrix type groups and non-plastic material groups. The three major groups, however, have to be split up into subgroups. This grouping is given in the Tables 1 and 2, with the same letters denoting ceramics made from a very similar raw material, hence most probably produced at a single pottery centre. The subgroups are indicated as numbers after the letter and may be different pottery workshops within a centre. The groups with low Mg, Cr and Ni are denoted by letters A to F and J. The groups with medium Cr and Ni are I, G and H. The groups with very high Cr and Ni are named K and EM.

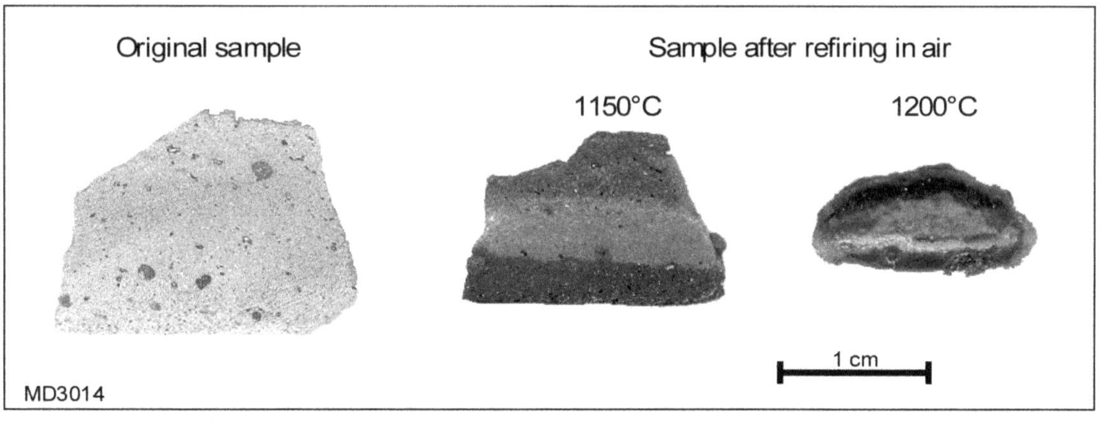

Fig. 1 – Cut sections of an altered Greco-Italic amphorae (sample MD 3014) before and after refiring in air.

sample	group	SiO_2	TiO_2	Al_2O_3	Fe_2O_3	MnO	MgO	CaO	Na_2O	K_2O	P_2O_5	V	Cr	Ni	(Cu)	Zn	Rb	Sr	Y	Zr	(Nb)	Ba	(La	Ce	Pb	Th)	l.o.i.	total
no.		% by weight										ppm															%	%
Dressel 1A																												
MD 3008	**A-2**	61,03	0,88	18,25	6,94	0,145	3,01	4,75	1,13	2,61	1,27	89	130	74	31	106	94	132	47	230	14	454	56	85	28	17	4,95	99,40
Lamboglia 2																												
MD 3007	**A-3**	57,86	0,85	17,48	6,98	0,104	3,61	8,55	1,25	2,73	0,59	106	149	69	27	108	99	225	39	190	16	465	58	78	23	15	1,12	98,87
Greco-Italic Type Will b																												
MD 3017	**A-1**	58,80	0,92	19,30	7,54	0,116	3,53	5,22	1,15	2,63	0,79	111	163	77	37	113	103	165	39	194	17	406	44	91	35	27	2,73	99,54
Magna Graecia Type MGS V																												
MD 3015	**A-4**	57,81	0,90	18,11	7,24	0,134	3,16	8,26	1,14	2,42	0,84	112	170	88	34	92	94	279	44	225	16	454	49	98	23	18	2,75	98,96
Greco-Italic																												
MD 3010	**A-1**	59,47	0,93	19,25	7,46	0,124	3,33	5,09	1,12	2,63	0,62	117	158	78	35	109	98	157	40	182	18	443	39	94	25	20	3,01	99,70
MD 3012	**A-1**	60,13	0,91	18,86	7,36	0,134	3,36	4,86	1,20	2,70	0,50	119	159	79	31	108	115	170	43	193	19	444	64	90	28	17	2,04	99,64
MD 3016	**C-1**	55,62	0,85	17,92	7,30	0,152	3,08	9,69	0,68	2,17	2,54	87	162	78	42	118	82	252	38	169	17	401	35	84	27	23	5,36	98,51
MD 3014	**D-1**	53,17	0,76	16,01	6,38	0,127	2,93	15,10	0,62	2,21	2,68	61	110	63	32	117	76	260	40	179	16	491	51	70	22	15	9,34	97,44
MD 3009	**E-1**	53,91	0,89	19,61	7,60	0,162	2,56	10,95	0,46	2,37	1,48	127	168	86	39	108	79	365	44	191	19	495	51	93	<5	11	6,15	99,47
MD 3013	**J-1**	58,81	0,80	16,47	6,38	0,096	3,24	10,26	1,12	1,62	1,21	115	132	56	14	52	53	195	39	197	15	361	49	92	17	16	2,96	98,79
Magna Graecia Type MGS II																												
MD 3011	**J-2**	60,31	0,76	16,03	6,12	0,101	2,68	10,26	0,65	1,94	1,15	93	143	64	<5	26	41	190	35	190	14	286	59	86	10	13	3,34	98,40
Solocha I/ l'épave de la Tour d'Angello sim.																												
MD 3018	**C-2**	52,12	0,83	17,65	7,04	0,145	3,22	12,10	0,69	2,31	3,89	77	147	65	46	146	75	260	39	181	15	595	34	69	28	18	8,42	98,05
Type Rhodos 1																												
MD 3004	**H-1**	57,60	0,80	14,30	6,78	0,095	4,26	11,81	1,57	2,53	0,25	171	226	71	30	74	93	351	33	151	10	331	37	43	11	11	2,82	100,08
MD 3006	**K-1**	65,15	0,72	13,08	6,70	0,154	4,62	6,04	1,69	1,65	0,21	112	551	277	35	71	63	208	31	162	<10	720	27	34	16	11	5,73	100,27

Table 1 – Results of chemical analysis by WD-XRF on amphorae.

sample	group	SiO_2	TiO_2	Al_2O_3	Fe_2O_3	MnO	MgO	CaO	Na_2O	K_2O	P_2O_5	V	Cr	Ni	(Cu)	Zn	Rb	Sr	Y	Zr	(Nb)	Ba	(La	Ce	Pb	Th)	l.o.i.	total
no.		% by weight										ppm															%	%
local clay																												
MD 3485	--	65,55	0,70	13,66	6,21	0,176	2,17	7,42	0,91	2,22	0,98	111	281	144	88	138	125	133	34	157	8	291	43	67	77	11	9,16	100,89
Adriatic coast?																												
MD 3030	**A-5**	59,33	0,87	18,50	7,05	0,096	3,46	5,57	1,28	2,43	1,41	106	143	70	40	102	101	304	42	239	15	582	56	96	28	22	2,34	100,31
MD 3029	**B-1**	57,45	0,90	18,87	7,29	0,086	2,11	7,51	1,07	2,31	2,41	116	144	59	na	119	90	316	na	136	na	540	na	47	44	na	3,97	97,26
MD 3026	**F-1**	48,69	0,95	20,19	8,51	0,160	4,50	10,34	0,70	2,30	3,65	85	182	82	54	104	71	358	46	178	18	780	56	87	32	21	5,28	98,85
MD 3025	**I-1**	52,14	0,93	18,01	7,78	0,120	3,58	10,01	1,01	2,22	4,21	134	218	129	na	121	66	538	na	217	na	609	na	86	na	na	5,15	98,85
MD 3028	**G-1**	52,12	0,80	16,34	6,93	0,090	3,69	15,19	1,10	2,13	1,61	120	203	104	29	84	62	514	35	167	15	421	38	68	21	15	4,33	98,64
East-Mediterranean?																												
MD 3022	**EM-1**	49,74	0,82	15,83	9,65	0,136	7,28	12,02	0,77	2,60	1,16	170	497	442	77	131	80	439	31	136	13	499	44	58	37	11	1,11	98,63
MD 3020	**EM-2**	51,14	0,81	15,31	9,35	0,127	7,00	12,43	1,20	2,40	0,23	157	500	411	42	118	77	294	32	137	13	403	30	56	30	12	0,33	98,88
MD 3024	**EM-2**	51,83	0,82	15,59	9,30	0,129	7,00	11,30	1,02	2,49	0,54	150	512	414	35	111	74	434	29	141	13	427	33	62	27	13	0,91	98,96
MD 3021	**EM-3**	49,75	0,90	17,29	10,48	0,159	7,39	10,14	0,71	2,02	1,18	159	541	477	44	88	65	252	41	144	14	414	47	73	24	17	2,52	98,55
MD 3027	**EM-3**	50,34	0,85	16,67	9,99	0,134	7,40	11,02	0,83	2,10	0,67	155	530	454	44	116	64	281	33	137	13	414	42	50	28	15	1,56	98,87
MD 3023	**EM-4**	46,84	0,79	16,09	9,42	0,170	6,89	14,00	0,76	2,28	2,76	137	470	414	83	147	70	931	31	156	11	769	50	65	98	16	2,14	97,78

Table 2 – Results of chemical analysis by WD-XRF on a sample of local clay and on Gnathia-type ware.

A local clay sample (MD3485) taken from the sediments of the excavation in Risan does not match any of the groups chemically. It is particularly very different from the low Cr-Ni groups and from the very high Cr-Ni groups and excludes a local production for these groups of pottery. This can also be concluded from the geology of the region of Risan where sediments could only be derived from weathering of limestone and dolomite. For the intermediate group a local production could be considered. To confirm this hypothesis, however, further studies are needed.

Amphorae with low Mg, Cr and Ni

The scattergram of Ni versus Cr (Fig. 2) illustrates that 12 of the amphorae analysed belong to the first major group characterised by low Mg, Cr and Ni. Calcium contents range between 5 and 15% CaO, potassium is below 2.7% K_2O, and Rb below ca. 100 ppm which is unusually low. In thin sections (Fig. 3) all show a more or less similar very fine (silty) to fine (< 0.2 mm) fabric with some larger grains of quartz or chert, clay or carbonate aggregates. Typical rock fragments or *e.g.* volcanic glass are missing. MD3010 and MD3012 is characterised by frequent quartz and pale mica with very little feldspar, and rare hornblende. MD3008, with an optically inactive matrix, contains quartz, chert, mica and, besides very little K-feldspar and plagioclase, rare and very tiny grains of epidote and ortho-pyroxene. In the optically inactive matrix of MD3007, besides quartz and pale mica, some inclusions of plagioclase and hornblende, and one grain of microcline were detected. The matrix of MD3018 is characterized by recarbonised calcite. Larger inclusions are chert and aggregates of calcite. MD3011 shows an isotropic yellow matrix in which, however, remains of microfossils can be detected. Because of the high firing temperature mica is rare. MD3013, which is very similar to MD3011 chemically, in refiring and in the colour of the matrix, does not show any remains of microfossils. It contains little mica and, besides quartz, chert and some intergrowth of quartz and feldspar, also a few grains of garnet, microcline, hornblende and ortho-pyroxene.

For comparison of amphorae from the western coast of Italy mainly the reference groups analysed by Thierrin-Michael (1992) were used. Her later publications (Thierrin-Michael 2000 and 2003; Thierrin-Michael *et al.* 2004; Thierrin-Michael *et al.* 2005) presented further data. Excluding the North-Etrurian amphorae from the region of Rosignano, petrographically all Italian amphorae analysed by her are characterized by their contents of volcanic material (Thierrin-Michael 2000, p. 226). This includes fragments of volcanic rocks with an ophitic structure, typical minerals, such as sanidine, clino-pyroxene, volcanic plagioclases and volcanic glass which we did not find in our thin sections even as very small fragments. Also, the comparison of chemical data did not show any agreement with any of her reference groups nor with data taken from our own data bank which includes published

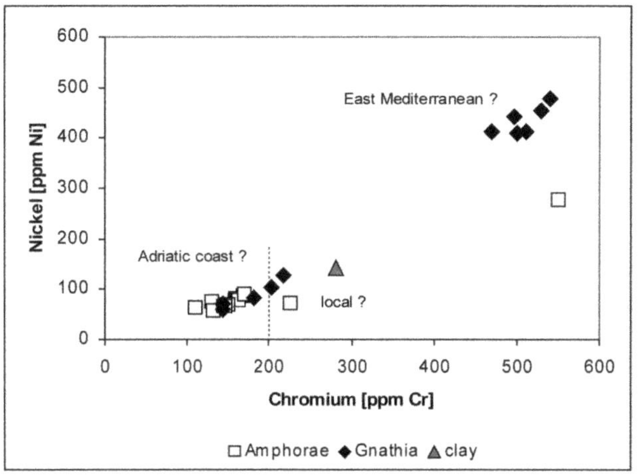

Fig. 2 – Variation of Chromium and Nickel in Gnathia-type ware, amphorae and a local clay sample from Risan.

and unpublished analyses of amphorae (Daszkiewicz *et al.* 1997; Daszkiewicz *et al.* 2000; Daszkiewicz *et al.* 2006). We also tried to compare our chemical and petrographic data with those of various authors (Slusallek *et al.* 1983; Jones 1986; Empereur and Picon 1986; Hesnard *et al.* 1989; Whitbread 1995), however, without any success.

Chemical data for amphorae from the east or west Adriatic coast were not available and therefore could not be compared. The geology of this area is different from that of the west Italian coast. Clay or ceramics from the Adriatic coasts should not show any volcanic inclusions. Probably the Risan amphorae can be attributed to a provenance somewhere within this region. Four amphorae petrographically analysed by Sondi and Slovenec (2003) seem to show some similarity to our amphorae. The data given in our paper leave open the possibility for later more detailed interpretations and secure attributions.

Amphorae with high Cr and Ni

The two amphorae of type Rhodos 1 differ from each other and from all other analysed amphorae. The first of them (sample MD3004) is characterised by inclusions of up to 1 mm from andesitic volcanic rocks, *e.g.* clino-pyroxene (Fig. 4, top left), typical rock fragments, volcanic plagioclase, but no volcanic glass. Quartz and occasional mica mainly belong to the silt and fine sand fraction. This amphora could not come from the Adriatic coasts, nor could it belong to the Italian reference groups analysed by Thierrin-Michael. Further studies will be made to check if the east Sicilian coast or other volcanic areas could be candidates for the provenance of this amphora.

The other amphora of type Rhodos 1 (MD3006) is chemically similar in its Mg, Cr and Ni values to amphorae from Galafone in the region of Rosignano (Thierrin-Michael *et al.* 2004, p. 242). However, it differs significantly in other element concentrations and its markedly different non-plastic particle composition precludes this possibility.

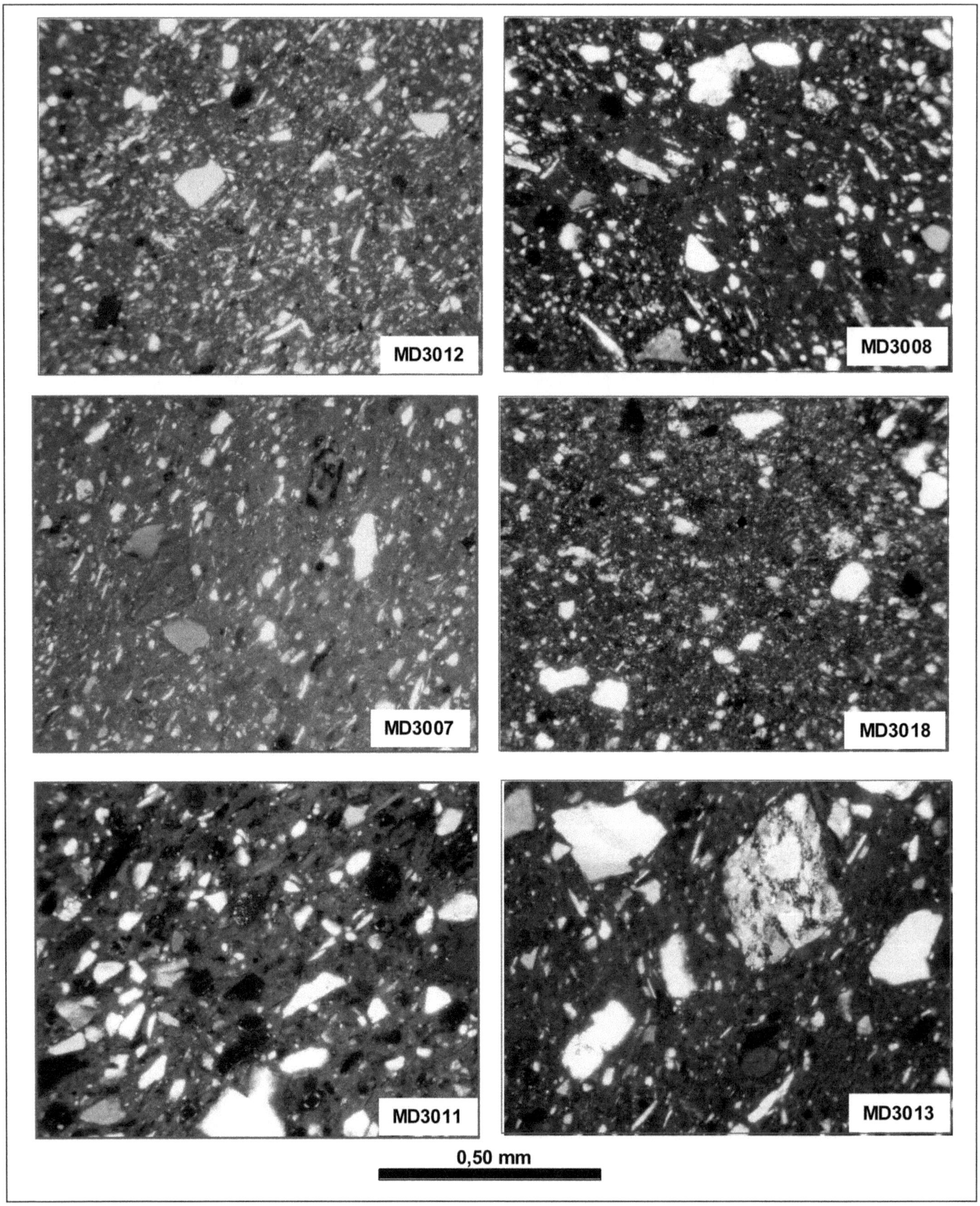

Fig. 3 – Thin section photomicrographs of six amphorae with low concentrations of chromium and nickel (XPL).

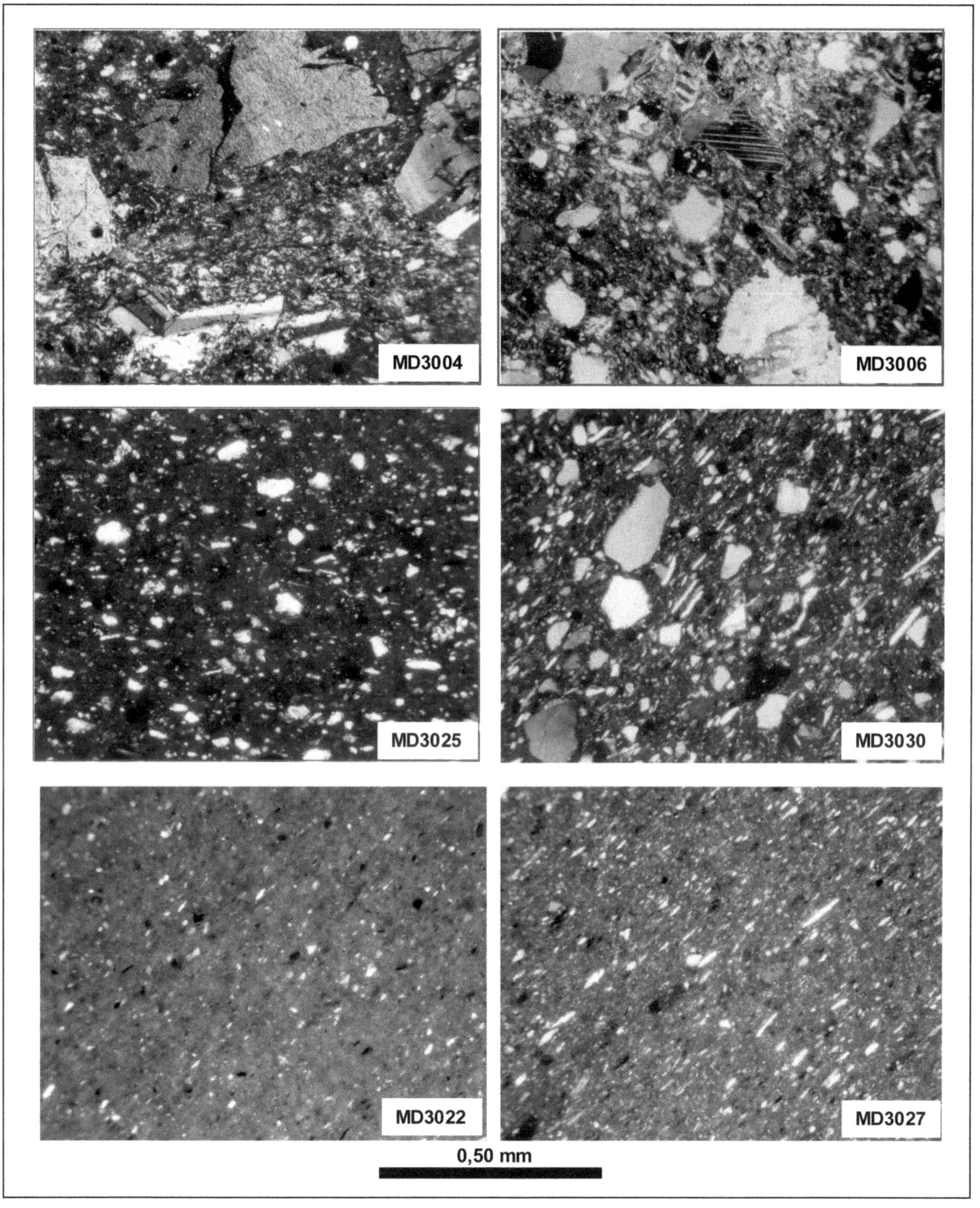

Fig. 4 – Thin section photomicrographs of two amphorae with high concentrations of chromium and nickel, and of four samples of Gnathia-type pottery (XPL).

Other regions with ophiolitic rocks, *e.g.* in the north of Albania and the sediments at the coastal plain or many regions in the eastern Mediterranean, could be possibilities for the origin of this amphora. The composition seems to be very different from that of amphorae from Rhodes, as established by analysis of Rhodian amphorae,[5] since the Mg concentration is much higher and Ni is about equal to Cr. In the thin section, quartz grains of up to 0.4 mm and primary calcite are dominant. The latter indicates a very low firing temperature. Other minerals in this amphora are serpentinite, plagioclase, chert and a few fragments of fine sandstone, pale mica and rare tiny crystals of epidote, ortho-pyroxene, and garnet.

Gnathia-type pottery

Gnathia-type pottery falls into several distinct groups (Table 2): five vessels which differ from each other and one homogeneous group of six vessels made from a raw material high in Mg, Cr and Ni, suggesting a provenance from the eastern Mediterranean Basin or from another region with ophiolitic rocks. Thin sections were made of two samples of the coarse group supposed to be local (*Fig. 4*, middle left and right). They contain quartz, pale and dark mica and decomposed calcite (MD3025). In the other sample (MD3030), besides quartz, pale and dark mica, occasional chert, plagioclase, K-feldspar and hornblende are visible in silt and fine sand fractions.

Because of the limited availability of local clay in Risan, and because of the difference from the composition of a local clay sample, local production of the analysed Gnathia-type ware at the site was not considered likely at the present stage of the research. On the other hand, a provenance from production centres somewhere along the Adriatic coasts seems probable for the five samples with low Cr and Ni. This is supported by the fact that one of these samples (MD3030) in chemical analysis, MGR-analysis and thin section is similar to the amphorae of group A. The chemical composition of all analysed samples, however, differs significantly from published data relating to Gnathia-type ware from Taranto (Hatcher *et al.* 1980; Jones 1986) and, in this first stage in the interpretation, no chemical composition analogies with other types of pottery were found.

The group with high Cr and Ni contents represents very fine pottery with no non-plastic particles larger than silt size (Fig. 4, bottom left and right). Particles consist of predominant quartz, very little serpentinite, mica and feldspar and rare hornblende in an isotropic matrix. Judging by the significant differences from the Rhodos 1 amphora (MD3006) they do not come from the same region. An Attic provenance can also be excluded. For geological reasons its provenance from a production centre in the eastern Mediterranean (or from an area in the north of Albania) is hypothesized.

CONCLUSIONS

1. The provenance of both amphorae and Gnathia-type pottery found in Risan remains an open issue. None of the analysed pottery fragments could be attributed to a known reference group. Workshops and production regions could be distinguished. The compositional groups are clear from three independent methods and data are given here for further interpretation. The Adriatic coast of Italy and/or Croatia, Montenegro or Albania for one section of the amphorae and the eastern Mediterranean Basin for another section of the imported amphorae were likely production areas. Analysis of kiln pottery and/or ceramic raw materials from hypothesized production sites within the region is essential, as the criterion of quality may not necessarily be a reliable indicator of local/regional provenance for some of the Gnathia-type wares. Also, analyses of other ceramics from Risan could be helpful in resolving the problems of provenance (Baranowski *et al.* 2006).

2. The analysed amphorae belong to several groups. One of the groups or production areas, determined according to chemical similarities (group A), includes the types Dressel 1, Lamboglia 2, Magna Graecia MGS II and two Greco-Italic amphorae as well as one vessel of Gnathia-type pottery. These and the other analysed amphorae, with the exception of the Rhodos 1 type vessels, were probably made in a region on the west or east Adriatic coast. A provenance from the Tyrrhenian coast in Italy is unlikely both for chemical and petrographic reasons. One of the Rhodos 1 amphorae, yet not attributed, must have been produced in a volcanic area. The other example, in view of its very high Mg, Cr and Ni values, comes from a region with ophiolitic rocks. However, for this sample a provenance from Rhodes seems unlikely because of its different chemical composition.

3. A large part of the Gnathia-type pottery samples analysed represent products which were very probably imported from workshops in the eastern Mediterranean Basin, though not from the same production centre where the Rhodos 1 type amphora was made. This chemically homogeneous group comprises five bowls and one skyphos. Some of the other analysed sherds, representing a bowl, a skyphos and a plate, were imported to Risan from unknown production centres, probably from the Adriatic coast like the larger part of the amphorae. Comparison with published data for Gnathia-type pottery from Taranto showed no agreement.

5 – Data for comparison of amphorae type Rhodes IIc were kindly provided by M. Picon, Lyon.

REFERENCES

Baranowski M., Bobryk E., Daszkiewicz M., Schneider G. 2006, "Vergleich makroskopisch bestimmter Keramikwaren und chemischer Gruppierung am Beispiel römischer und spätantiker Gebrauchskeramik aus Novae und Risinium", *Archäometrie und Denkmalpflege: Kurzberichte 2006, Jahrestagung in Stuttgart*, p. 160-162.

Beltrán Lloris M. 1970, *Ánforas romanas en España*, Zaragoza.

Bernardini M. 1961, *Vasi dello stile di Gnathia vasi a sernice nera*, Bari.

Curti F. 1998, *La céramique de Gnathia du Musée d'Art et d'Histoire de Genève*, Genève.

Daszkiewicz M., Bobryk E. 2001, "Pottery manufacturing technology detected from ceramic properties (K-H method)", *Archäometrie und Denkmalpflege: Kurzberichte 2001*, Köln, p. 113-115.

Daszkiewicz M., Schneider G. 2001, "Klassifizierung von Keramik durch Nachbrennen von Scherben", *Zeitschrift für Schweizerische Archäologie und Kunstgeschichte* 58, p. 25-32.

Daszkiewicz M., Meyza H., Schneider G. 1997, "A preliminary study of amphorae of the Mau XXVII/XXVIII type", *Polish archaeology in the Mediterranean VIII, Reports 1996*, Warsaw, p. 132-138.

Daszkiewcz M., Bobryk E., Dyczek P, Schneider G. 2000, "Chemical and mineralogical composition of Roman amphorae from Novae and some other sites in Bulgaria: first results", *Novensies* 12, p. 23-41.

Daszkiewicz M., Schneider G., Bobryk E. 2006, "Studies of pottery production in Selinunt (Italy)", *Archäometrie und Denkmalpflege: Kurzberichte 2006*, Stuttgart, p. 144-146.

Dyczek P. 2005, Rhizon, 2001–2003, "Preliminary Report on the excavations of the Center for archaeological research – Novae, Warsaw University", *Archeologia* LV, p. 101-118.

Empereur, J-Y., Picon M., 1986, "Des ateliers d'amphores à Naxos et à Paros", *Bulletin de correspondance hellénique* CX, p. 495-511 et p. 647-653.

Evans A. 1884, *Antiquarian Researches in Illiricum*, Westminster.

Forti L. 1965, *La ceramica di Gnathia, Monumenti Antichi della Magna Grecia II*, Napoli.

Green J.R. 1976, *Gnathia Pottery in the Akademisches Kunstmuseum Bonn*, Mainz.

Hatcher H., Hedges R.E.M., Pollard A.H., Kenrick P.M. 1980, "Analysis of Hellenistic and Roman fine pottery from Benghazi", *Archaeometry* 22, p. 133-152.

Hesnard A., Ricq M., Arthur P., Picon M., Tchernia A., 1989, "Aires de production des gréco-italiques et des Dr. 1", *in Amphores romaines et histoire économique: dix ans de recherche*, Collection de l'École française de Rome 114, p. 255-268.

Jones R.E. 1986, *Greek and Cypriot pottery, A review of scientific studies*, Fitch Laboratory Occasional Paper 1, British School at Athens, Athens.

Kovačevič V. 1997a, "Istraživanja arheoložskog lokaliteta 'Carine' u Risnu", *Glasnik Otdjeljenja umjetnosti Cernogorska akademija nauka i umjetnosti* 16, p. 17-51.

Kovačevič V. 1997b, "Rezultati istraživanja arheoložskog lokaliteta 'Carine' u Risnu", *Glasnik srbskog arheoložskog družestva* 14, p. 18-40.

Panella C. 1998, "Anfore e archeologia subaquea", *in* G. Volpe (ed.), *Archeologia Subaquea*, Firenze.

Peacock D.P., Williams D.F. 1986, *Amphorae and the Roman economy*, London.

Py M. 1993, Mélanges d'histoire et d'archéologie de Lattara. *Dictionnaire des céramiques antiques (VIIème s. av. n.è -VIIème s. de n.è.) en Méditerranée nord-occidentale*, Lattara 6.

Sciallano M., Sibella P. 1991, *Amphores: comment les identifier?*, Aix-en-Provence.

Slusallek K., Burmester A., Börker C. 1983, "Neutronenaktivierungs-analytische Untersuchungen an gestempelten griechischen Amphorenhenkeln: Erste Ergebnisse", *Berliner Beiträge zur Archäometrie* 8, p. 261-276.

Sondi I., Slovenec D. 2003, "The mineralogical characterisation of the Lamboglia 2 Roman-age amphorae from the Central Adriatic (Croatia)", *Archaeometry* 45, p. 251-262.

Thierrin-Michael G. 1992, *Römische Weinamphoren, Mineralogische und chemische Untersuchungen zur Klärung ihrer Herkunft und Herstellungsweise*, Dissertation Nr. 977, Universität Freiburg Schweiz.

Thierrin-Michael G. 2000, "Amphores gréco-italiques de l'oppidum de Pech-Maho (IIIe siècle av. J.-C.): détermination d'origine par l'étude pétrographique", *in* L. Rivet (ed.), *SFECAG, Actes du congrès de Libourne 2000*, Marseille, p. 225-231.

Thierrin-Michael G. 2003, "Classification des amphores vinaires italiques par l'examen macroscopique des pâtes: possibilités et limites", *in* L. Rivet (ed.), *SFECAG, Actes du congrès de St Romain-en-Gal 2003*, Marseille, p. 319-323.

Thierrin-Michael G., Menchelli S., Del Rio A., Cherubini L., Pasquinucci M. 2004, "Les amphores de l'Ager Pisanus et Volaterranus: production et distribution vers le Nord à la lumières des analyses", *in* L. Rivet (ed.), *SFECAG, Actes du congrès de Vallauris 2004*, Marseille, p. 237-244.

Thierrin-Michael G., Maza G., Thirion-Merle, V. 2005, "Banquets celtiques: comparaison des amphores à vin de trois sites", *in* L. Rivet (ed.), *SFECAG, Actes du congrès de Blois 2005*, Marseille, p. 741-752.

Ujes D. 1999, "Rhizon (Montenegrin Coast), a trading and staple town on the crossroads of Mediterranean and inland Balkan routes", *Numismatica e antichità classiche* 28, p. 203-220.

Vikić B., Damaevski V. 1998, "Apulske vaze Gnathia stila u arheološkom muzeju u Zagrebu", *Vjesnik arheološkog muzeja Zagreb* 3, p. 97-131.

Whitbread I.K. 1995, *Greek transport amphorae, a petrological and archaeological study*, Fitch Laboratory Occasional Paper 4, British School at Athens, Athens.

Will E.L. 1982, "Greco-Italic Amphoras", *Hesperia* 51, p. 338-356.

Zeest I.B. 1960, *Keramičeskaja tara Bospora*, Moskva.

TILES FROM THE LYON AREA IN THE 2nd CENTURY BC: LOCAL PRODUCTS OR IMPORTS?

N. CANTIN[1], A. DESBAT[1], A. SCHMITT[1]

[1] CNRS-UMR5138 "Archéométrie et Archéologie" Maison de l'Orient et de la Méditerranée, Université Lyon 2, 7 rue Raulin 69365 LYON Cedex 7, France, nadia.cantin@mom.fr

ABSTRACT

Tegulae have been used for roofing construction since the 2nd century BC in Lyon (France). The first discovery was made on the Rue du Souvenir site. Petrographic and chemical analyses have been carried out on tile samples from this site. Samples from other sites located in Lyon (of the same age or more recent) and the Rhône valley have been studied for comparison. The acid rock inclusions characterising all the studied fabrics favour the hypothesis of an early local tile production, while import from Tyrrhenian areas can be excluded. The observed variability in texture and composition suggests the existence of possibly several workshops in the local area.

INTRODUCTION

The archaeological site of Rue du Souvenir in Lyon, France (Fig. 1) was one of the first to reveal tiles (*tegulae*) dating from the second century BC (Desbat and Plassot 2000; Desbat 2003).

This discovery seemed to constitute an unique case in Gaul. Two hypotheses for the origin of these tiles can be proposed. A local origin would show the early introduction of Mediterranean building techniques. Importation, perhaps from Italy because of the high quantity of Italic amphorae found alongside the tiles (Thierrin-Michael and Maza 2002), or from southern Gaul, would be proof of trade (perhaps over a long distance) in building materials,

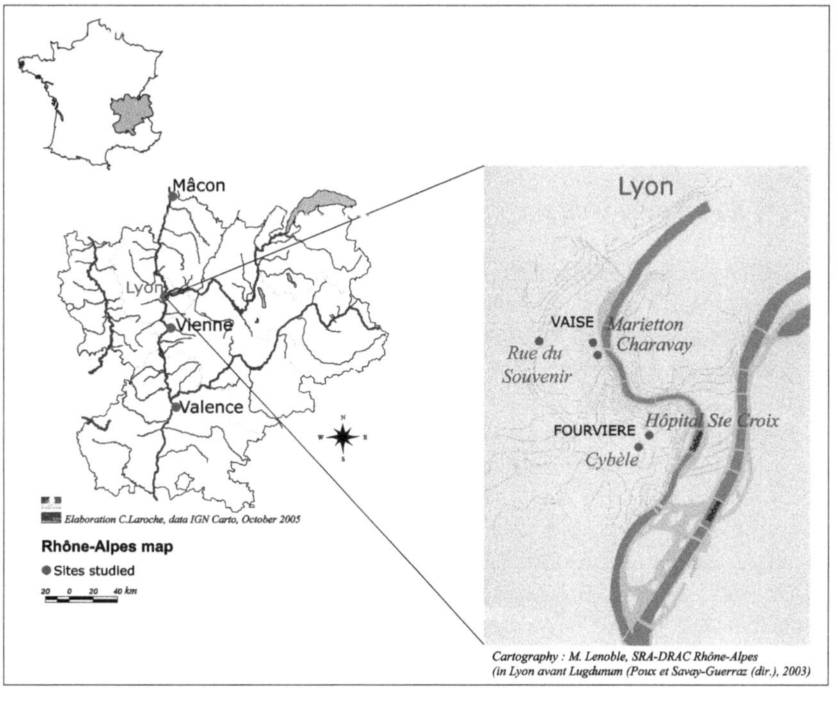

Fig. 1 – Geographic location of the sites presented in this study.

not suspected until now. Indeed, many architectural elements in Lyon seem to confirm the hypothesis of Roman influence before its foundation as a Roman colony in 43 BC. Petrographic and chemical analyses were carried out on tiles from the Rue du Souvenir to find out if they are local products or imports. The study was then extended to tiles from other sites in Lyon and in the Lyonese region, coming from contexts either contemporaneous with, or more recent than, those of the Rue du Souvenir. The latter were considered in order to assess the continuity of the choice of the raw materials used to manufacture the tiles.

1. SAMPLING AND METHODS OF ANALYSIS

Over the last fifteen years, many excavations have been carried out in Vaise, a district of Lyon, and have revealed the existence of tiles since the second century BC.

The site of "Rue du Souvenir" in Vaise in Lyon has been excavated three times, first in 1992, then in 1998 and 1999. This site, identified as an *emporium*, was situated near the Saône river. The tile fragments sampled from the site of Rue du Souvenir come from a roofing construction and are of three types : *tegulae*, *imbrices* and ridge tiles. The *tegulae* have various edge thicknesses, but all have an internal rounded profile as described by A. Desbat (2003). This is a peculiarity of early *tegulae* especially those found before the Augustan period in Lyon.

Tile samples from other sites discovered later in Vaise and in the area surrounding Lyon were studied archaeometrically for comparison. They are the following:

— The sites of ZAC Charavay (Lascoux and Gay 2003) and Marietton (Lyon, district of Vaise) possess tiles quite similar in shape and dimension to those found in Rue du Souvenir. These two sites have been dated to the 2nd century BC.

— We have also chosen samples from more recent sites like Hôpital Sainte-Croix (Lyon, district of Fourvière) (Monin and Maza 2003) dated to the 1st century BC and the so-called sanctuary of Cybèle (Lyon, district of Fourvière) (Desbat 1998). These two sites provided *tegulae* with angular profiles similar to those known from the Augustan period.

— South of Lyon, the use of tiles was suggested on two sites for the period under consideration : Vienne and Valence (Fig. 1). The site "25-31 rue de Bourgogne" in Vienne was excavated in 1987, and is described as a five period site (Durand 2000). We sampled material from the 4th period, which corresponds to a dwelling place constructed at the end of the 2nd century BC, and from the 5th period, dated to 40 BC. Finally, tiles discovered in levels dated to the beginning of the 1st century BC were sampled in Valence.

— More recently, excavations carried out in 2004-2005 at the site "Sennecé-les-Mâcon" near Mâcon, north of Lyon (Fig. 1), have indicated the use of *tegulae* and *imbrices* which were found in ditches dating to La Tène D (Barthelemy *et al.* in press). Macroscopic similarities with materials found in "Rue du Souvenir" were observed.

A total of 109 tiles coming from these sites were analysed, including 29 representative samples from the Rue du Souvenir (Table 1). Six samples of local clays from the Vaise plain were also examined. We used petrographic analyses to identify the raw material employed at the different sites and then we tried to identify their geological origin. Chemical analyses (WD-XRF) were also used to verify the coherence of our petrographic classifications.

2. RESULTS

2.1. Rue du Souvenir

Tiles from Rue du Souvenir have heterogeneous pastes, but have as a common feature many argillaceous inclusions *(argillaceous rock fragments (ARF), clay pellets*), whose nature is discussed below. Four compositional groups were distinguished by petrographic analysis (1, 2, 3, 4) corresponding to two chemical groups (A and B).

2.1.1. Petrographic observations

All the tiles of Rue du Souvenir have heterogeneous compositions with many inclusions, with minerals as well as rock fragments. These inclusions can have a very large diameter (up to 2 mm across).

Group 1 (n=3): C.3055, C.3057, C.3059

The fine fraction (50 to 100 μm) is composed of micas and discrete quartz grains. The middle fraction (400 to 700 μm) and the coarse fraction (about 1 mm to 2 mm) are composed of rock fragments (granitic fragments and cataclastic rocks) and discrete minerals (quartz, potassium and plagioclase feldspar, biotite, hornblende) (Fig. 2a). There is a very small quantity of secondary calcite.

Group 2 (n=9): C.3045, C.3049, C.3052, C.3053, C.3054, C.3056, C.3061, C.3250, C.3251

This is the most important group identified by petrography. The fine fraction (60 to 100 μm) is very abundant with much quartz (Fig. 2b). The middle fraction (around 500 μm) and the coarse fraction (1 mm) contain granitic rock fragments, discrete minerals and also fragments of detrital rocks. In addition to this we observed to a lesser degree rounded bioclastic limestone and fossil shells. If we compare this group with group 1, there is a lot of cryptocristalline calcite (corresponding to secondary calcite).

Group 3 (n= 8): C.3043, C.3044, C.3046, C.3050, C.3051, C.3063, C.3082, C.3249

Compared to group 2, there is less of the fine fraction (50 to 80 μm) composed of quartz, feldspar and some micas. The middle fraction (400 to 700 μm) and the coarse fraction (around 1 mm) are both composed of quartz, potassium and plagioclase feldspars, and granite

Provenance	Number of samples	Chemical sample id.	Petrographic sample id.	Material
Lyon, rue du Souvenir	29	SOV 1 to 21 SOV 30 to 32 SOV 78 to 82	C 3043 to C 3063 C 3082 to C 3084 C 3249 to C 3253	tile tile tile
Lyon, Villa Montel, rue du Bourbonnais, Vaise	6	ASC201 to 206	C 3264 to C 3269	clay
Lyon, ZAC Charavay, us2289	11	TUI 1 to 11	C 3212 to C 3222	tile
Lyon, Rue Marietton, fouilles 1995, us103	10	TUI 12 to 21	C 3223 to C 3232	tile
Lyon, Hôpital Sainte-Croix	8	TUI 23 to 25 TUI 30 to 31 TUI 34 TUI 36 to TUI 37	C 3234 to C 3236 C 3241 to C 3242 C 3245 C 3247 to C 3248	tile tile tile tile
Lyon, Cybèle	17	TUI 66 to 82	C 3420 to C 3436	tile
Mâcon, ZAC Sennecé-les-Mâcon	12	TUI 62 to 65 TUI 83 to 90	C 3416 to C 3419 C 3573 to C 3580	tile tile
Vienne, 25-31 rue de Bourgogne	14	TUI 48 to 61	C 3290 to C 3303	tile
Valence	6/8	TUI 38 to 42 not analysed TUI 45 not analysed	C 3254 to C 3258 C 3259 C 3261 C 3262	tile tile tile tile

Table 1 – List of the samples used in this article.

fragments (Fig. 2c). We noted in particular amphibole granite and micaschist. There are also a few calcareous compounds. The main difference from group 1 lies in the small proportion of micas in the fine fraction.

Group 4 (n=6): C.3047, C.3048, C.3058, C.3060, C.3252, C.3253

This group is characterized by pastes with a very significant proportion of inclusions. The composition is very close to that of group 2 but with some particular features. They contain vitreous inclusions due to clay that has been vitrified during the firing (Fig. 2d). Bioclastic limestone and fossils have also been identified in this group. As in group 2, there is also a great deal of secondary calcite in this group of tiles

Sometimes the differences between the groups 2 and 4 are not very clear. Perhaps the distinction is not necessary or pertinent.

Marginals

The two samples C.3062 and C.3084 are very different from the other tiles discovered in the site. In fact, the pastes are very homogeneous and the inclusions are very well sorted. They are composed only of fine inclusions (quartz and feldspar) (diameter between 50 to 100 μm).

The question concerning the tile making was: did potters add grog in their preparation of clay for these *tegulae*? Some tile fragments have many red, orange or black grains that were first interpreted as grog (to the naked eye). However, under the microscope the presence of grog was not easily recognisable (see Whitbread 1986 and Cuomo di Caprio and Vaughan 1993). These authors have noted the convergence of several criteria to determine the presence of grog. Although some criteria were observed, others were not, so that we could not positively identify these particles as grog. Therefore we would prefer to call them "argillaceous inclusions", in this case.

2.1.2. Chemical results

The pastes of all the tiles investigated from the Rue du Souvenir are Ca-poor. The analysis of their compositions reveals two main groups, A and B, and three marginal samples (SOV 4, SOV 20 and 32) (Table 2). SOV 20 and 32 are very rich in SiO_2 and their compositions differ completely from the other two groups. They have also been identified by petrography (cf. *supra*). The sample SOV 4 was considered as a marginal of group B and its composition was not included in the mean of the group. Differences were also noted in Sr and Ba concentrations.

Samples in group A contain more Fe_2O_3, TiO_2, K_2O, Al_2O_3, MnO, Rb, Zn, Cr, Ni, Ba and V and less CaO, SiO_2, Zr and Sr than samples in group B.

These differences can be illustrated in binary diagrams (Fig. 3). The graph 3a (Fe_2O_3/Al_2O_3) indicates the separation between the two groups and that Fe_2O_3 is indeed correlated with Al_2O_3. The figure 3b shows an inverse correlation SiO_2/Al_2O_3 that is due to the significant proportion of siliceous inclusions (minerals and rocks). These differences suggest that two types of raw materials were employed to produce the two groups.

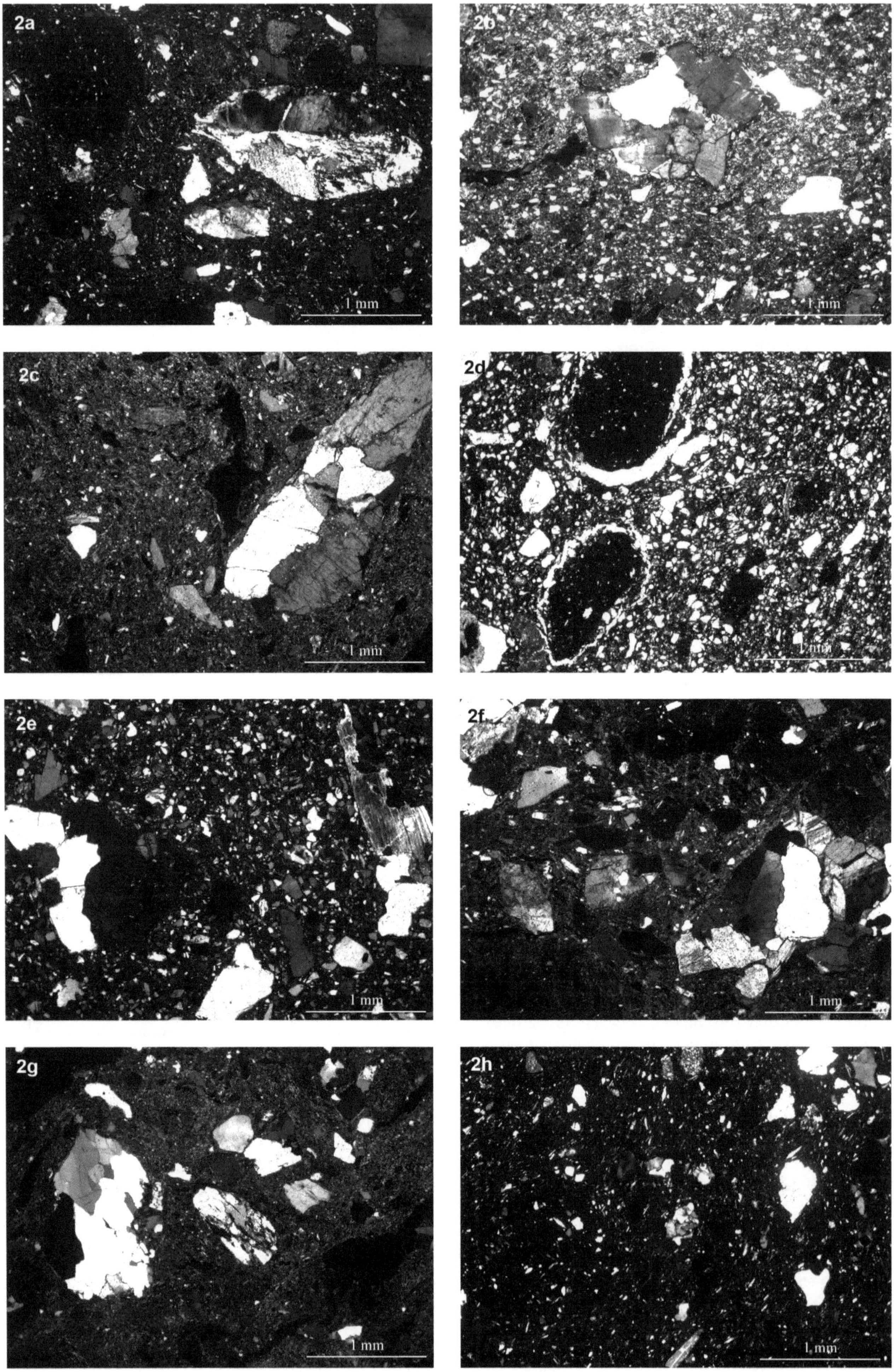

Fig. 2 – Photos of thin sections of tiles from Rue du Souvenir (2a, 2b, 2c, 2d), clay from Vaise (2e), tiles from Mâcon (2f), from Vienne (2g) and from Valence (2h). Photo 2a is representative of group 1, 2b of group 2, 2c of group 3, 2d of group 4. XPL.

		major elements (weight oxide %)										minor and traces elements (ppm)													
		CaO	Fe_2O_3	TiO_2	K_2O	SiO_2	Al_2O_3	MgO	MnO	Na_2O	P_2O_5	Zr	Sr	Rb	Zn	Cr	Ni	La	Ba	V	Ce	Y	Th	Pb	Cu
Lyon, Rue du Souvenir																									
group A (n=11)	m	2,03	7,54	0,942	3,18	63,12	19,54	1,94	0,0898	1,01	0,40	205	157	204	131	102	65	53	800	121	100	42	14	42	35
	σ	0,45	0,40	0,077	0,31	1,09	0,71	0,22	0,0131	0,20	0,22	15	37	21	6	6	6	4	99	12	7	3	2	4	5
	σ%	22	5	8	10	2	4	11	15	20	55	7	24	10	5	6	9	7	12	10	7	7	13	9	14
group B (n=15)	m	4,19	5,83	0,749	2,46	67,59	16,29	1,28	0,0621	0,89	0,47	286	165	142	107	88	56	50	638	98	98	41	15	37	28
	σ	1,48	0,62	0,064	0,17	1,32	0,92	0,13	0,0174	0,12	0,23	21	50	14	11	13	9	6	105	10	5	3	3	5	5
	σ%	35	11	9	7	2	6	10	28	13	48	7	30	10	10	14	15	12	16	10	5	7	19	13	19
SOV 4		3,42	6,86	0,769	2,74	65,19	17,91	1,46	0,0688	0,74	0,59	209	294	174	126	92	66	52	1013	104	93	39	16	35	35
SOV 20		1,57	4,31	0,794	2,23	75,66	12,93	1,03	0,0681	1,08	0,17	385	105	128	78	71	43	53	451	77	104	38	14	38	19
SOV 32		2,44	4,68	0,738	2,27	73,41	13,83	1,03	0,0711	1,10	0,26	392	129	133	86	72	53	52	547	70	101	39	16	33	27
Lyon, clays from Vaise																									
ASC201		0,96	3,72	0,746	2,34	77,63	12,19	0,98	0,0935	1,08	0,13	325	104	122	71	68	41	47	444	65	89	35	13	24	22
ASC202		1,08	7,03	0,813	2,48	71,46	14,57	1,33	0,0454	0,80	0,25	284	98	145	93	88	51	47	433	91	98	36	12	34	24
ASC203		1,39	8,48	0,832	2,27	69,54	15,04	1,35	0,0494	0,58	0,30	329	104	144	100	91	53	56	461	98	103	38	14	32	24
ASC205		1,10	5,18	0,791	2,41	73,44	14,51	1,28	0,0665	0,88	0,20	304	102	142	92	82	54	46	449	92	96	40	12	34	23
ASC206		1,31	8,20	0,982	3,16	63,35	19,97	1,99	0,1232	0,46	0,27	159	106	223	155	112	73	61	630	130	105	41	15	34	40
Lyon, Charavay																									
group A (n=4)	m	2,50	7,44	0,853	2,99	64,14	18,71	1,69	0,1000	0,92	0,50	202	127	188	137	110	64	52	703	119	99	44	16	41	30
	σ%	33	5	8	10	1	3	15	17	20	43	5	10	15	5	9	11	4	8	10	4	5	12	7	7
group B (n=7)	m	5,58	5,57	0,708	2,76	66,65	15,62	1,21	0,0818	1,04	0,60	279	153	143	109	89	51	53	723	89	93	41	15	40	25
	σ%	11	8	6	4	2	7	12	29	11	41	5	12	13	9	14	9	5	16	11	5	5	15	11	13
Lyon, Marietton																									
group B (n=5)	m	4,20	6,42	0,785	2,47	66,04	16,78	1,35	0,0756	0,85	0,83	279	135	148	122	98	61	57	766	104	101	43	13	49	27
	σ%	48	19	12	9	4	12	16	26	19	40	18	14	17	17	12	17	5	15	20	3	8	12	18	14
group 5 (n=5)	m	2,06	6,23	0,811	3,38	65,43	17,64	1,45	0,0944	1,39	1,29	179	170	186	117	88	54	48	1063	87	88	38	11	44	27
	σ%	11	4	2	2	1	2	9	10	4	32	3	9	2	1	4	1	7	10	14	4	4	10	12	4
Lyon, Hôpital Sainte-Croix																									
(n=6)	m	1,22	6,10	0,831	3,52	67,02	17,80	1,66	0,0888	1,40	0,18	237	122	199	120	89	50	60	687	93	99	40	14	44	23
	σ%	13	5	2	6	2	5	6	7	14	11	17	8	5	13	5	11	11	12	16	7	8	17	16	28
TUI 30		3,08	7,35	0,795	2,91	64,34	18,30	1,52	0,1184	0,77	0,63	201	134	162	133	104	65	56	784	105	92	41	13	46	32
TUI 37		0,95	4,84	0,658	3,53	71,58	15,42	1,18	0,0445	1,48	0,15	242	105	184	84	69	35	50	556	71	91	32	15	48	16
Lyon, Cybèle																									
(n=17)	m	1,05	6,93	0,909	3,73	64,40	19,19	1,86	0,0899	1,38	0,24	252	116	226	125	95	54	61	723	101	108	46	17	39	30
	σ%	23	8	5	7	3	8	5	26	11	11	10	7	7	14	9	13	18	17	12	16	14	18	14	15
Mâcon																									
(n=10)	m	1,58	7,03	0,896	3,15	64,17	18,79	1,78	0,1095	1,32	0,98	207	140	191	125	104	57	51	778	92	96	44	14	42	30
	σ%	14	5	6	10	2	5	17	27	14	84	19	16	15	7	5	9	7	18	15	8	4	17	13	17
TUI 88		0,60	8,57	0,870	4,89	59,29	20,88	3,29	0,0320	0,27	1,13	209	131	292	99	118	61	51	526	132	90	32	17	15	34
TUI 89		1,46	9,00	1,099	3,61	54,92	23,23	1,91	0,0947	1,04	3,37	214	158	226	164	142	75	62	1191	126	104	52	18	37	44
Vienne																									
(n=12)	m	1,53	7,21	0,946	3,71	63,39	19,45	1,95	0,0972	1,24	0,26	211	127	221	136	106	61	58	804	113	102	43	13	50	40
	σ%	24	13	13	7	4	9	17	14	16	17	14	11	7	11	13	12	20	7	24	9	11	19	17	25
TUI 56		1,60	4,60	0,654	3,23	71,91	14,65	1,28	0,0571	1,62	0,24	219	130	172	85	82	41	41	512	82	98	32	15	47	19
TUI 58		1,31	5,63	0,713	3,68	68,73	16,50	1,45	0,0773	1,52	0,23	161	131	198	95	76	44	47	673	74	84	33	10	43	29
Valence																									
(n=6)	m	1,36	7,39	0,985	3,57	62,77	20,28	1,95	0,1155	1,17	0,21	247	124	205	136	106	64	63	806	113	114	47	17	43	32
	σ%	18	11	5	3	3	4	14	10	11	10	15	10	6	6	12	8	13	7	13	11	7	12	6	12

Table 2 – Results of WD-XRF analysis of tiles from Lyon and the Rhône valley.
(n = number of samples; m = mean ; σ = standard deviation ; σ% = relative standard deviation).

To conclude, the chemical results show a good agreement with those derived from petrography. The petrographic sets 1 and 3 correspond to the composition of group A whereas group B contains the sets 2 and 4. Indeed, certain differences observed by petrography do not show up in the chemical compositions.

2.1.3. Comparison with local clays

The principal question was to identify the origin of the raw materials employed. As the results above do not indicate a foreign origin for the tiles, which was the first archaeological hypothesis, we needed to compare their pastes with locally available clays.

Lyon is the point of convergence of several geological areas and offers different possibilities for clay sources. On the west side, we can find weathered clays of crystalline and metamorphic rocks. These clays can be residual or can have been moved. Morainic clays and loess are very common to the east and north east of Lyon. Alluvial deposits of the Rhône or Saône rivers can also be employed as raw materials. The ancient urban development of the city is responsible for a failure to recognize certain zones. Prospection in the city is practically impossible now, except during archaeological excavations. So, most of our clay sampling in Lyon today is linked to archaeological intervention and carried out with the help of geomorphologists, as was the case during the last excavations carried out in Vaise. Indeed, the clays presented here (ASC201 to 206) were sampled in this area. The petrographic composition of Vaise clays varies a lot but is close to petrographic group 2 (Fig. 2e). They contain red nodules but no calcareous fossils. Their chemical compositions (Table 2 and Fig. 3) indicate variations and differences too. In spite of this, figure 3b shows the same

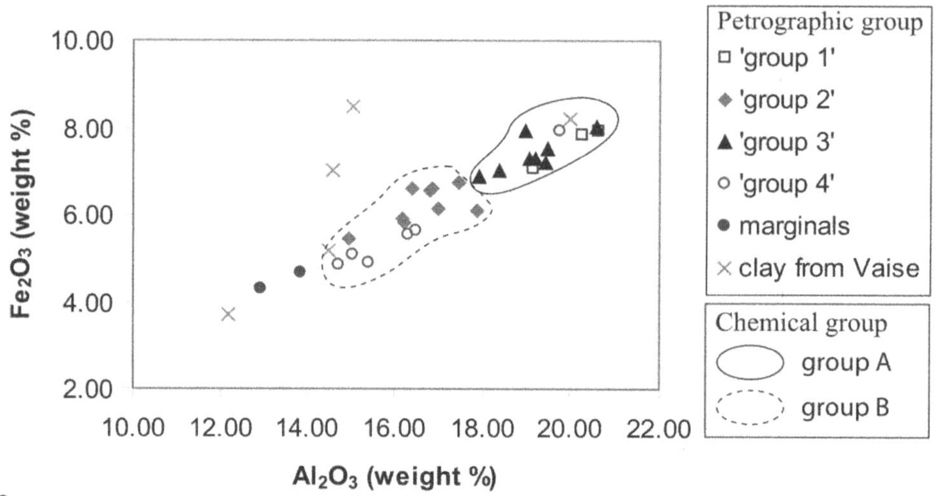

3a

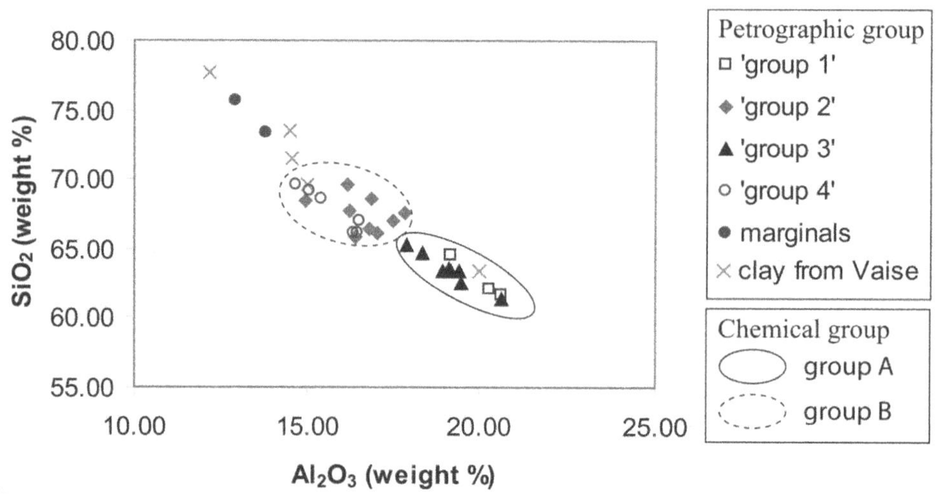

3b

Fig. 3 – Binary diagrams showing the separation between chemical groups and correlation with petrographic groups in Rue du Souvenir, Lyon.

inverse correlation as that demonstrated for the tiles, due to the important natural siliceous coarse fraction.

2.1.4. Conclusion

Four petrographic groups, corresponding to two chemical groups, were identified. Some distinctions made by petrographic examination between group 1 and 3, and groups 2 and 4, were not observed in their geochemical compositions.

Thus, two main groups of composition can be hypothesized for tiles from Rue du Souvenir, and may relate to the use of two different clays.

Even if we have not identified truly discriminant non-plastic inclusions, the nature of pastes is in agreement with the composition of local raw materials. They may come from weathered clay derived from granitic (normal and amphibole granite) and cataclastic rocks (A) and granitic sands (colluvium) (B) (similar in composition to the clays of Vaise). In addition, potters probably did not use temper but only the natural raw material.

2.2. Sites in Lyon

First, we looked at tiles from contemporary sites (Charavay, Marietton) and then we examined tiles from more recent sites (Hôpital Sainte-Croix, Cybèle) to assess the continuity of the choice of raw materials employed.

Charavay

This site shows many similarities with Rue du Souvenir. Seven samples are close to those of petrographic group 2, whereas four other tiles have the same composition as group 3 identified at Rue du Souvenir. The chemical analysis indicates the same distinction between a group A and a group B (Table 2). Tiles discovered in Charavay have the same compositions as those found in Rue du Souvenir. Therefore, they may have been produced from the same raw materials and may have come from the same workshop.

Marietton

Half of the samples examined for Marietton are similar to the petrographic group 2. The others form a new petrographic group, named 5. This last group is very similar to Souvenir-group 3. Indeed, the pastes have the same mineralogy but show granulometric differences (the coarse fraction is more common and inclusions have larger diameters—up to 3 mm across). Chemical analysis also shows two main composition groups. One can be compared to group B (identified at Rue du Souvenir) (Table 2) but shows greater deviation from the mean than the others, comparable to group B for instance in iron, aluminium and magnesium. The other group called "5" (as the corresponding petrographic group) has an intermediate composition between group A and B. It is quite close to the chemical group A but shows differences for instance in iron, titanium, aluminium, zirconium. Chemical results have smaller deviations from the mean, which indicates an homogeneous group compared to group Marietton (B).

Hôpital Sainte-Croix

The pastes have heterogeneous compositions with micaceous clays with non-plastic inclusions such as granite, gneiss. Fabrics are different from those of Rue du Souvenir (Table 2) except for one case (TUI 30) which has a petrographic and chemical composition much closer to Souvenir petrographic group 3 (corresponding to the chemical group A). The raw materials employed for the production of these tiles may come from another, but similar deposit.

Cybèle

The tiles found in Cybèle also have mica-rich compositions. Petrographically they are close to those of Hôpital Sainte-Croix, but chemically they are not. They differ from Rue du Souvenir samples too. Chemical compositions differ mainly from Rue du Souvenir ones in minor and trace element concentrations (Table 2). The sampling was carried out in several very well dated levels but there is no clear relationship between composition and date. This may suggest a continuity in the choice of raw materials for the periods investigated (40 BC to 30 BC).

To conclude, the three sites dating from the second century BC, located in Vaise (Rue du Souvenir, Marietton, Charavay) have typologic, petrographic and chemical similarities. The convergence of these resemblances is an argument in favour of the hypothesis of a common local origin for these tiles. However, the two more recent sites (first century BC), located in Fourvière (Hôpital Sainte-Croix and Cybèle), show differences from the earlier ones. This may indicate either a change in the choice of raw materials or that another workshop located somewhere else in Lyon supplied these sites.

There is a certain diversity in the local raw materials employed to produce ceramic building materials in Lyon during the second and the first centuries BC. But until now we have not observed a real continuity in the choice of raw materials employed for these tiles, and we are also short of data to allow us to understand how these building materials were produced. Moreover, for this period tile workshops are unknown and there are few sites where tiles have been found.

2.3. Mâcon

Three petrographic groups were distinguished. In the main group, pastes have many inclusions spread over several granulometric classes. Discrete minerals are mostly found in the fine fraction (50 to 80 μm) and in the middle fraction (80 to 250 μm); rock fragments are more often observed in the coarse fraction (diameter up to 2 mm across). We noted the presence of quartz, potassium feldspar, muscovite and in lesser proportion plagioclase feldspar and biotite. Granitic rock fragments predominate over cataclastic rocks and rare micaschists. This composition is comparable to Souvenir-group 3 but with granulometric differences (Fig. 2f). In the second group, tiles contain vitreous argillaceous inclusions, as described for Rue du Souvenir. Amphibole has also been identified as an accessory phase.

Pastes classified in the third group have a very abundant fine fraction (50 μm), composed of quartz, feldspar and a few muscovite inclusions. The coarse fraction is not abundant and is composed of granitic fragments. Argillaceous inclusions are very abundant in this group.

The chemical analysis confirms that tiles from Mâcon are Ca-poor with a quite significant quantity of Fe_2O_3 (between 7 and 8%). These compositions are very close to those of Souvenir-group A (Table 2).

We conclude that raw materials used in Lyon and Mâcon are very similar petrographically and chemically. In the two cases clay from weathered granite has probably been employed. But it is difficult for us to determine if tiles found on these two sites come from the same workshop. Both hypotheses are possible: the use of local raw materials and local production, or importation from Lyon.

2.4. Vienne

Two petrographic groups were identified, showing similarities to Souvenir-groups 1 and 3 (Fig. 2g). Indeed, discrete minerals such as quartz, feldspar, some mica, amphibole, with granitic and cataclastic rocks are the main inclusions in these tile pastes. Differences from Rue du Souvenir are observed because of the proportion of inclusions that seems larger in Vienne.

The chemical analysis shows quite homogeneous compositions (Table 2). Two samples are not included in the mean of the group (TUI 56 and TUI 58). TUI 56 presents differences in several chemical elements (Fe_2O_3, TiO_2, SiO_2, Al_2O_3, MnO, Rb, Zn, Ni) that can be explained by petrographic observations. Indeed, a high quantity of granitic fragments were observed in this sample. TUI 58 also differs from the group in several elements (TiO_2, SiO_2, Al_2O_3, MnO, Zr, Zn, Cr). This sample is comparable to group A and is characterised by a higher proportion of inclusions.

Even so, similarities with tiles discovered in Rue du Souvenir (group A) and also with Mâcon can be observed. The nature of the non-plastic inclusions involves similarities of composition, and as these inclusions can be found at the three sites it is difficult to say if there was one or many workshops producing these tiles. But until now, very few cases of tiles for this period have been reported.

2.5. Valence

The pastes of tiles sampled in Valence are very micaceous (Fig. 2h), the main petrographic difference from the samples from Rue du Souvenir. The Valence tiles were probably made with a clay from altered mica schists. Nevertheless, chemical analysis (Table 2) are close to Souvenir-group A with some differences in MnO and Zr. But thanks to the petrographic observations, we can suggest that the raw materials employed to produce tiles in Valence are not the same as those used in Lyon, Rue du Souvenir, Vienne and Mâcon even if their chemical compositions are not very different.

3. CONCLUSIONS

Tiles discovered at three sites in Lyon dating from the second century BC (Rue du Souvenir, Charavay, Marietton) were made with similar raw materials. As we do not know any production structure for the period and the area considered, we can only argue from geological criteria.

The tiles from the more recent sites studied in Lyon seem to have been produced with different raw materials than the earlier ones. This might indicate an evolution in the choice of raw materials and their processing.

We have to develop studies in Lyon in order to understand better which raw materials were employed for ceramic building materials and where they came from during the Gallo-Roman period. Presently, we do not know of any production structure in Lyon during the Gallo-Roman period. There is no tile workshop identified before the 12th century AD.

Analysis established similar compositions between Lyon, Mâcon and Vienne. Clays derived from the alteration of granites and some metamorphic rocks were used. These types of weathered clays are known in the eastern limit of the Central Massif from the south of Mâcon to the south of Valence. The immature state of the sediment identified in the tiles indicates that fluvial and fluvio-glacial clays were not used to produce these tiles. A local production in the Lyon area can be hypothesized on the basis of these arguments. In addition, the hypothesis of an Italic origin can be rejected although granites and metamorphic rocks exist in northern Italy; they do not seem to be used frequently to produce tiles. The workshops known in Italy are usually situated in a more volcanic regions.

This study has been limited to the Lyon area. In the future, it would be interesting to extend our research area, for example to the "Narbonnaise province". However, this would not change our finding that geological conditions are in favour of a local production (in the case of Lyon).

ACKNOWLEDGEMENTS

We are grateful to the archaeologists who provided samples D. Barthélémy, C. Ramponi (INRAP) for Mâcon, M. Zannettacci for Vienne, and to Anne Mc Birney for her help with translation.

REFERENCES

BARTHELEMY D., CANTIN N., RAMPONI C., VIDEAU G., *in press*, "L'habitat de Sennecé-lès-Mâcon (Saône-et-Loire) : nouvel exemple de tuiles en contexte laténien" *in Actes du colloque de l'AFEAF (may 2006)*.

CUOMO DI CAPRIO N, VAUGHAN S.J. 1993, "An experimental study in distinguishing grog from argillaceaous inclusions in ceramic thin sections", *Archeomaterials* 7 (1), p. 21-40.

DESBAT A. 1998, "Nouvelles recherches sur l'emplacement du prétendu sanctuaire lyonnais de Cybèle. Premiers résultats", *Gallia* 55, p. 237-277.

DESBAT A. 2003, "Les tuiles en céramique", *in* M. Poux and H. Savay-Guerraz (dir.), *Lyon avant Lugdunum*, catalogue d'exposition, p. 136-139.

DESBAT A., PLASSOT E. 2000, "Le site de la rue du Souvenir à Lyon", *in* V. Guichard, S. Sievers and O.H. Urban (dir.), *Les processus d'urbanisation à l'âge du Fer, colloque des 8-11 juin 1998*, Bibracte 4, p. 189-190.

DURAND V. 2000, "Vienne Gauloise", *in* V. Guichard, S. Sievers and O.H. Urban (dir.), *Les processus d'urbanisation à l'âge du Fer, colloque des 8-11 juin 1998*, Bibracte 4, p. 191-195.

LASCOUX J.-P., GAY J.-P. 2003, "Charavay: l'occupation de la fin du second âge du fer", *in* M. Poux and H. Savay-Guerraz (dir.), *Lyon avant Lugdunum*, catalogue d'exposition, p. 121.

MONIN M., MAZA G. 2003, "Les fossés de l'Hôpital Sainte-Croix", *in* M. Poux and H. Savay-Guerraz (dir.), *Lyon avant Lugdunum*, catalogue d'exposition, p. 106-107.

THIERRIN-MICHAEL G. MAZA G. 2002, "Amphores gréco-italiques et Dressel 1: signification des pâtes marbrées", *in SFECAG, actes du congrès de Bayeux, 2002*, p. 457-466.

WHITBREAD I.K. 1986, "The characterisation of argillaceous inclusions in ceramic thin sections", *Archaeometry* 28 (1), p. 79-88.

LYON AMPHORAE IN THE NORTH: STUDIES IN DISTRIBUTION, CHRONOLOGY, TYPOLOGY AND PETROLOGY

PATRICK MONSIEUR[1], PAUL DE PAEPE[2], CHRISTINE BRAET[3]

[1] Doctor-Assistant, Section of Classical Archaeology, Ghent University
[2] Professor, Laboratory for Mineralogy, Petrology and Micropedology, Ghent University
[3] Master of Arts, graphics

INTRODUCTION[1]

Over some twenty years Lyon amphorae have become more and more well identified, thanks to interdisciplinary studies carried out in Lyon and Augst. The combined results proceeding from the investigations on the production sites of *Lugdunum* and on highly valued consumption sites such as *Augusta Raurica* and *Vindonissa* remain the basis for all research on Lyon amphorae.[2] Nevertheless, besides new work that is needed on the complex typology and epigraphy, the study of Lyon amphorae on consumption sites calls for more attention. It is surprising that Lyon amphorae are ill-represented in the publications of the archaeological material of excavated sites in the North, especially *Gallia Belgica* and *Germania Inferior*.

Apparently the excellent publications on the production and consumption sites in and around Lyon have had little impact. Admittedly, if one is not acquainted with the material, Lyon amphorae can be confusing: when in fragments the fabric at first sight leads them to be identified as mortars, dolia or simply as coarse ware or storage vessels, whereas the diagnostic elements of the typology lead many scholars to classify a good part of them as Baetican fish sauce amphorae, the bulk of the export of Lyon amphorae being imitations of these types. Yet, research in military installations such as Vechten or Köln-Alteburg, and in civil settlements such as Titelberg, Braives, Tongeren, Kruishoutem or Velzeke[3] proves that the distribution of Lyon amphorae to the North was appreciable. This study concentrates on the evidence yielded by Velzeke, ancient *Fel(i)ciacum**, a vicus in *Gallia Belgica* (*civitas Nerviorum*).[4]

Situated at an important road junction Velzeke was a thriving vicus already in Julio-Claudian and Flavian times. At its floruit in the 2nd century AD, it possessed stone buildings and at least two sanctuaries. The settlement most probably grew out of an Augustan military outpost installed about 10/5 BC. In 172/174 AD there was an important decline probably due to a raid of infiltrating *Chauci* pirates. The vicus came to an end in 265/268 AD as a result of the disturbances during the reign of Postumus. The amphora material is considerable and diverse, indicating an important marketplace of the romanized *Nervii*. It is our aim to present the potential of some new material, related, whenever possible, to the chronological data from the contexts and to the contribution of petrography in its identification. The variety of the typology, and even more, the quantity of the Velzeke material is remarkable in terms of its distribution to the North. Special attention is paid to some rare types such as the small fish-sauce amphora Lyonnaise 4b and the large fish-sauce type Lyonnaise 6, also called Dressel 16 similis to distinguish it from the South-Gaulish Dressel 16.

1 – We sincerely thank the editorial team for their remarks and thorough criticism of the first draft of this text, making it more easily readable and avoiding some weakly founded allegations.
2 – *Lugdunum*: Dangréaux and Desbat 1987-1988; Dangréaux *et al.* 1992; Desbat 1987; Desbat and Dangréaux 1997; Desbat and Schmitt 1997; Schmitt 1988; *ead.* 1993; *ead.* 1998. *Augusta Raurica* and *Vindonissa*: Bohn 1926; Ettlinger 1977; Martin-Kilcher 1992; *ead.* 1994a; *ead.* 1994b.
3 – Vechten (the Netherlands): Van der Werff 2004. Köln-Alteburg and Mainz (Germany): Hanel 2003; Ehmig 2004. Titelberg (Luxemburg): Laubenheimer and Hénon 1998. Braives, Kruishoutem, Velzeke and Tongeren (Belgium): Brulet *et al.* 1992; Monsieur and Braeckman 1995; Monsieur 2001; *id.* 2005. The amphora material of the *municipium* of Tongeren (ancient *Atuatuca Tungrorum*) is still mostly unpublished and partly under study by the first author. See also distribution patterns in Desbat and Dangréaux 1997, p. 90-94 and fig. 33.
4 – Velzeke lies in the province of East-Flanders, at some 25 km south of Ghent, Belgium; the results concerning the Lyon amphorae are part of a Ph.D. presented in April 2005 at Ghent University, see Monsieur 2005.

THE AMPHORA MATERIAL OF VELZEKE: ARCHAEOLOGICAL DATA AND HYPOTHESES

(Fig. 1, Fig. 2)

Armand Desbat and his collaborators were able to distinguish at least seven main types of Lyon amphorae (Lyonnaises 1-7) used for the transport of wine, fish sauces and olives. Four main types are divided into two subtypes (Lyonnaises 2a-b, 3a-b, 4a-b and 7a-b). They consist of imitations of Italic, Greek, Baetican and South-Gaulish types and of unique creations.[5] Not all of these main- and/ or subtypes do occur in Velzeke, but at least 3 of the main types (Lyonnaises 3, 4 and 5) were traced. Two subtypes were recognized (Lyonnaises 3b and 4b). A rim fragment probably points to a Lyonnaise 7b, as does a handle fragment with the stamp BAS F. The conserved size of a shoulder with handle fragment makes the choice of an attribution to a Lyonnaise 6 or a Lyonnaise 2b difficult even if the chronology of the context would incline to the first identification. Finally there is the upper part of another Lyonnaise 6 type—or should we call it more conveniently a Dressel 16 similis—that remains problematic concerning its origin.

As may be expected, the Lyonnaise 3b subtype that contained fish-sauce is, if not overwhelmingly so, at least the most common type among the diagnostic fragments in Velzeke. Different rim- and handle profiles make it clear that this subtype is liable to a further subdivision of a wider range of new subtypes. By contrast, the Lyonnaise 4b and the Lyonnaise 6 types, both also manufactured for the transport of fish-sauce, turned out to be rare. The latter type especially seems extremely rare, even in Lyon.[6] So the Velzeke examples can be considered of great importance, not only with regard to their distribution but also in their typology and the epigraphy: one Lyonnaise 3b type with the stamp DAGOV; a fractional type 4b, is nearly completely preserved in profile while the neck of the Lyonnaise 6 type bears a remarkable *titulus pictus*. This painted inscription [LX]XII / C I C / ACIT[I ? is probably only comparable to one preserved in Augst.[7] The rim fragment of a Lyonnaise 5 undoubtedly belongs to a flat-bottomed wine amphora imitating the Baetican Dressel 28, the Tarraconese Oberaden 74 or the Gauloise 2-3 types. Finally, the rim fragment and the stamped handle BAS F belonging most probably to a Lyonnaise 7b, also known as the Augst 21 type, may be evidence for import of *defrutum* and/or olives.[8]

In fact, the Lyon amphorae in Velzeke represent mostly the so-called second generation, that is to say the Lyon amphorae produced in the second half of the 1st century AD. Contexts in Lyon and in Augst[9] give dating clues to the greater part of the Lyonnaise types, which is partly confirmed by the Velzeke contexts. A substantial number of the Lyon amphorae were retrieved from settlement contexts in the centre of the vicus (see Table 1 below). Unfortunately during the reign of Trajan this quarter was the scene of a thorough reorganization involving the destruction of much of the evidence of the 1st century AD. Indeed, one part of this residential zone was reused for the construction of a sanctuary, another part apparently for loam extraction and elsewhere, wooden houses were replaced by stone buildings. Nevertheless, a substantial number of refuse pits in the former yards that were related to the dwelling-houses of the 1st century AD escaped the later transformations and remain reliable contexts. These pits date from the early Tiberian to the late Flavian periods, now and then mingled with residual material of the Augustan age. Some of the pits have been cut by more recent ones, causing chronological contamination.

METHODOLOGY

When calculating the minimum number of individual amphorae every information that made a distinction possible was taken into account: diagnostic fragments, wall fragments if they came from different contexts, differences in fabric. The total number of all amphora sherds of Mediterranean origin[10] recovered in Velzeke is 4871, representing a minimum of 527 individuals. The total number of Lyon amphora sherds is 250, representing a minimum number of 45 individuals. Wall sherds form the bulk. Since it may be considered that most of the Lyon imports must have been fish-sauce amphorae, wall sherds that could be used for quantification were also involved in this identification. In this way most probably up to some 40 individuals represent fish-sauce amphorae.[11]

Without taking chronology into account, the minimum number of individual Lyon amphorae comes to 8,5% of the total of imported amphorae in Velzeke. For the period 50/60-80/90 AD, the material coming from four well preserved refuse pits brings Lyon amphorae to 20% of the imports. These four contexts also show that fish-

5 – Dangréaux *et al.* 1992; Desbat and Dangréaux 1997, p. 73-79 and fig. 22-23 (typology), 88-90 (contents). Cf. Martin-Kilcher 1994a, p. 380-382 (Gruppe 6), 385-392 (Gruppe 8), 410-427 (Gruppen 10 and 11).

6 – Or not recognized: a Lyonnaise 6 discovered at the Place Bellecour in Lyon was erroneously identified as a Gaulish Dressel 2-4 (cf. Lyonnaise 2b): Burnouf and Laubenheimer 1998, p. 184 and fig. 12, 1.

7 – Martin-Kilcher 1994, p. 426-427 and fig. 190: MVR(iae) /FL[OS]/ (figure probably lost)/L(or P?) AVR[ ]. As there seem to have existed a sort of standard formula, we consider as lost the first two lines of the Velzeke example in which the sort of fish-sauce and its quality must have been mentioned, whereas in the missing third line on the Augst example most probably a figure was noted (a measure LXX as is also the case on the Lyonnaise 3b). The last line or last two lines give the names of the merchants, cf. Martin-Kilcher 1994a, p. 420 sq. and table p. 421.

8 – Martin-Kilcher 1994a, p. 391-392; Monsieur 2001.

9 – Desbat and Dangréaux 1997, p. 73-79; Martin-Kilcher 1994a, p. 382 and table fig. 157, p. 391 and table fig. 165, p. 414-415 and table fig. 183.

10 – Because of identification problems, which are related to the state of knowledge of locally and regionally made amphorae, these were not taken into consideration.

11 – As was noted in the previous paragraph only one diagnostic rim fragment could be securely identified as a wine amphora, and the identification of two other diagnostic fragments remains unclear.

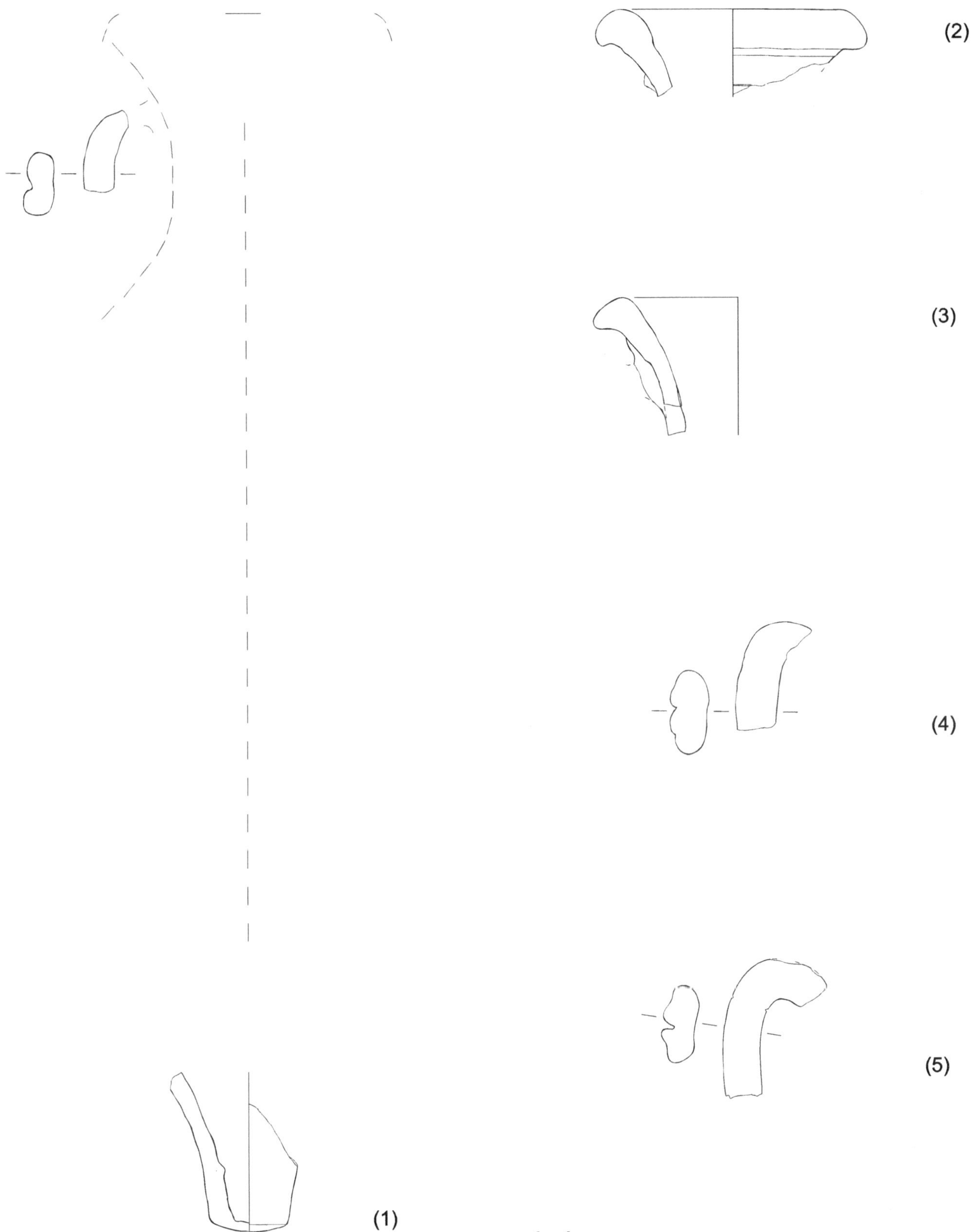

Fig. 1 – Lyon amphorae from Velzeke (Gallia Belgica): typology:
(1) Lyonnaise 3b (nr 161): (2) Lyonnaise 3b (nr 158); (3) Lyonnaise 3b (nr 159); (4) Lyonnaise 3b (nr 165); (5) Lyonnaise 3b (nr 164).

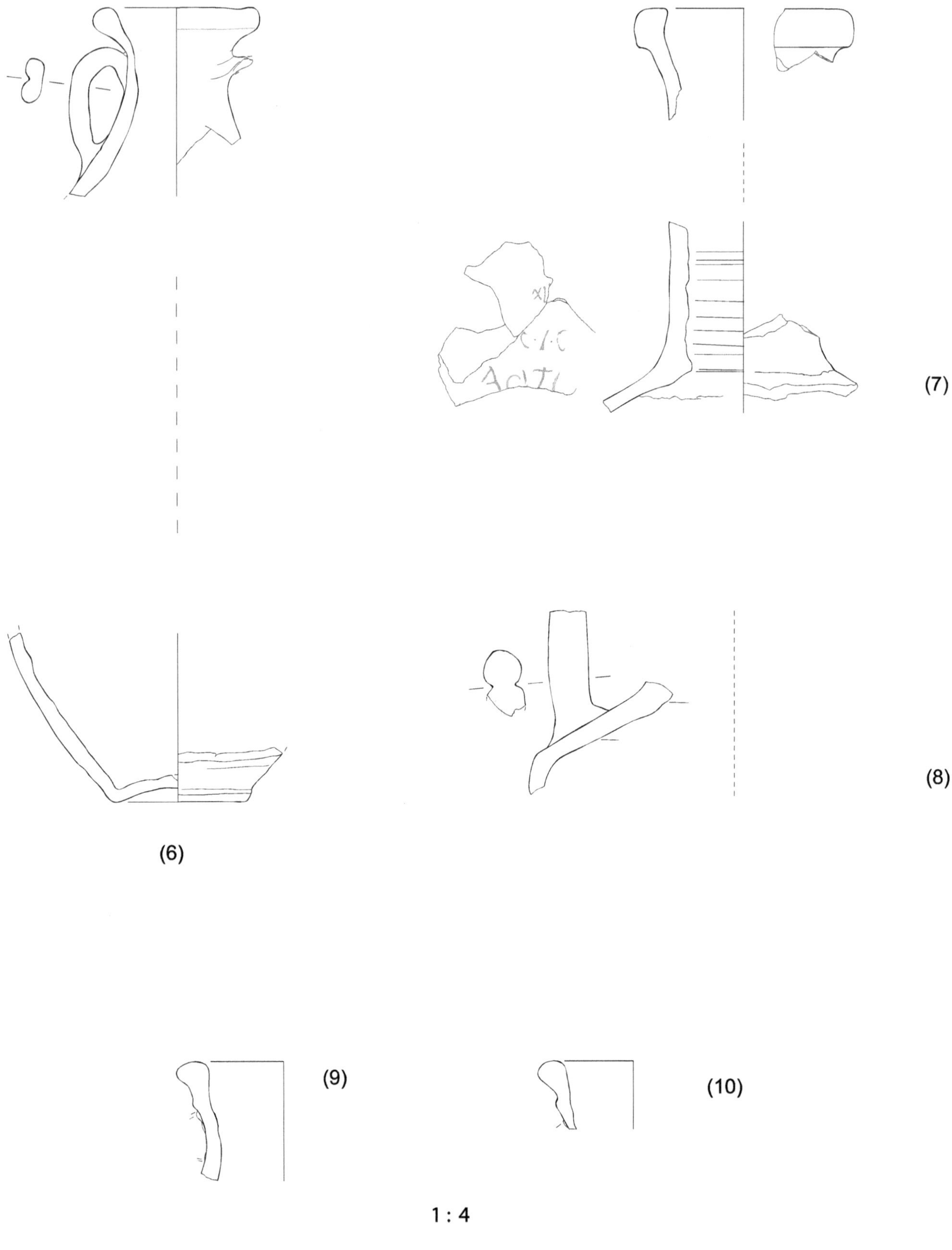

Fig. 2 – *Lyon amphorae from Velzeke (Gallia Belgica): typology:*
(6) Lyonnaise 4b (nr 167); (7) Lyonnaise 6 (nr 168); (8) Lyonnaise 6? 2b? (nr 169); (9) Lyonnaise 5 (nr 154); (10) Lyonnaise 7b? (nr 155).

sauce was imported in the proportion of approximately 60% in amphorae from Baetica, and 40% in amphorae from Lyon.[12]

These figures remain very approximate because only the greater part of the amphorae found in Velzeke was studied; only a small part of the site was excavated; there is a difference in the quality of old and new excavations; it is not known if all the material in the older excavation was kept; etc.

To determine the provenance of the Lyonese amphorae, we first proceeded to a classification based on a combination of typology, technical characteristics and visual fabric characteristics. Two intact, or nearly intact Lyonnaise 3b, one with stamp DAGOV, were a useful starting point, as were different diagnostic fragments which could be compared with wall sherds macroscopically. Different groups of recognizable and more doubtful Lyon fabrics were formed. The choice of the samples for petrographic analysis was motivated by the idea of comparing typologically securely identified diagnostic fragments with wall sherds of a doubtful Lyon fabric, and thus to confirm their provenance.[13] The choice was also influenced by fragments recovered from dated contexts. This gave rise to three fabric groups, of which the first two groups point to a Lyon provenance because of their components of granitic origin. The second group contains more mica. The third fabric group, with only one sample, seems to have a doubtful origin (Lyonnaise 6 or 2b type, S51) With regard to the second fabric group, two samples (an undefined wall sherd, S16, and a Lyonnaise 6 type) can be distinguished on the basis of several variations. The choice of only 12 samples was dictated by a lack of finances.[14] For the same reason, no chemical analysis was carried out. It is our wish that in the near future more sampling could be done and that the results may also be compared with those of chemical analysis. In this light it would be useful to extend the sampling to neighbouring groups of already identified or presumed Baetican and Tarraconese provenance.

PETROLOGICAL ANALYSIS (Fig. 3, Fig. 4)

Ten amphorae and two mortars assigned on visual and typological grounds to Lyonese production centres were examined in thin section under a petrological microscope to find the full range of mineral and rock fragments present in their fired clay matrix (Table 1). Petrological analysis provides an alternative approach to characterising ancient ceramics. It may also help to determine the likely parent outcrops of their raw materials and their place of manufacture. Thin sections for the present study were made available by J. Jurčeka of the Laboratory of Mineralogy and Petrology of Ghent University. To ensure the quality of the thin sections, all sherds were impregnated with polyester.[15] On the basis of the aplastic inclusions, three fabrics were recognized.

Fabric 1 (S17, S49, S55, S85, S90). Five specimens are very similar in fabric, forming a first group. They contain scattered sand-sized mineral and rock fragments with a relatively wide range of particle sizes set in a matrix of brown fired clay. The average grain size of the aplastic inclusions is about 0.3-0.4 millimetres but much larger grains (up to 3 mm across) are dispersed throughout. They show all gradations of roundness. The fabric has a pale-brown or dark-brown matrix which is almost isotropic under crossed nicols and may be obscured by the presence of finely divided organic matter. Discrete mineral grains consist principally of quartz and feldspars. Biotite flakes are less common. They are usually dark reddish-brown, highly oxidized, or heavily iron-stained. Muscovite, minerals of the epidote family, and zircon are accessory constituents. Quartz grains are sometimes polycrystalline, and strain extinction is a common characteristic. Three types of feldspar have been identified in thin section: Carlsbad-twinned orthoclase (microperthite), microcline and plagioclase are twinned according to the albite law. Among the feldspars, potassium-rich varieties predominate. Some orthoclase crystals are cloudy in plane polarized light, and most plagioclase feldspars have been considerably altered by sericitization. Microscopic examination reveals that most rock fragments enclosed in fabric 1 are of granitic origin and composed of quartz and potassium-feldspar, and intergrowths of these two components. Minor constituents include plagioclase feldspar, strongly oxidized biotite and opaque iron ore. In addition, a few lithic inclusions deriving from quartz mica schist, chert, sandstone, and possibly felsic porphyritic lava have also been observed. The lava particles are silicic (rhyolite?) and carry phenocrysts of potash feldspar and corroded quartz. The thin sections of S55 and S85 contain a maximum of two pieces of lava; in the other amphorae volcanic fragments are completely lacking. The matrix of S49 and S90 is characterized by the presence of a varying quantity of slender reddish-brown biotite crystals. The biotite content of these two vessels is, however, insignificant when compared to what we noted in most samples classified into fabric 2.

12 – This is of course no evidence that it should be generalized to the whole of the site of Velzeke.

13 – Two mortar wall sherds were also examined and could be classified into the two fabric groups that were confirmed as Lyonese by typology and petrography.

14 – We are grateful to the Administration of the Province of East-Flanders for their financial support.

15 – TRA polyester (Farbwerke Hoechst AG, Frankfurt-Hoechst, D.R.Germany).

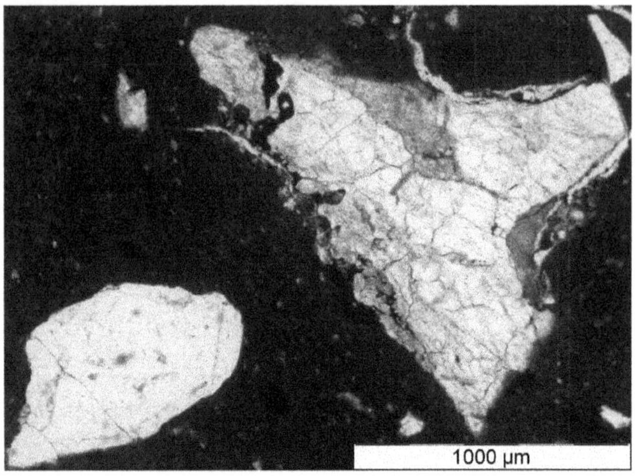

Fig. 3 – Lyon amphorae and mortars from Velzeke (Gallia Belgica): fabrics:
(1) S17 (fabric 1): angular granitic rock fragment and subrounded quartz grain in a dark-brown matrix (plane polarized light); (2) S16 (fabric 2): muscovite-bearing granitic inclusion in a dark-brown matrix (plane polarized light).

Fabric 1

Snr	*Catnr*	**Type amph/mort**	*Fragment*	*Type context*	*Date context*
S17	-	mortarium	wall	refuse pit 5/settlement	60-80
S49	158	Lyonnaise 3b	complete rim	arable layer	disturbed
S55	165	Lyonnaise 3b	handle+bottom	refuse pit 55/settlement	40-60
S85	159	Lyonnaise 3b	rim	construction layer/settlem	disturbed
S90	161	Lyonnaise 3b	handle+spike	ref pit 346 (+residual?)/sett	20/30-60/70

Fabric 2

Snr	*Catnr*	**Type amph/mort**	*Fragment*	*Type context*	*Date context*
S16	-	undefined	wall	refuse pit 2 (+cut)/settlem	15/30-30/50
S30	168	Lyonnaise 6	upper part	filling layers/loam extract	disturbed
S36	164	Lyonnaise 3b	handle	construction layer/settlem	disturbed
S61	-	mortarium	wall	refuse pit 56/settlement	70/80-100
S84	167	Lyonnaise 4b	most profile	filling layers/loam extract	disturbed
S93	154	Lyonnaise 5	rim	refuse pit 5/settlement	60-80

Fabric 3

Snr	*Catnr*	*Type amphora*	*Fragment*	*Type context*	*Date context*
S51	169	Lyonnaise 6? 2b?	handle+shoulder	refuse pit 5/settlement	60-80

Table 1 – Amphorae and mortars sampled for petrography, presented by fabric.

Fabric 2 (S16, S30, S36, S61, S84, S93). This fabric is closely related to fabric 1 as it also contains many mineral and rock fragments of granitic origin. The only consistent compositional differences from the latter are a relatively high mica content and the fact that many aplastics fall into silt grade particle sizes. Reddish-brown biotite and muscovite are widespread and large flecks are readily visible with the unaided eye on the outer surface of some ceramics. The micaceous nature of this ware, in combination with the presence of much granitic debris, favours an origin in or close to outcrops of granitic and metamorphic rocks. The mica content of S16 is impressive. The commonest components of the non plastic bodies included in this amphora are quartz, potash feldspar and less commonly plagioclase feldspar, biotite, muscovite and iron ore. The majority of the quartz grains are silt-sized and angular in shape, with a scatter of larger grains about 0.3 mm across. Feldspar crystals are angular or subangular and usually fall within the range covered by the fine and medium sand grade categories. Red-brown biotite and muscovite flakes are very conspicuous under the microscope, of varying sizes and range up to 1 mm in length in rare instances. Biotite is the predominating mica. Epidote minerals and zircon occur as accessories. Besides unsorted fragments of granite (ranging up to 3.1 mm across), S16 also contains small pieces of mica schist, gneiss and aplite. S30, S36, S61, S84 and S93 are less micaceous and often characterized by much finer textured non plastic inclusions. As a consequence, their rock fragment content is lower and sometimes almost nil. In some sherds the micas are found exclusively in the matrix of fired clay. Rare grains of tourmaline, garnet and clinopyroxene, and cherty inclusions were also noted. In addition, S30 also contains dispersed silicified tests of foraminifera and two fragments of felsic porphyritic lava, probably rhyolitic in composition.

Fabric 3 (S51). Only one sample is classified in this fabric. The mineral fragments characteristic of this fabric have about the same composition as those found in the amphorae described above, although in this case feldspars and micas (mainly biotite) occur in much smaller quantities. Quartz is predominant; most grains are silt-sized and angular in shape. Those of sand size grade are on average about 0.3 mm across but range up to 1.3 mm. Rock fragments are small, rare and composed exclusively of chert. Due to the lack of specific mineral and rock inclusions, it is impossible to say much about the parent rocks from which the temper of S51 was derived. On the available evidence, a source in a granitic terrane is within the bounds of possibility.

As mentioned earlier, the description of fabric 1 leaves little doubt as to the granitic origin of the non plastic inclusions, while fabric 2 has an admixture of silt- and sand-sized grains derived from a geological context dominated by granitic and micaceous metamorphic rocks. The tempering materials of both fabrics can be matched petrologically with the solid and drift geology of the Lyon region. Production within the Lyon area is therefore plausible, although other source regions cannot be disregarded. Lyon lies at the confluence of the Rhône and Saône rivers. The eastern edge of the *Massif Central* is hardly a few kilometers away from the westernmost quarters of this city. The central part of France has a complex and varied lithology. Its basement consists of metamorphic rocks intruded by hercynian granite and covered by Palaeozoic, Mesozoic and Cenozoic sedimentary and volcanic rocks. The part of the *Massif Central* facing the city of Lyon mainly consists of the granitic and metamorphic rocks (mica schists, gneisses and migmatites) of the basement. Detritus of these rocks is carried by brooks and small rivers to the valleys of the Saône and the Rhône, where it is deposited and was used in the past for pottery-making purposes.

A. Schmitt examined the mineralogical and petrographic properties of the aplastic inclusions, as well as the bulk chemical composition and trace element content of Roman amphorae found at Lyon and thought

Fig. 4 – Lyon amphorae from Velzeke (Gallia Belgica): fabrics:
(1) S30 (fabric 2): granitic rock fragments composed of potassium-feldspars and quartz, and a rare skeleton of microfossil enclosed in a dark-brown matrix (plane polarized light);
(2) S84 (fabric 2): large inclusions of Carlsbad-twinned orthoclase in a well-oriented clay matrix (crossed polarized light);
(3) S85 (fabric 1): quartz grains, granitic rock fragments and porphyritic rhyolitic lava in a very dark-brown matrix (plane polarized light).

to come from local pottery manufacturing centres. On the same occasion, this author also analysed alluvial sands collected in the Lyon region, in particular along the Azergues, the Saône and the Rhône. Schmitt's data make it clear that the Roman amphorae from Velzeke supposed to come from the Lyon region may indeed have been produced in this part of France, in particular in the Rhône valley close to the confluence with the Saône or south of Lyon.[16] The alluvial deposits outcropping in this area are poor in sand grains with rhyolitic composition, while those occurring along the Azergues and the Saône have high percentages of rhyolitic elements. These volcanic detrital grains are known to originate in Upper Visean strata exposed north of Lyon in the Beaujolais region.[17] The commonplace nature of the aplastic inclusions taken in fabric 3 implies that speculation about the origin of amphora S51 is extremely hazardous although, on the basis of the available petrological data, it may also be a Lyonese product.

DISCUSSION AND CONCLUSIONS

The first and second fabric groups may be considered of Lyon provenance, judging from a combination of typological identification and petrological analysis. Nevertheless, if the first fabric group looks more homogeneous than the second, from a typological as well as from a petrological point of view, the second group shows a great diversity in types and in fabric composition. Fabric group 1 contains only Lyonnaise 3b fish-sauce amphorae, the most frequent imported amphora type, and a mortar. Fabric group 2 contains at least four different amphora types and a mortar, and is more micaceous in its fabric. This group is also distinguished by some variations in samples S16 and S30. The first is a typologically undefined wall fragment from a Tiberian context, remarkable in its higher content of mica, that could be considered uncommon in Lyon productions. The second contains some rare foraminifera which do not normally occur in Lyon fabrics, making its origin doubtful. At any rate, we consider the typology of S30, a Lyonnaise 6 type or an imitation of the South-Gaulish Dressel 16, secure considering the rim and neck type and the epigraphy. The high, everted neck and the massive rim type leave little doubt, as does the *titulus pictus*. When considering the unit of measure LXX and the initials of the *tria nomina* of the merchant, this painted inscription can be added to the range of standard formulae on Lyonnaise 3b.[18] And Lyonnaise 3b and 6 have in common that they were designed and used to transport fish-sauce. The existence of different types could possibly be explained by the quality and the sort of fish-sauces (cf. also the fractional type Lyonnaise 4b, S84) these amphorae carried, or as mentioned in the *tituli picti* by their origins, *Antipolis* and *Hispania*. Apart from that, this typological observation presents an even better explanation for the sources of imitation of these two amphora types.[19] Finally, while the typology (Lyonnaise 6 or 2b, S51) and composition of fabric 3 does not rule out a Lyonese origin, the provenance remains doubtful because of the common nature of the inclusions and faint evidence for a granitic source.

There are different observations to be made on the chronology and the typology of the Lyon amphorae discovered in Velzeke, and some of them are related to the results of the petrological analysis, even with limited sampling.

The dating of the Velzeke contexts does look attractive but should be handled with care since it may lead to misinterpretations. Indeed, dating of the rim fragment of a Lyonnaise 5 wine-amphora (S93) would fit better in the first half of the 1st century AD than is suggested by the late Neronian and early Flavian date of refuse pit 5. The fragment may be residual, or the context could prove that Lyonnaise 5 continued to be produced in the second half of the 1st century AD. Likewise, the shoulder with handle fragment from the same context could be residual if it is interpreted as a Lyonnaise 2b wine-amphora[20] of the first half of the 1st century AD, instead of the later Lyonnaise 6 fish-sauce container (S53). In fragments these two types are easily confused. One could argue that the distribution of Lyonnaise 2b is considered to be very limited, but this is also thought of Lyonnaise 6. And the latter type seems to be present in Velzeke (S30). Moreover this problem of typological attribution is enhanced by the fact that fabric 3 is of doubtful Lyonese provenance. Returning to the chronology, the fragments of the Lyonnaise 6 type (S30), as well as the fractional Lyonnaise 4b (S84) were unfortunately found in disturbed layers filling an enormous excavation (probably for loam extraction) during the 2nd century AD. Finally, with the exception of one undefined wall fragment from a Tiberian deposit (possibly slightly contaminated by a Claudian superposition) there is for the moment no clear evidence of Lyon amphorae in Velzeke dating from the first half of the 1st century AD.

16 – For fabric 2 rich in mica compare with Schmitt 1993, p. 56, tab. 1 and 1998, p. 300-301 where the maximum mica content of the pottery samples is 17 percent volume.

17 – Sider *et al.* 1988.

18 – See note 7. The figures LXXI and probably LXXII are also seen. The names of the merchants read in *tituli picti* on Baetican fish-sauce amphora are not abbreviated.

19 – On the problem of the origin of the fish-sauce in these amphorae and its bulk transport to Lyon in caskets and/or dolia see Desbat 1991; Martin-Kilcher 1994a, p. 485-487; Desbat and Dangréaux 1997, p. 90-94; cf. also Schmitt 1998, p. 308 with further bibliography.

20 – Imitating Italic and Tarraconese Dressel 2-4 examples.

REFERENCES

BOHN O. 1926, "Pinselaufschriften auf Amphoren aus Augst und Windisch", *Anzeiger für schweizerische Altertumskunde, NF* 28, p. 197-212.

BRULET R., LAUBENHEIMER F., VILVORDER F. 1992, "Les amphores de Braives, un vicus de Gaule Belgique", *in Les amphores en Gaule: production et circulation: table ronde internationale, Metz, 4-6 octobre 1990. 1*, Paris, p. 172-177.

BURNOUF J., LAUBENHEIMER F. 1998, "Des vides sanitaires, Place Bellecour à Lyon", *in Les amphores en Gaule: production et circulation: table ronde internationale, Metz, 4-6 octobre 1990. 2*, Paris, p. 175-192.

DANGREAUX B., DESBAT A. 1987-1988, "Les amphores du dépotoir flavien du Bas-de-Loyasse à Lyon", *Gallia* 45, p. 115-153.

DANGREAUX B., DESBAT A., PICON M., SCHMITT A. 1992, "La production d'amphores à Lyon", in *Les amphores en Gaule: production et circulation: table ronde internationale, Metz, 4-6 octobre 1990. 1*, Paris, p. 37-50.

DESBAT A. 1987, "Note sur la production d'amphores à Lyon au début de l'Empire", *in SFECAG, Actes du congrès de Caen*, p. 159-166.

DESBAT A. 1991, "Un bouchon de bois du 1er siècle apr. J.-C. recueilli dans la Saône à Lyon et la question du tonneau à l'époque romaine", *Gallia* 48, p. 314-344.

DESBAT A., DANGREAUX B. 1997, "La production d'amphores à Lyon", *Gallia* 54, p. 73-117.

DESBAT A., SCHMITT A. 1997, "L'atelier de la Manutention", *Gallia* 54, p. 45-49.

DESBAT A., PICON M., SCHMITT A. 1994, "Sur l'origine de quelques types d'amphores d'Augst", *in* S. Martin-Kilcher, *Die römischen Amphoren aus Augst und Kaiseraugst. 3, Archäologische und naturwissenschaftliche Tonbestimmungen. Beiträge von Gisela Thierrin–Michael sowie von Armand Desbat, Maurice Picon, Anne Schmitt und Katalog und Tafeln (Gruppen 2-24)*, Augst, Forschungen in Augst, Band 7/3, p. 649-656.

DE PAEPE P. 2005, "Petrologisch onderzoek van Romeinse amforen afkomstig van Velzeke en Kruishoutem", *in* P. Monsieur, *Mediterrane amforen in de* Civitates Nerviorum et Menapiorum. *Identificatie, epigrafie, chronologie, kwantificering en interpretatie van de vondsten in Velzeke en Kruishoutem (Oost-Vlaanderen, België)*. Deel 1b, Ph.D. Ghent University, p.1-26.

EHMIG U. 2004, *Die römischen Amphoren aus Mainz*, Frankfurt, Frankfurter Archäologische Schriften 5.

ETTLINGER E. 1977, "Aspects of Amphora-Typology seen from the North", in *Méthodes classiques et méthodes formelles dans l'étude des amphores, actes du colloque de Rome, 27-29 mai 1974*, Rome, Collection de l'École française de Rome 32, p. 9-16.

HANEL N. 2003, "Die Amphoren der Ausgrabungen 1998 im Hauptstützpunkt der Classis Germanica 'Alteburg' bei der Colonia Claudia Ara Agrippinensium: ein Vobericht", *Rei Cretariae Romanae Fautorum. Acta* 37, p. 361-364.

LAUBENHEIMER F., HENON M. 1998, "Les amphores du Titelberg (Luxembourg)", *in Les amphores en Gaule: production et circulation: table ronde internationale, Metz, 4-6 octobre 1990. 2*, Paris, p. 107-142.

MARTIN-KILCHER S. 1992, "Les amphores de Gaule romaine : leur présence à Augusta Rauricum", *in Les amphores en Gaule: production et circulation: table ronde internationale, Metz, 4-6 octobre 1990. 1*. Paris, p. 157-161.

MARTIN-KILCHER S. 1994a, *Die römischen Amphoren aus Augst und Kaiseraugst. Ein Beitrag zur römischen Handels- und Kulturgeschichte. 2, Die Amphoren für Wein, Fischsauce, Südfrüchte (Gruppen 2-24) und Gesamtauswertung (mit einem Beitrag von Markus Schaub)*, Augst, Forschungen in Augst, Band 7/2.

MARTIN-KILCHER S. 1994b, *Die römischen Amphoren aus Augst und Kaiseraugst. 3, Archäologische und naturwissenschaftliche Tonbestimmungen. Beiträge von Gisela Thierrin–Michael sowie von Armand Desbat, Maurice Picon, Anne Schmitt und Katalog und Tafeln (Gruppen 2-24)*, Augst, Forschungen in Augst, Band 7/3.

MONSIEUR P. 2001, "L'amphore gauloise *Augst 21*: une grande inconnue", *in Belgian Archaeology in a European Setting II. Studies in Honor of Jozef Mertens*, Acta Archaeologica Lovaniensia 13, Leuven, p. 171-184.

MONSIEUR P. 2005, *Mediterrane amforen in de* Civitates Nerviorum et Menapiorum. *Identificatie, epigrafie, chronologie, kwantificering en interpretatie van de vondsten in Velzeke en Kruishoutem (Oost-Vlaanderen, België)*. Ph.D., Ghent University.

MONSIEUR P., BRAECKMAN K. 1995, "Romeinse amforen in de *vicus* Velzeke", *Archeologische Kroniek van Zuid-Oost-Vlaanderen* 4, p. 289-316.

SCHMITT A. 1988, "Les productions d'amphores dans la moyenne vallée du Rhône: mise au point sur quelques types", *SFECAG, Actes du congrès d'Orange*, p. 25-33.

SCHMITT A. 1993, "Apports et limites de la pétrographie quantitative: application au cas des amphores de Lyon", *Revue d'archéométrie* 17, p. 51-63.

SCHMITT A. 1998. "Amphorae from Lyons: Petrographic and chemical arguments", *Archaeometry* 40, p. 293-310.

SIDER H., GAGNY C., MOUTERDE R., PAREL J.-L., GUYOU A., KERRIEN Y., FLEURY R. 1988, *Carte géologique de la France (1/50.000), feuille Amplepuis (673)*, BRGM, Orléans.

VAN DER WERFF J.H. 2004, "Amphoras from Vechten (Excavations 1946-1947)", *in Archaeology in confrontation. Aspects of Roman military presence in the Northwest. Studies in honour of Prof. Em. Hugo Thoen*, Ghent, Archaeological Reports Ghent University 2, p. 287-304.

ARCHAEOMETRIC CHARACTERISATION OF ROMAN WINE AMPHORAE FROM BARCELONA (SPAIN)

V. MARTÍNEZ FERRERAS[1], J. BUXEDA I GARRIGÓS[1], J.M. GURT I ESPARRAGUERA[1], V. KILIKOGLOU[2]

[1] ERAUB, Dept. de Prehistòria, Història Antiga i Arqueologia, Universitat de Barcelona, C/ Montalegre, 6-8, 08001, Barcelona (Spain) (vmartinez@ub.edu; jbuxeda@ub.edu; jmgurt@ub.edu)

[2] Institute of Materials Science, NCSR Demokritos, 15310, Agia Paraskevi, Attiki, Greece (kilikog@ims.demokritos.gr)

INTRODUCTION

Particularly from the end of the 2nd century to the end of the 1st century BC, the coastal Catalan region witnessed a deep change in the settlement pattern, with the progressive abandonment of the Iberian villages, favoring newly created Roman cities, like *Barcino*, established during the Augustan period (Fig. 1). These towns were placed in rural areas whose activity was then based on a surplus production, especially of wine, intended for widespread distribution in a large complex trade-network, using locally produced amphorae. However, the role of the new Roman towns seems to be not only amphora production centers, but also redistribution centers for their hinterlands. Thus, along the lines of this Romanization process, in the Catalan region, Dressel 1 and Laietana 1 amphora types were already being produced from the second quarter of the 1st c. BC. In contrast, the first evidence of Roman wine amphorae production in ancient *Barcino* (Barcelona) is the Pascual 1 amphora from the end of the 1st c. BC, most probably related to the foundation of the town.

During the 70s, archaeological studies characterized the Roman wine amphorae as a unique regional group, called *Tarraconense Amphorae*, based on formal typologies and macroscopic descriptions. This led to the belief in a single provenance because of their common characteristics, derived from the presence of an igneous geological formation in the central part of the Catalan coast. Archaeometric studies started in the 80s and continued in the 90s. In general, little work was done, on a limited number of samples. Even so, all the results pointed to the possibility of differentiating several productions within the so called *Tarraconenses Amphorae* (Buxeda *et al.* 2002a, and references therein). This was to be expected since archaeologically up to 60 possible production centers in that region are known (Miró 1988; Revilla 1995; Tremoleda 2000). Therefore, a large program has been established in order to characterize a significant number of kiln sites across this region, as well as several reception centers. The final objective is to understand the production systems and to determine the distribution areas (Buxeda *et al.* 2004; Martínez *et al.* 2005; Vila *et al.* 2005; Casas and Martínez 2006).

Within this region, one important place is *Barcino*. Here, recent archaeological excavations have brought to light new evidence on amphora production (Fig. 1). On the one hand, the *Princesa* site (from now on PR) corresponds to an amphora workshop. It includes a ceramic kiln structure, ceramic dumps and some Pascual 1 amphorae standing upside down in a row. The workshop is dated between the 2nd half of the 1st c. BC and the beginning of the 1st c. AD. The production includes Pascual 1 and Dressel 2-4 amphorae, fine coarse pottery, *tegulae* and *pondii*. On the other hand, another possible workshop has been located at *Av. Cambó-Mercat de Santa Caterina* (from now on CM), about 100 m from PR workshop. Several archaeological remains can be considered as evidence for Pascual 1 and Dressel 2-4 amphora production, including the presence of

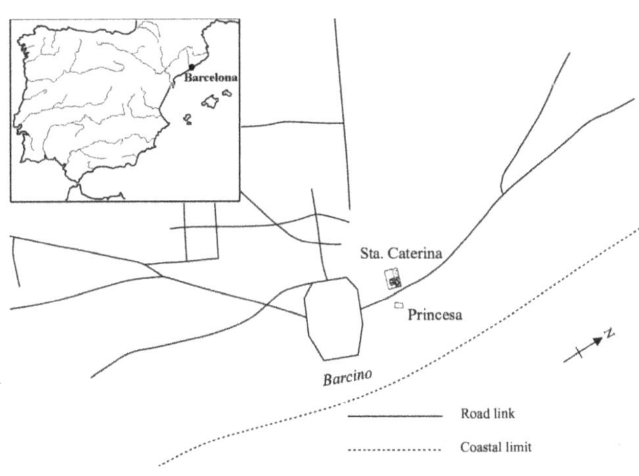

Fig. 1 – Map indicating the geographical location of Barcino (Barcelona) and their territory in the Roman period, with the emplacement of Princesa and Av. Cambó-Mercat de Santa Caterina sites.

overfired fragments buried in a dump. However, no kiln structure has been found so far. Moreover, this site is also a possible production centre of fine coarse pottery and building materials. Similarly to PR, it is dated between the end of the 1st c. BC and the 1st century AD. It is important to highlight that both possible production centres were located in the rural area of Roman *Barcino*, very close to the most important communication route at this period, the Via Augusta, and to the port complex (Aguelo *et al.* 2005; Casas and Martínez 2006).

The objective of this paper is the definition of the reference group (from now on RG) of PR, and, if CM is demonstrated to be such a production centre, also to define the RG of CM, or the paste reference compositional units (from now on PRCU) if it turns out not to be a production center (Buxeda *et al.* 1995). Moreover, it is also intended to draw conclusions on technological aspects of the amphorae production.

MATERIALS AND TECHNIQUES

To date, 42 Pascual 1 amphorae (CSC053-CSC092), 2 *pondii* (CSC093 and CSC094) and 5 clay samples from the kiln structure (CSC098-CSC102) have been characterized from the PR kiln site. As well, 52 Pascual 1 amphorae (CSC001-CSC052) and 3 clay samples (CSC095-CSC097) from the CM site have also been characterized (Fig. 2).

The 102 samples have been analyzed using X-Ray Fluorescence (XRF) to determine the chemical concentrations for the provenance study and X-Ray Diffraction analysis (XRD) to determine mineralogical compositions in order to understand technological characteristics. Mechanical properties of 24 samples, 12 for each site, have also been measured, and the results will be published elsewhere. The analytical routine for XRF and XRD analysis has been described at length elsewhere (Hein *et al.* 2002; Tsantini *et al.* 2005).

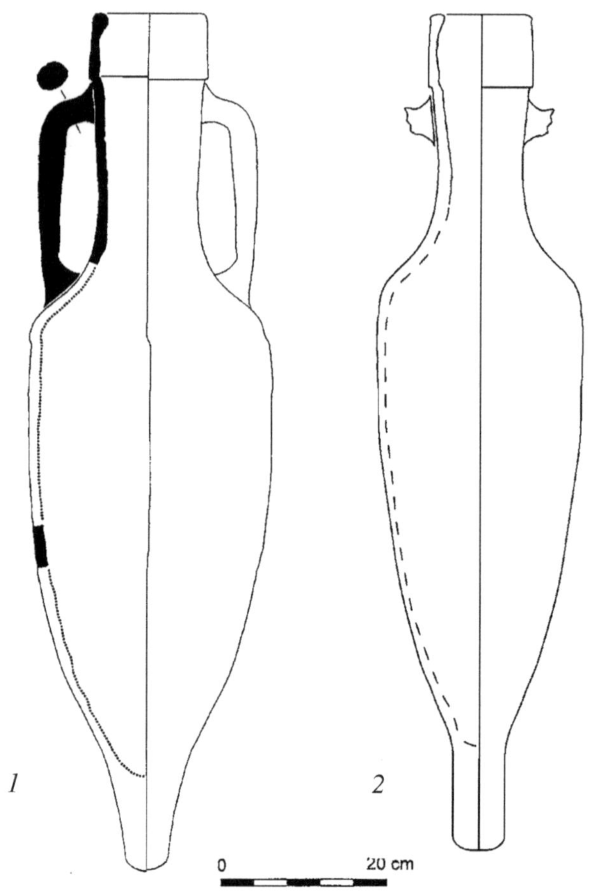

Fig. 2 – (1) Pascual 1 amphora from Av. Cambó-Mercat Sta. Caterina site; (2) Pascual 1 amphora from Princesa site.

ANALYTICAL RESULTS AND DISCUSSION

For the statistical treatment, Aitchison's observations on compositional data were taken into consideration (Aitchison 1986; Buxeda 1999; Buxeda and Kilikoglou 2003). Raw data were logratio transformed on the following elements Fe_2O_3, Al_2O_3, MnO, TiO_2, MgO, CaO, Na_2O, K_2O, Ba, Rb, Nb, Zr, Y, Sr, Ce, Ga, V, Zn, Ni and Cr, using SiO_2 as a divisor. The total variance of the variation matrix was found to be important (vt =1.4862), indicating the existence of significant chemical differences. The maximum contribution to the chemical variability is provided by the relative concentrations of the following elements: MgO, CaO, Na_2O, K_2O, MnO, Sr, Rb, Ba, Ga, and Zr.

The results are summarized in the dendrogram (Fig. 3) resulting from the cluster analysis performed with S-Plus 2000 (MathSoft 1999), on the previous subcomposition, using the square mean euclidean distance and the centroid clustering algorithm. The study of the dendrogram reveals a relatively complex structure formed by four major chemical groups. The 44 amphorae and *pondii* from the PR site cluster together forming a homogeneous chemical group. Four clays relating to different parts of the kiln structure (CSC099-CSC102) show a chemical composition close to PR ceramics. However, the individual CSC098, corresponding to the external wall of the same kiln, exhibits clear significant chemical differences. In the CM case, the chemical composition of the 55 individuals shows a more complex structure. The CM1 group is formed by thirteen individuals from CM. Another thirty individuals cluster together in CM2. Finally, three individuals are clustered in group CM3. Other CM individuals that cannot be associated with these groups are amphorae CSC004, CSC006 and CSC010, and the three CM clays. The CM individual CSC049 is a special case, because it shows chemical similarities with the PR kiln site.

The chemical differences between these chemical groups (Table 1) are clear in the biplot of the principal components analysis performed using S-Plus 2000, on

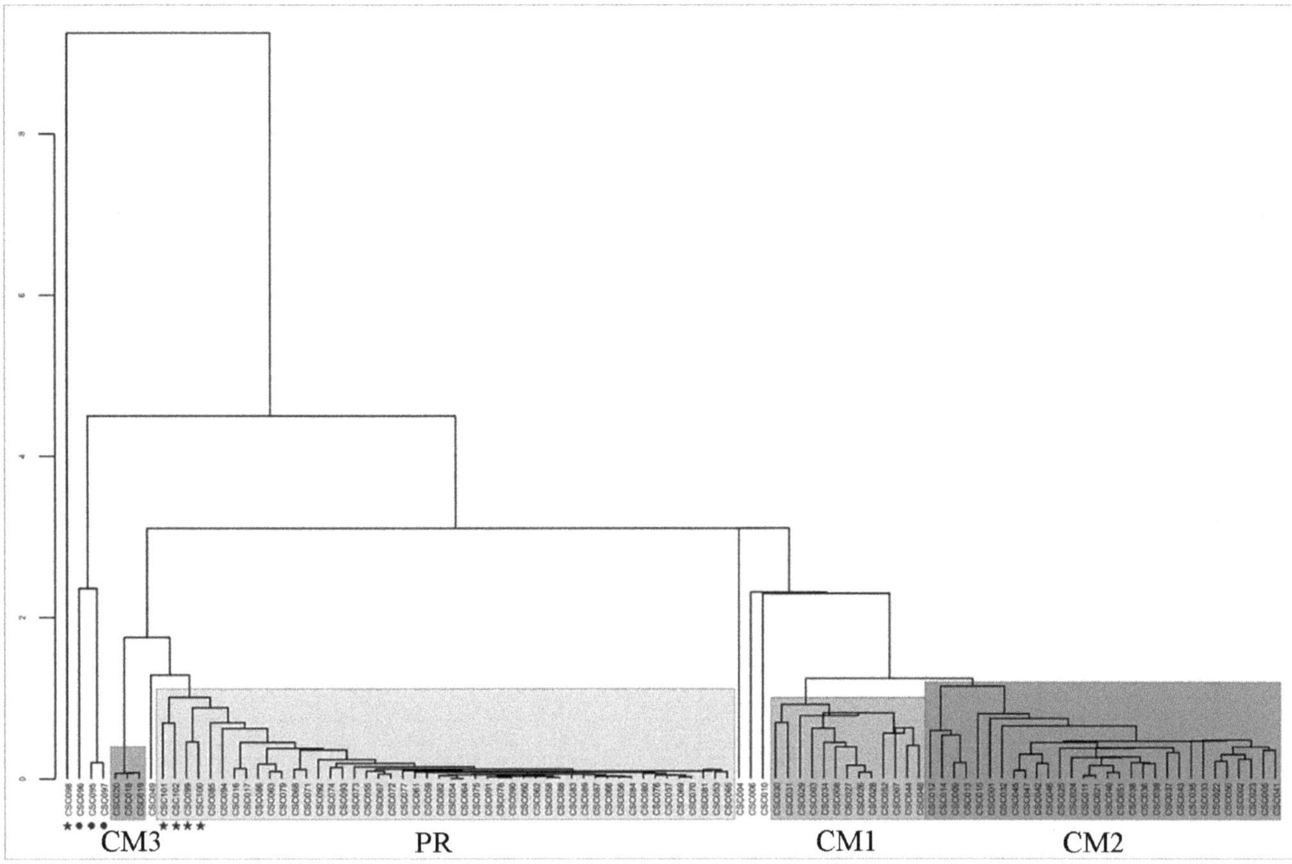

Fig. 3 – Dendrogram resulting from the cluster analysis performed on the logratio transformed subcomposition Fe_2O_3, Al_2O_3, MnO, TiO_2, MgO, CaO, Na_2O, K_2O, Ba, Rb, Nb, Zr, Y, Sr, Ce, Ga, V, Zn, Ni and Cr, using SiO_2 as divisor. All but one PR individuals are grouped in cluster PR. Clusters CM1, CM2 and CM3 include individuals from CM. Dark circle: CM clays; Dark star: PR kiln struct. clays.

the covariance matrix of the previous logratio transformed subcomposition (Fig. 4). It shows the first and second principal components, which account for 70.5% of the chemical variation. The PR group exhibits higher Rb and Zr relative concentrations, while the clay sample CSC098 appears isolated mainly due to its very low CaO relative content (0.93% in normalized concentrations). This PR group can be considered as quite homogeneous, although the four clays in that cluster show relatively small chemical differences. Therefore, the RG composition finally defined is based only on the 44 amphorae and *pondii* individuals. In the CM case, clear differences can be observed. On the one hand, the CM1 group has been separated from the rest because it exhibits an elevated relative content in CaO and MgO. On the other hand, the CM2 group is separated because it presents also elevated relative concentration of MgO and CaO, but it is different from CM1 because of the higher Na_2O and Sr, and the lower K_2O relative contents. As for the clay sample CSC098, group CM3 appears isolated due to its low CaO content. Finally, it has to be noted that there is no direct chemical similarity between the CM ceramic groups and the CM clays, contrary to what was observed at PR.

From a technological point of view, all the analyzed ceramic individuals, except those of the CM3 group and CSC098 PR clay sample, are calcareous (Fig. 5). Moreover, the study of the mineralogical data has enabled us to identify some different fabrics (Buxeda *et al.* 1995), according to the association of different crystalline phases with different firing temperatures. The results are summarized in Table 2.

For PR calcareous amphorae, four different fabrics can be differentiated, the estimated equivalent firing temperature (EFT) of which vary from below 800-850 °C up to over 1000 °C (Table 2). The 6 individuals from PR-F1 show only primary phases, related to a low-fired fabric (< 800-850 °C). The fabric PR-F2 shows primary phases together with the formation of firing phases like gehlenite and the initial formation of pyroxene, indicating a well-fired fabric. The estimated EFT is around 850-950 °C. Four amphorae individuals are grouped in PR-F3, showing the almost total decomposition of illite-muscovite (the 10Å peak is no longer identifiable) and development of higher firing-phases (plagioclase, gehlenite and pyroxene). The EFT can be estimated at around 950-1000 °C. Finally, a large number of individuals can be related to an overfired

	PR (n=44)		CM1 (n=13)		CM2 (n=30)		CM3 (n=3)	
	m	sd	m	sd	m	sd	m	sd
Fe_2O_3 (%)	5,82	0,25	5,13	0,31	5.09	0,26	6,74	0,19
Al_2O_3 (%)	15,99	0,44	14,28	0,60	14.01	0,55	17,25	0,45
MnO (%)	0,08	0,01	0,09	0,03	0.08	0,01	0,09	0,00
TiO_2 (%)	0,75	0,02	0,66	0,03	0.65	0,03	0,84	0,01
MgO (%)	1,94	0,18	4,93	1,64	5,39	0,86	1,45	0,03
CaO (%)	8,11	1,38	11,66	2,16	12,59	1,85	2,66	0,22
Na_2O (%)	0,80	0,11	0,83	0,29	1,91	0,43	0,78	0,02
K_2O (%)	3,37	0,15	3,39	0,59	2,14	0,30	3,02	0,05
SiO_2 (%)	62,99	1,26	58,87	2,36	57,98	1,59	67,01	0,53
Ba (ppm)	564	65	680	121	619	140	642	66
Rb (ppm)	130	6	100	16	67	16	127	3
Nb (ppm)	17	1	16	1	16	1	20	0
Zr (ppm)	225	20	191	19	185	17	270	7
Y (ppm)	30	2	26	2	25	1	31	0
Sr (ppm)	156	18	300	69	414	229	124	12
Ce (ppm)	76	10	63	6	61	9	83	3
Ga (ppm)	20	1	18	1	17	1	22	1
V (ppm)	85	7	74	7	74	6	109	2
Zn (ppm)	96	4	81	6	76	9	108	4
Ni (ppm)	27	2	29	2	29	2	38	2
Cr (ppm)	79	4	76	4	73	4	96	4

Table 1 – Mean and Standard deviation for the four chemical groups defined.

fabric, PR-F4, which shows maximum development of the firing phases and the total decomposition of illite-muscovite. This fabric accounts for 25 overfired individuals.

In the CM case study, 3 different fabrics can be established for the CM1 group, whose firing temperatures varies in the range 850/950 up to > 1000 °C. As can be observed in Table 2, CM1-F1 and CM1-F2 correspond to well-fired fabrics (850-950 °C and 950-1000 °C). The differences between them are related to the disappearance of the illite-muscovite 10Å peak in the case of fabric CM1-F2. Finally, CM1-F3 shows an overfired fabric (> 1000 °C) with the presence of firing and secondary phases such as plagioclase, pyroxene, gehlenite, forsterite and analcime. Furthermore, XRD data for the 34 individuals from CM2 reveal their overfired nature. All these individuals are characterized by the total decomposition of illite-muscovite and, especially, the important presence of analcime. The significant crystallization of this secondary Na-zeolite must be the explanation for the high relative values of Na_2O, and, indirectly, is also related to the low relative values of K_2O, which are distinctive for this group. Its formation has been reported several times in overfired calcareous pottery and it is possibly related to the decomposition of the glassy phase, with the leaching of potassium, and the crystallization of analcime on the silicate material left by this alteration (Buxeda *et al.* 2002b; Schwedt *et al.* 2006). In all these samples, calcite corresponds, at least partly, to a secondary phase (Cau *et al.* 2002). Finally, the 3 low calcareous individuals corresponding to the chemical group CM3 show a low-fired fabric (estimated EFT < 800-850 °C) in which unclear firing phases are observed.

CONCLUSIONS

The evidence discussed above proves the existence of a Roman wine amphorae production in the area of ancient *Barcino*. Archaeometric analyses show that the PR site can be considered as a production centre, due to the homogeneity of its amphorae production. A unique RG product can, therefore, be defined for the PR workshop, that is a calcareous amphora, fired most probably around the 950-1000 °C range, although a significant number of overfired individuals occur in the kiln dump.

In contrast, amphorae from the CM site show a complex chemical structure, indicating possible differences in provenance, as well as technological differences between low calcareous and calcareous amphorae, and the

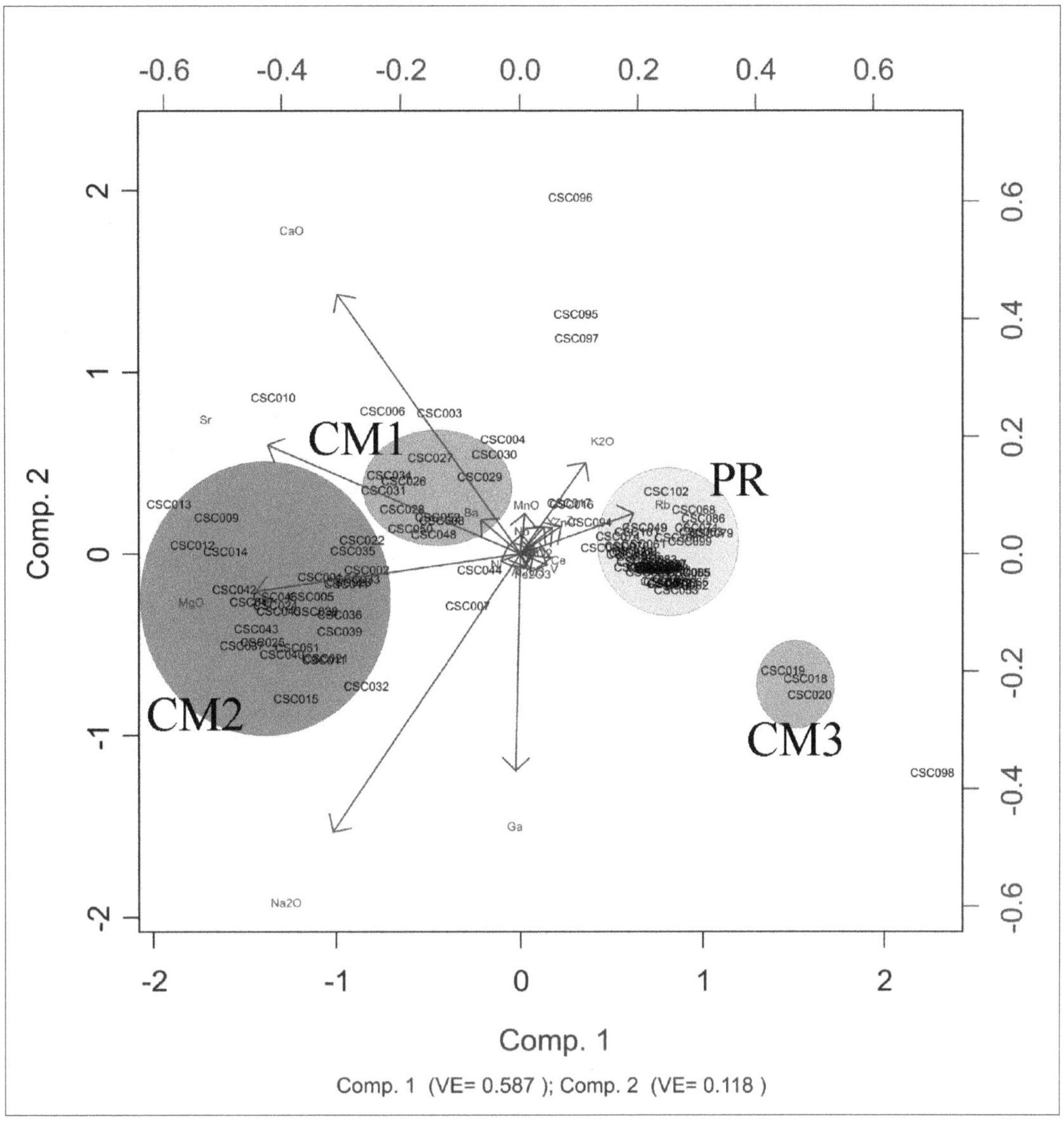

Fig. 4 – Biplot of the first and second principal component after a Principal Component Analysis performed on the logratio transformed subcomposition: Fe_2O_3, Al_2O_3, MnO, TiO_2, MgO, CaO, Na_2O, K_2O, SiO_2, Ba, Rb, Nb, Zr, Y, Sr, Ce, Ga, V, Zn, Ni and Cr, using SiO_2 as divisor.

presence of extensive secondary processes in an important number of individuals. For these reasons and, contrary to archaeological interpretations, we would propose a different role for the CM site, where only some pottery dumps have been found. From the archaeological and archaeometric results, we infer that at the CM site we find amphorae produced in the territory of *Barcino* and, possibly, in other closer areas, pointing to the role of the city not only in consumption, but also in redistribution. In this way, it is interesting to note that the individual CSC049 found in CM seems to be associated with the PR production center.

The clay samples from the kiln structures of PR workshop show a similar chemical composition to the paste used for amphora production. Even so, the CSC098 clay sample corresponds to a very different low calcareous material. This fact demonstrates that at least part of the building material used in the kiln construction is not the same as that used in the amphora paste preparation. With regard to the CM case study, the sampled clay sediments exhibit significant chemical differences in relation to the whole ceramic material. This fact could indicate that none of these amphorae were manufactured at this site, unless very different raw materials or clay recipes were used. However, the existing chemical groups at CM site, and the large chemical variability that they exhibit, seem to reinforce the hypothesis that those amphorae were produced somewhere else.

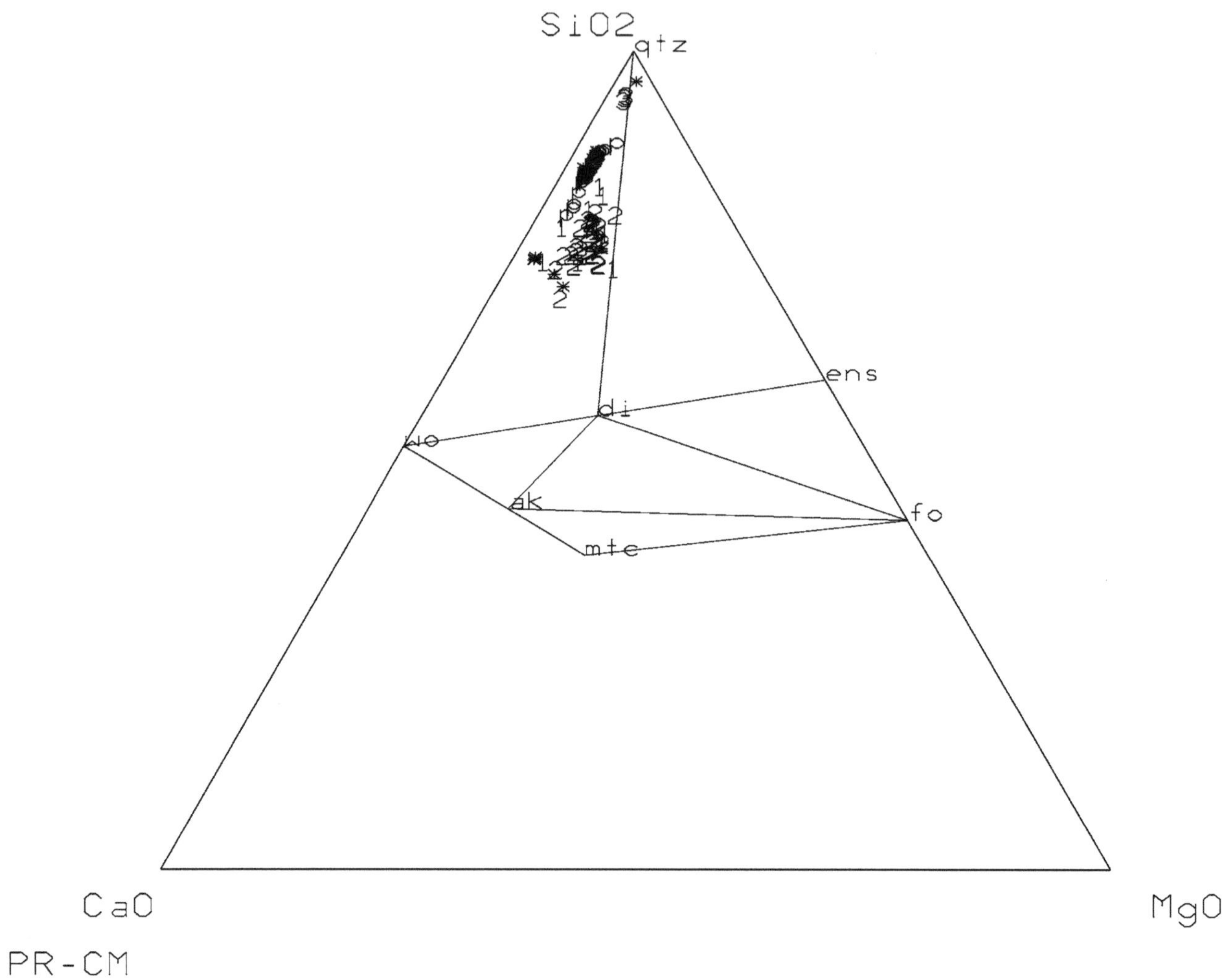

*Fig. 5 – Ternary diagram of the CaO-MgO-SiO$_2$ system. p: PR individuals; 1: CM1 individuals; 2: CM2 individuals; 3: CM3 individuals; *: ungrouped individuals.*

RG/PCRU	FABRIC	EFT	MINERALOGICAL PHASES
PR	PR - F1 (n=6)	< 800-850 °C	qtz, ill, kfs, pg, cal
	PR - F2 (n=7)	850-950 °C	qtz, ill, kfs, pg, cal, px, hm
	PR - F3 (n=4)	950-1000 °C	qtz, ill, kfs, pg, cal, px, hm
	PR - F4 (n=25)	> 1000 °C	qtz, kfs, pg, cal, px, hm, gh
CM1	CM1 - F1 (n=3)	850-950 °C	qtz, ill, kfs, pg, cal, px, hm, gh
	CM1 - F2 (n=6)	950-1000 °C	qtz, kfs, pg, cal, px, hm, gh, fo
	CM1 - F3 (n=4)	> 1000 °C	qtz, kfs, pg, cal, px, hm, gh, fo, anl
CM2	CM2 - F1 (n=34)	> 1000 °C	qtz, kfs, pg, cal, px, hm, gh, fo, anl
CM3	CM3 - F1 (n=3)	800-850 °C	qtz, ill, kfs, pg, cal, hm

Table 2 – Table indicating the fabrics corresponding to PR, CM1, CM2 and CM3 amphorae. qtz: quartz; ill: illite-muscovite; kfs: alkaline feldspar; pg: plagioclase; cal: calcite; px: pyroxene; hm: hematite; gh: gehlenite; fo: forsterite; anl: analcime.

All these data prove that *Barcino* played an important and diversified role in Roman wine amphorae production and distribution, because it was not just a production centre, but also a consumption and redistribution centre. In that sense, it is clear that for a full understanding of this first period of the romanization process, it is important to gain a deep insight into the strategic position of *Barcino*. For that reason, more sites and more individuals from Barcelona are now under study. We hope they will serve to complement this first investigation.

ACKNOWLEDGEMENTS

This study is included in the project PRODIFAN, Estudio arqueométrico y arqueológico de la producción y difusión de ánforas del nordeste peninsular durante los S. I a.C.- I d.C. financed by the Dirección General de Investigación del Ministerio de Ciencia y Tecnología (Spain) and the ERDF (European Union). Verònica Martínez Ferreras has a Formació en la Recerca i la Docència predoctoral fellowship of the Universitat de Barcelona. The XRF and XRD analysis were performed at the Serveis Cientificotècnics of the Universitat de Barcelona.

REFERENCES

AITCHISON J. 1986, *The Statistical Analysis of Compositional Data*, Chapman and Hall, London.

AGUELO J., HUERTAS P., PUIG, F. 2005, "Santa Caterina de Barcelona: assaig d'ocupació i evolució", *Quaderns d'arqueologia i d'història de la Ciutat de Barcelona* 1, p. 11-43.

BUXEDA I GARRIGÓS J. 1999, "Alteration and contamination of archaeological ceramics: the perturbation problem", *Journal of archaeological sciences* 26, p. 295-313.

BUXEDA I GARRIGÓS J., CAU ONTIVEROS M.A., GURT I ESPARRAGUERA J.M., TUSET I BERTRAN F., 1995, "Análisis tradicional y análisis arqueométrica en el estudio de las cerámicas comunes de época romana", *in Ceràmica comuna romana d'època alto-imperial a la Península Ibèrica. Estat de la qüestió*, Monografies Emporitanes VIII, Conjunt Monumental d'Empúries, Empúries, p. 39-60.

BUXEDA I GARRIGÓS J., COMAS I SOLÀ M., GURT I ESPARRAGUERA J.M. 2002a, "Roman amphorae production in *Baetulo* (Badalona, Catalonia). Evidence of Pascual 1", in V. Kilikoglou, A. Hein and Y. Maniatis (eds.), *Modern trends in scientific studies on ancient ceramics,* BAR International Series S1011, Oxford, p. 277-285.

BUXEDA I GARRIGÓS J., MOMMSEN H., TSOLAKIDOU A. 2002b, "Alterations of Na-, K- and Rb-concentrations in Mycenaean pottery and a proposed explanation using X-ray diffraction", *Archaeometry* 44, p. 187-198.

BUXEDA I GARRIGÓS J., KILIKOGLOU V. 2003, "Total variation as a measure of variability in chemical data sets", *in* L. van Zelst (ed.), *Patterns and process. A Festschrift in honor of Dr. Edward V. Sayre*, Suitland, p. 185-198.

BUXEDA I GARRIGÓS J., MARTÍNEZ FERRERAS V., VILA I SOCÍAS L. 2004, "Caracterització arqueomètrica de les àmfores Pascual 1 del derelicte Culip VIII", *in Culip VIII i les àmfores Haltern 70*, Monografies del CASC, 5, Museu d'Arqueologia de Catalunya i Centre d'Arqueologia Subaquàtica de Catalunya, Girona, p. 167-173.

CASAS J., MARTÍNEZ FERRERAS V. 2006, "El taller ceràmic d'època romana del carrer Princesa de Barcelona. Estudi arqueològic de les restes i estudi arqueomètric del material ceràmic", *Quaderns d'arqueologia i d'història de la Ciutat de Barcelona* 2, p. 36-59.

CAU ONTIVEROS M.A., DAY P., MONTANA G. 2002, "Secondary calcite in archaeological ceramics: evaluation of alteration and contamination processes by thin section", in V. Kilikoglou, A. Hein, Y. Maniatis, (eds.), *Modern trends in scientific studies on ancient ceramics*, BAR International Series S1011, Oxford, p. 9-18.

HEIN A., TSOLAKIDOU A., ILIOPOULOS I., MOMMSEN H., BUXEDA J., MONTANA G., KILIKOGLOU V. 2002, "Standardisation of elemental analytical techniques applied to provenance studies of the archaeological ceramics: an inter laboratory calibration study", *Analyst* 127, p. 542-553.

MARTÍNEZ FERRERAS V., BUXEDA I GARRIGÓS J., MARTÍN I MENÉNDEZ A. 2005, "L'évolution des premières amphores romaines produites à Cabrera de Mar (Catalogne) d'après leur caractérisation archéométrique", *SFECAG, actes du colloque de Blois*, p. 391-402.

MATHSOFT 1999, *S-PLUS 2000. User's Guide*, Data Analysis Products Division, MathSoft, Seatle.

MIRÓ J. 1988, *La producción de ánforas romanas en Catalunya. Un estudio sobre el comercio del vino de la Tarraconense (siglos I A.C.-I D.C.)*, BAR International Series 473, Oxford.

REVILLA CALVO V. 1995, *Producción cerámica, viticultura y propiedad rural en Hispania Tarraconensis (siglos I A.C.-III D.C.)*, Cuadernos de Arqueología 8, Edicions Servei del Llibre L'Estaquirot, Barcelona.

SCHWEDT A., MOMMSEN H., ZACHARIAS N., BUXEDA I GARRIGÓS J., 2006, "Analcime crystallization and compositional profiles: comparing approaches to detect post-depositional alterations in archaeological pottery", *Archaeometry* 48 (2), p. 237-251.

TREMOLEDA I TRILLA J. 2000, *Industria y artesanado cerámico de época romana en el nordeste de Cataluña, Época augustea y altoimperial*, BAR International Series 835, Oxford.

TSANTINI E., BUXEDA J., MADRID M., GURT J.M., MIÑARRO M. 2005, "Amphora production in the pre-Roman northeast of the Iberian Peninsula and evidence of trade with the Balearic Islands", *in* I. Prudencio, M.I. Dias and J.C. Waerenborgh (eds.), *Understanding people through their pottery*, Instituto Tecnológico e Nuclear, Lisbon, Portugal, Travalhos de Arqueología 42, p. 279-286.

VILA I SOCÍAS L., BUXEDA I GARRIGÓS J., TREMOLEDA I TRILLA J. 2005, "Caracterització arqueomètrica de la producció d'àmfores Pascual 1 i Dressel 7-11 del taller romà de Llafranc (Palafrugell, Catalunya)", *SFECAG, actes du colloque de Blois*, p. 761-769.

A LATE ROMAN POTTERY AND BRICK FACTORY IN SICILY (SANTA VENERA AL POZZO)

SUSANNA AMARI

Superintendence of Catania, Via Vitaliano Brancati 13, 95030 Nicolosi, Italy
amarisusanna@tiscali.it

1. INTRODUCTION

The archaeological area of Santa Venera al Pozzo is in the territory of Catania, 14 km north of the city and 4 km northwest of the ancient port of Capo Mulini. This area was well-known as a Roman spa[1] and is still today as in ancient times a wonderful natural reserve with water springs, and plants, trees and flowers of Mediterranean scrub. The modern church, which is in use, is located a few metres east of the well where St. Venera was martyred.

2. THE EXCAVATIONS

At St. Venera al Pozzo, excavation work was undertaken by the Superintendence of Syracuse (Spigo 1980-81; Spigo 1984-85) and Catania (Branciforti 1999) and then, from 1980 to 1997, work was concentrated on the public buildings: the baths and the temple. New excavations carried out by the Superintendence of Catania in the years 1999-2001, have been very useful in clarifying the location of the built-up area of the Greek period, *Akis*, and the Roman *statio Acium* with the public baths. The excavations have also provided information about the earlier occupation of the site that dates back to the late Copper Age, 3rd millennium BC. At the end of the project, a little museum was inaugurated inside the archaeological area in a 19th century building restored for this purpose.[2] A new research programme is being carried out, in ancient *Akis* and *Acium*.[3]

3. FUNCTION AND ROLE OF THE SITE

The site, identified with the ancient Roman *statio Acium*, is listed in the Antonine Itinerary along the road *Pompeia* between Messina and Catania (Cicero, *In Verrem* II, 5, 66, 169; Cuntz 1990, I, 12). Because of this, we can talk today about the *mansio Acium* which has a long history (Branciforti 2002; Branciforti, Pagnano, Amari, *et al.* 2006). Little is known about the economic and social context of *Acium* in the countryside. Few of the regional towns nearby have had significant modern excavations. The Greek and Roman rural cult-centre,[4] listed in the ancient itinerary along the extensive network of roads of the Roman period, continued to be used well into the late 19th century. On the basis of modern local sources we know that the spa building, the open market and the church at St. Venera al Pozzo were still frequented in medieval times, and the modern site was a rural cult-centre until two centuries ago. At the end of the 19th century only the warm sulphureous water springs (22 °C) were piped to a modern spa building in Acireale (the present-day spa of Santa Venera) and only the church remained in use at St. Venera al Pozzo.[5] The feast-day is on 11th January.

1 – Paternò Castello Principe di Biscari 1781, p. 22-26; Houel 1782-1787; Pagnano 2001, p. 72-74, 154-155.
2 – Branciforti 2002; Branciforti *et al.* 2006; Amari (in press).
3 – For the bilingual inscription (in Greek and Latin) found at Casalotto in a granary near S. Venera al Pozzo, see Libertini 1922, p. 493-494, fig. 2; Wilson 1990, p. 191, fig. 159a; p. 292 and no. 80 for the references; p. 314, fig. 267. For the *Acium* tile-stamp found not a long way from the factory, see Wilson 1979, p. 37, n. 57; Amari (in press).
4 – The Demeter and Kore cult is attested from the Greek period, from the end of the 5th to the first half of the 4th century BC by the presence of little terracotta statues, *ex voto* figures on the site. The religious function is already suggested by the discovery of a significant temple building , and other material as well as the marble head of Minerva, see Wilson 1990, p. 212, 226-228, 291.
5 – The place was an important destination for pilgrims. A free fair was held here in 15th c. AD and continued until 1647, see Branciforti 2002; Branciforti *et al.* 2006; Amari (in press).

4. THE BRICK AND TILE PRODUCTION, FEATURES AND DIFFUSION

The late Roman factory is located on the northeast side of the archaeological area of the Superintendence of Catania, 160 m north of the ancient spa. Many rooms for the manufacture of pottery and bricks, warehouses, three round kilns of vertical type have been found in the factory. The kilns also produced amphorae and common ceramics for table and larder. We have been able to date the kilns by coins of Constantine I found buried under the combustion chamber. The kilns were used until the mid 5th century AD. The attacks of the Vandals on the western Mediterranean coasts may have caused the end of the activity of the factory (Branciforti 2002; Branciforti *et al.* 2006; Amari in press) and the slackening-off of the economic life of the site that served trade and industries. Some of the rooms of the Roman factory were fully excavated in 2000. A new research programme has just started at the Roman factory of *Acium* and a considerable quantity of pottery, amphorae, brick and tiles is coming to light in the deposit rooms. These confirm activity of intensive production and demand for materials from a local market.

This paper concerns the ceramics used as building material which were produced in the pottery and tile kilns, where we have found many bricks and tiles of good quality and many shapes, used for the building and tiling of walls, columns, arches, floors, roofs, and *suspensurae* with ring-bricks. In the above paragraph, I have not mentioned anything about the tools, the brick types in everyday usage in the factory, or the manufacture and firing processes. Even if we did not have wasters, the dimensions of the *Acium* bricks which were different from Italian ones, confirm that they were locally manufactured and that the island used only bricks and tiles made in Sicily during the whole period of the Roman Empire. We have knowledge of several of the brick and tile industries of the region in this period of the late Roman Empire (Wilson 1990, p. 268-270). In the course of the 4th and 5th c. AD, at *Acium* and in Sicily as throughout the whole of the Roman Empire, regional types of brick and tiles were produced.

This analysis focuses on the features of the materials, to research their diffusion and commercialization in the surrounding territory and elsewhere. In this paper I present a complete typology of the bricks and tiles found inside the factory and some found in the ancient baths and in the neighbouring area. An important long study has been carried out to describe the materials using metrological and physical properties. The brick and tile features gathered together in the *table of data*, nos. 1-33, permit us to make comparisons with other productions and gain some idea of the diversity of measurements and variation in fabric. The presence elsewhere of these tiles may indicate an organised trade. The finds show that the upper sides of the majority of the bricks from *Acium* have marks. They usually seem to have been made with the fingers (a single groove, double lines or double curves, wavy lines, single or triple X) or cut (stars and herringbone motifs) when the clay was not too dry (leather hard). We do not know the significance of the marks, but we probably would not be wrong to think of these as aids to gripping the lime mortar better and distinguishing between batches of wares or the products of specific workshops.

The pastes of the bricks and the tiles found at *Acium* are red (Munsell 1992, 2.5YR, 6/8 or 5YR, 6/6) occurring again in the ring-shaped bricks, or reddish orange (Munsell 1992, 5YR, 6/6, 6/8 or 7/8) with the last two most frequent in all products, and orange (Munsell 1992, 7.5YR, 7/4 pale orange or 7/6 and 7/8) even with many inclusions between less than 1 mm and more than 2 mm in size. Only further analysis will identify the presumably volcanic rock inclusions. The variation in fabric, with or without mica and gravels on the underside, suggests that the paste production and drying may have been carried out in more than one way even for the same type, according to the bulk of the wares.

The majority of the brick and tile production has beige or pale pink colour coating. A portable LNS PIXE-alpha system[6] was used to analyse it. This work was carried out in the LANDIS laboratories of the National Institute of Nuclear Physics at the University of Catania. The PIXE analysis is non-destructive and is a surface technique. The system allows the detection of elements as light as Na. PIXE-alpha analysis revealed that the light colour coat is rich in calcium, $CaCO_3$. There are no traces of sodium (Na) and chlorine (Cl) in the coating of the *Acium* tiles. There is not the whitening of surfaces by sodium, widely used in the North African tiles productions. Other examples were chosen for analysis because they are indicators of production techniques (wasters, overfired) or because they are very common tiles in their paste colour and inclusions and similar to the majority of the tiles found. More detailed petrological analysis will be conducted on examples nos. 2 and 5 (*imbrices*), 10 and 11 (frg. of flat-*tegulae*), 18 (frg. of ring brick), 23 (square brick), 33 (rhomboid brick). A programme of thin sectioning on powdered samples should be carried out on the *Acium* brick and tile examples and wasters to identify the physical and chemical features of this fabric, in order to determine whether the fabrics that have been thought to be the same are in fact so. This will be the subject of the next publication.

As mentioned above, the kiln fired-brick and tiles were produced for walling, columns, paving and roofing. For roof tiles the most widespread types in Sicily were the curving tiles which overlapped in the "Laconian" system of roofing, and the rectangular flat *tegulae* with the semicircular *imbrices* covering the adjoining flanges between two *tegulae*. Sicilian tiles are generally much thinner than bricks. *Acium* tiles are usually between 3 and 1 cm thick. In the factory both *tegula* and *imbrex* are found. Three different types of roof-tiles can be distinguished

6 – Pappalardo *et al.* 2003.

inside the workshop. The first type represents the fabric of Sicilian origin and is the latest product of the factory, in the first half of the 5th c. AD. It is decorated with finger-streaked lines on its upper surface and/or with "combed lines". The latter motif is on the tiles discovered in the Roman villa at Heraclea Minoa (Agrigento).[7] Wasters of this type with "combed lines" were also found in a circular brick-kiln at Nunziata (a district in the modern province of Catania, close to Naxos).[8] The samples from *Acium* in this paper, no. 1, Fig. 1-2, no. 2 and waster no. 3, Fig. 3, found in room A of the factory, consist of three curved roof-tiles with simple edges and straight and wavy finger-streaked lines and/or combed lines made when the paste was not dry. In any case, the finger decoration examples are a singular *Acium* production, different from similar contemporaneous types manufactured in western and eastern Sicily, in the territory of Catania. The *Acium* type retains the exclusive factory method of marking the bricks and tiles with wide finger marks. The workshop, strongly rooted in its tradition, did not adopt only the new "combed lines". This may also be an experimental product of the factory. The second scheme of roofing is well represented by semicircular profile tiles (*imbrices*), of the second type, and flat-*tegulae*, all with finger marks. The *imbrices*, no. 4, Fig. 4, and no. 5, Fig. 5, found inside the big kiln, do or do not have a rudimentary flange at the end on the edge, and two characteristic straight streaked lines made by fingers on the upper surface. The third type of roof-tile is the flat *tegulae* nos. 6 and 7, Fig. 6-7, found in room H where there is one little kiln, and no. 8, found inside the big kiln. They have rectangular shapes (48 cm in width)[9] with a flanged edge rounded in profile, 4.5 cm in width, on two parallel sides, finger marks on the upper surface along the inner part of the high edge, maybe to help the lime mortar grip better, and at the centre they have finger marks, generally parallel lines, X and double V, Fig. 6 and 9a. They are covered by a buff pink or beige wash. In this paper, the shape of the flat-*tegula* is the type with a curved side, cut perhaps before firing, example no. 9, Fig. 8, found inside the big kiln. The type is well represented *in situ* because one little kiln is built with these tiles. Examples nos. 10 and 11, Fig. 9b, were chosen for the analysis, the first because it has the features of the majority of the tiles, in the colour of its paste, and the second because it is an overfired waster. At Santa Venera al Pozzo the flat-*tegulae* type is also used as a double wall in the swimming pools with mosaics, to keep the temperature constant inside. This was also the function of one special type of square or rectangular brick, 3 cm thick, with four feet on one face. No complete examples have been found in the factory, only flat-*tegula* fragments with conical feet and with many marks imprinted before firing, maybe to help grip the lime mortar better. Examples no. 12 found in room B and no. 13, Fig. 10a, found in the south of the baths, have a herringbone motif[10] cut into the circular face before firing. These imprints are also present on the upper surface of the brick, thickness 3 cm, used to built the side door of the combustion chamber of the big kiln. No. 14, Fig. 11-12, found in room N, has a star motif (see Chavane 1975, no. 60, p. 27, pl. 8) cut into the circular face before firing. The last one, no. 15, Fig. 10b, found in room N, has an X motif (see Chavane 1975, no. 57, p. 27, pl. 8) imprinted onto the circular face before firing.

Sicilian bricks in the Roman period are thicker than the bricks in the other regions of Italy and the different thicknesses of bricks do not indicate different periods of manufacture (Wilson 1990, p. 15). So in this case in the late Roman period the *Acium* kilns did not fire standard size bricks as in central Italy. The second brick type recorded in this paper, the ring-shaped brick, was employed in the stacks of the *hypocaust* (*suspensurae*) but also for building columns. The hole in the centre of each brick may be present for ease of manufacture and drying before firing (Wilson 1990, p. 11). This ring brick type used for *suspensurae* is very common elsewhere in Sicily (Wilson 1979, p. 33, n. 5). Many *Acium* ring bricks are close in shape to the Sicilian production, but most have finger grooves imprinted before firing on the upper side roughly about 1.5 cm from the rounded edges, as shown in examples no. 16, Fig. 13 and no. 17, found in room B of the *Acium* factory. The second one has a rudimentary shape, maybe because of a defect in firing. This type, also found inside the big kiln, was used here to build the pillars in the combustion chamber. They are all wasters and overfired, as example no. 18, fragment of a ring brick diameter 25 cm, with rough sides and made lighter (white, yellow and pink) by firing and the greyish red colour paste with the same inclusions as the other examples and the lower side rough. Many other ring bricks with or without finger groove imprints and rounded edges are present inside the kilns and the rooms of the factory at St. Venera al Pozzo. They have a thickness of between 5.5 and 8 cm and apart from the 24/25.5 cm size, we have found five different diameters, 22,5 cm, 33 cm, 36/37 cm (also *in situ* in room D), 39/40 cm and 45 cm (also *in situ* employed as a column in the portico in the east of the big kiln). The complete examples, no. 19 found in room N, and no. 20, Fig. 14, found in room B, were part of columns built with circular bricks. The first has a finger groove imprint made before firing on the upper side around the edge and a layer of lime mortar on the sides and inside the hole. Another system for building columns involved the use of segments of brick as well as no. 18 and special brick. The type found in the archaeological area of Santa Venera al Pozzo, no. 21, Fig. 15a, consists in a complete brick of triangular shape with a circular side, a quarter of a circle. We rarely

7 – Wilson 1979, p. 23, pl. 2.VI.b, dated from early in the 5th century AD.

8 – Wilson 1990, p. 269-270.

9 – The Roman rectangular flat-*tegula* has different dimensions. In Italy, Ostia and Pompei have a type 48 cm in width and 72 or 59 in length, see Adam 1994, p. 229. The *Acium* type is more than 59 cm in length, up to about 72 cm, see *infra* the flat-*tegulae* employed in the tube.

10 – For the motif, see Chavane 1975, no. 59, p. 27, pl. 8.

Fig. 1 – Imbrex with finger-streaked lines.

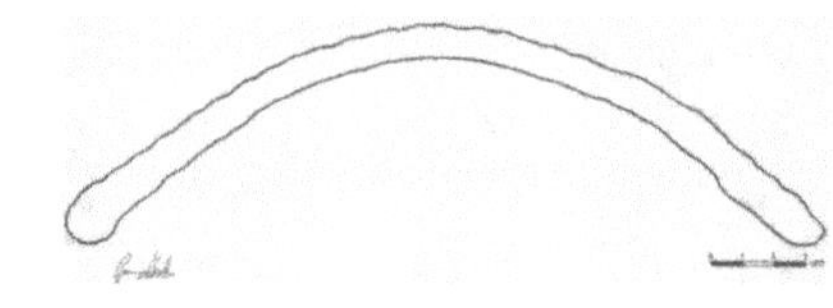

Fig. 2 – Profile of imbrex: no. SV 721.

Fig. 3 – Imbrex waster: no. SV 709.

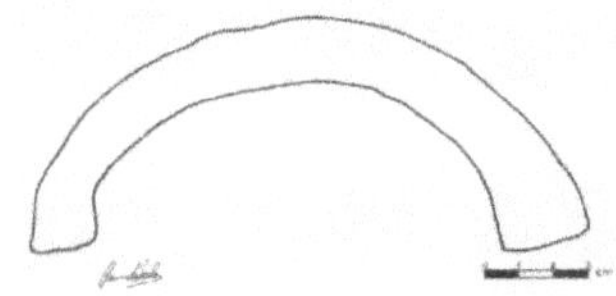

Fig. 4 – Profile of imbrex: no. SV 640.

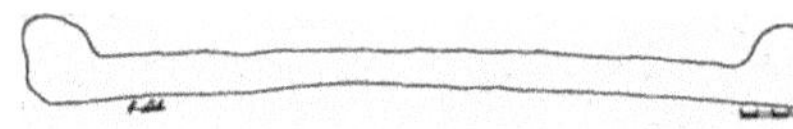

Fig. 7 – Profile of tegula: no. SV 732.

Fig. 5 – Imbrex no. SV 642.

Fig. 6 – Tegula no. SV 732.

Fig. 8 – Curved side tegula no. SV 502.

Fig. 9 – Tegulae: mark on SV 733 and over fired no. SV 838.

Fig. 10 – Tegulae mammatae: feet no. SV 228 and no. SV 649.

Fig. 11 – Tegula mammata no. SV 647.

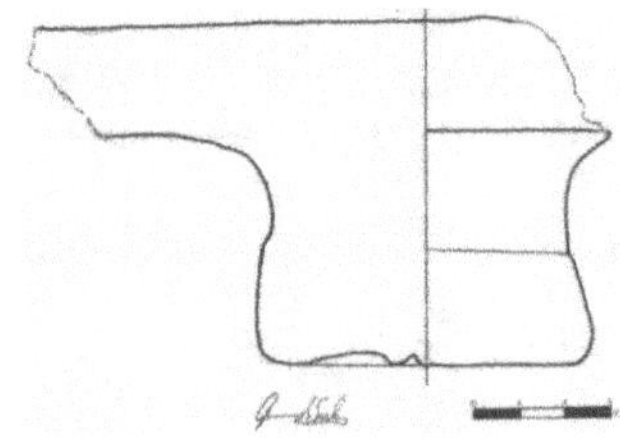

Fig. 12 – Profile of tegula: no. SV 647.

Fig. 13 – Ring brick no. SV 706.

Fig. 14 – Ring brick no. SV 839.

Fig. 15 – Quarter circle and triangular shape bricks.

Fig. 16 – Square brick no. SV 644.

Fig. 17 – Square brick no. SV 681.

Fig. 18 – Opus spicatum bricks.

Fig. 19 – Almond brick no. SV 691.

Fig. 20 – Rhomboid bricks: no. SV 663 and no. SV 646.

find walls entirely built in bricks in the territory of Catania, because of the presence of local basaltic stone. At *Acium*, apart from the kilns built entirely of bricks and tiles, bricks were used in the factory walls. The bricks were used like blocks of masonry. The first type presented here, *in situ*, was used in the south perimeter wall of the factory, and is a rectangular brick (52 cm in length, 34 cm in width and 7.5 cm thick). This type is related to the next type (48/49x32/33x6/7 cm) used in the factory. Another example of a small structure entirely built of bricks and tiles is the tube inside the factory. It is also *in situ* and has a floor partly laid of bricks (48/49 cm in length, 32/33 in width and 6/7 thickness) and in part in flat-*tegulae* (72/73 cm in length, 49/52 cm in width and 3/5.5 cm in thickness *tegula* and listel), and the walls in bricks (about the size of 48/49 cm in length, 32/33 cm in width and 6/7 cm thickness). There are many other types of bricks in the rooms used for manufacture and in the warehouses of the factory. We can find rectangular-shaped bricks of different sizes used in the rooms of the factory. One type used as a bench, is brick no. 22 which supports a *dolium* for water in room D with a clay tank built with flat-*tegulae* set upright. The other type widely manufactured was a square brick, with sides of 24 cm and 6 cm thickness, with reddish orange paste, rough and very hard, with black inclusions, some larger than 2 mm, some with mica. The lower side is granular. Like other ones, the complete examples in this paper, found in room N (warehouse), nos. 23, 24 and 25, Fig. 16, show the same X-shaped finger mark in a corner on the upper side. It is rare today to find a *pedale* size. One example, *in situ* in room E, has 30 cm sides and 6 cm thickness. The red paste is covered with a beige wash.

The other major production of kiln-fired bricks in *Acium* was for paving. Brick pavements were laid out either with square, or rectangular bricks, and sometimes in flat-*tegulae* as the example in the brick-tube, in some rather more specialized products, such as rhomboid, almond-shaped, triangular and blocks of soap-shaped bricks, this latest type being used in herringbone brickwork (*opus spicatum*). On the east side of the big kiln, on the opposite side of the *praefurnium*, the portico has an example of rectangular flat-*tegulae* pavement *in situ* (72/73 cm in length, 49/52 cm in width and 3/5.5 cm thickness *tegula* and listel) of the same type of floor of the brick tube mentioned above. The square bricks of 24 cm and 29 cm could be used everywhere for paving, as could another type of rather thin quadrangular brick, found in the factory. The complete example no. 26, Fig. 17, has a triple X finger print on the entire upper side which was made when the paste was not dry, maybe to allow the lime mortar to grip better and to distinguish one batch of wares. This mark recurs on different size bricks for paving found in Italy and in other regions of the Roman Empire.[11] The type of brick used in herringbone brickwork (*opus spicatum*) is also widespread

11 – *Settefinestre* 1985, p. 35, Type Tav. 2.3a-b; Green 1979, p. 363-373, pl. 20.1.

in Sicily.[12] Of the complete examples here, nos. 27, 28 and 29, the first was found in room A of the factory and the others in the baths, and they have the same features and paste, Fig. 18. The last three types of brick presented in this paper, rhomboid, almond, no. 30, Fig. 19, and no. 31 triangular shape, Fig. 15b, were used for wall tiling and for paving during the period of the Roman Empire.[13] The triangular shape brick is also well attested, in concrete walling, for the side tiling of arches. The rhomboid type also occurs in the back of the arches and in the architraves.[14] The complete examples, nos. 32 and 33, were one found in room N and the second in room D, Fig. 20.

5. CONCLUSIONS

During the Roman Empire the Sicilian brickyards and specially the Catanian brick industry did not undergo a huge expansion in production such as occurred in the brick industries of peninsular Italy (Wilson 1990, p. 268). In Catania, volcanic block was used as a facing material for concrete. The importance of the Santa Venera al Pozzo area during this period of the late Roman period is demonstrated by the discovery of the tile and brick yard. We do not know whether it operated as a municipal works.[15] The factory supplied the area near the workshop and the raw materials (water, clay and wood) were available close to the settlement. The activity of intensive production suggests a market demand for the products. Little is known about the neighbouring rural settlements: a granary and two ovens at Casalotto[16] and storage at Aciplatani (La Rosa 1972) are the only data we have. It is unknown what part *Acium* played in the exchange of its products with other territories. As well as in the *Acium* kilns, pottery, amphorae, and tiles were discovered in the late Roman kilns in Naxos, in eastern Sicily, north of Santa Venera al Pozzo (Pelagatti 1976-1977, p. 544-545; Fallico 1976-1977, p. 632-633; Wilson 1990, p. 269). The rich variety of their manufacture suggests a lively building and economic activity in Sicily in the late Roman period. The role played by bricks and tiles in the late antique economy would be better understood with a more detailed knowledge of forms and production sites. More quantified studies of closed deposits are very important. With more of this data available, we will be in a better position to ask questions about economic activity in the ancient world and the roles played internally by some regions and some types of material. In the future, the search, for the presence elsewhere of these building materials through comparison with other fabric series, would demonstrate the existence or non-existence of a brick and tile sea-trade in which the bricks and tiles would have served as accompanying goods in directed trade, or to fill empty merchant-ships after or before freight, or as ballast for sale (Gianfrotta and Pomey 1981, p. 222-225; Tomber 1987, p. 169-171; Amari in press). Though building materials are recorded as part of the excavation of settlements, we have little understanding of the potential significance, economically and socially, of the production and marketing of such materials. Though in many cases we see essentially local strategies of acquisition, building materials could have played a significant role in the scale and direction of economic networks. Considerable quantities of bricks and tiles have come to light in the Mediterranean sea as part of loads or entire cargoes.[17] Nothing is known about their Sicilian origin. In the sea, near the Catanian coast, tiles of red paste which do not seem to be products of *Acium* have also been found (Tortorici 2003, p. 287, Fig. 14-15). At St. Venera al Pozzo the three most visible industries archaeologically are pottery, amphorae and tile production. All have urban and rural manifestations in Sicily. Decorated styles found at *Acium* suggest a distinguishable production which give an indication of "trade zones" across the territory. Naturally, the largest Roman industry was agriculture. The island in the Roman period is characterized by intensively occupied and extensive rural landscapes related to expanded agricultural production, besides regional scale craft and industrial production of pottery, amphorae and tiles and others. This paper focuses attention on the study of the tiles produced in the kilns and on the typology of the forms and the marks, which may indicate whether or not they were a product intended for a local market only. If this was the distribution of these wares, only further work can show. The presence of some tile shipwrecks in the Sicilian sea and in the whole of the Mediterranean sea confirms the existence of a sea-trade in brick and tile. The discovery of this factory, a kiln-site, is of great importance for Sicilian studies. The closed road network of Sicily is especially important as it represents a key phase in which the economic, social and political geography of the Roman province was established and formalised and *Acium* is a rural town near the junction of four of the most important roads on the island in a large and complex network of local and regional routes in and out of Sicily. Furthermore, this network needs to be considered alongside the role of rivers in understanding the overall framework of communications and transport to the sea. We have some documented evidence for the road between St. Venera al Pozzo and Capo Mulini, an important port used from the prehistoric times (like the St. Venera al Pozzo area), placed

12 – Giuliani 1990, p. 152-160, fig. 7.4; *Settefinestre* 1985, p. 36-37, Type pl. 3.8a-b; Wilson 1990, p. 18, fig. 2, V, a.

13 – These types were used for decorative tiling on a wall in Pompei, see Adam 1994, p. 160, fig. 351. The type of flooring with rhomboid shape bricks is attested in the territory of Modena, see Calzolari and Giordani (eds.) 2001, p. 19, fig. 6.

14 – So in the Roman theatre of Ferento and in the Odeon of Catania, see Lugli 1957, p. 354, fig. 81.

15 – Wilson 1979, p. 37, n. 57, a tile stamped Ακιος is in the Accademia Zelantea collection in Acireale.

16 – Libertini 1922, p. 499, fig. 1. The ovens were built putting flat-*tegulae* together.

17 – Parker 1992, see Index and for layers and loads of bricks and tiles on ship-wrecks of the Roman period p. 52, 92-93, 107-108, 113, 115, 121, 132, 137, 169, 206, 226, 244, 247, 292, 295, 299-300, 320, 322, 326, 338-339, 360, 370, 394, 407, 414, 416, 450.

on the Reitana estuary, of the ancient river *Akis* (Tortorici 2003, p. 291-306, 318-325, fig. 21, 47). The river route was especially suitable for bricks and tiles, which were very heavy wares, and hard to transport by land. I believe that soon we will be able to understand more about the brick and tile trade and its organisation, production and distribution.

ACKNOWLEDGEMENTS

I am very grateful to Dr. Maria Grazia Branciforti, archaeologist and supervisor of the Superintendence of Catania, for the right to publish the entire content of my paper including the drawings and photos belonging to the Superintendence of Catania. My husband, Giovanni La Scala, produced the drawings and Giuseppe Barbagiovanni the photographs of the materials. I would like to thank them. I am indebted to Prof. Giuseppe Pappalardo, Dr. Lighea Pappalardo and to Dr. Paolo Romano for permission to present here the preliminary results of the archaeometrical analysis in progress, as part of collaboration between the LANDIS laboratories of the National Institute of Nuclear Physics at the University of Catania and the Superintendence of Catania. My deepest thanks go to all them.

Table of data

1. SV 721. Inv. no. 9107. 7 cm high, max. 30 cm long and 22 cm wide. 1/1.5 cm thick. Red colour paste (Munsell 1992, 5YR 6/6), with black rock inclusions. Restored.
2. SV 743. Inv. no. 9812. 10 cm high, max. 16.5 cm long and 28 cm wide. Red orange colour paste (Munsell 1992, 5YR 6/8), with black rock inclusions. Restored. Example analysed.
3. SV 709. Inv. no. 9106. 1.5 cm high, max. 6.5 cm long and 9 cm wide. Grey paste.
4. SV 640. Inv. no. 9813. 8 cm high, max. 35 cm long and 17 cm wide. Red orange colour paste (Munsell 1992, 5YR 7/8), with black rock inclusions. It is covered with a buff pink wash.
5. SV 642. Inv. no. 9814. 8 cm high, max. 37.5 cm long and 17.5 cm wide. Red orange colour paste (Munsell 1992, 5YR 7/8), with black rock inclusions. It is covered with a buff pink wash. Example analysed.
6. SV. 731. Inv. no. 9815. 3 and 5 cm thick, max. 51 cm long and 48 cm wide. Edge 4.5 cm wide. Red orange colour paste (Munsell 1992, 5YR 6/8), with black rock inclusions. It is covered with a beige wash.
7. SV. 732. Inv. no. 9816. 3 and 5 cm thick, max. 54.5 cm long and 48 cm wide. Edge 4.5 cm wide. Red orange colour paste (Munsell 1992, 5YR 6/8), with black rock inclusions. It is covered with a buff pink wash.
8. SV. 733. Inv. no. 9817. 3.5 and 6 cm thick, max. 39 cm long and 48 cm wide. Edge 4.5 cm wide. Beige grey colour paste (Munsell 1992, 10YR 7/3), with black rock inclusions. It is covered with a beige wash.
9. SV. 502. Inv. 9122. 2 and 5.5 cm high, 22.5 cm long. Edge 5 cm wide. Reddish orange colour paste (Munsell 1992, 5YR 6/8), with black rock inclusions. Complete.
10. SV. 836. Inv. no. 9818. 3 cm high, max. 22 cm long and 18 cm wide. Reddish orange colour paste (Munsell 1992, 5YR 6/8), with black rock inclusions more than 2 mm long and mica.
11. SV. 838. Inv. no. 9819. 3.5 and 6.5 cm high, max. 20 cm long and 17 cm wide. Very hard red colour paste (Munsell 1992, 2.5YR 6/8), with black rock inclusions. Gravels are on the under side.
12. SV. 27. Inv. no. 9099. 5.5 cm high and 9 cm in diameter. Red orange colour paste (Munsell 1992, 5YR 7/8), with black rock inclusions.
13. SV. 228. Inv. no. 9100. 6.5 cm high and 8.5 cm in diameter. Red orange colour paste (Munsell 1992, 5YR 7/8), with black rock inclusions. It is covered with a beige wash.
14. SV. 647. Inv. no. 9098. 7.9 cm high and 7.6 cm in diameter. Reddish orange paste (Munsell 1992, 5YR 6/8) with black inclusions. It is covered with a beige wash.
15. SV. 649. Inv. no. 9101. 3 cm high and 7 cm in diameter. Beige orange paste (Munsell 1992, 7.5YR 7/4) with black inclusions and mica. It is covered with a beige wash.
16. SV. 706. Inv. no. 9124. 6.5 cm thick and 24.5 cm in diameter. Coarse red fabric (Munsell 1992, 2.5YR, 6/8) very hard, with black inclusions. Complete.
17. SV. 707. Inv. no. 9125. 6 cm thick and 25.5 in diameter. Coarse red fabric (Munsell 1992, 2.5YR, 6/8) very hard, with black inclusions. It is covered with a beige wash on the upper side. Complete.
18. SV. 837. Inv. no. 9820. 5.5/5.9 cm thick and max. 13.8 cm wide. Red colour paste (Munsell 1992, 2.5YR, 6/8). Example analysed.
19. SV. 648. Inv. no. 9821. 5.5 cm thick, 33 cm in diameter, grey and orange paste.
20. SV. 839. Inv. no. 9822. 7 cm thick, 39 cm in diameter, red in fabric (Munsell 1992, 2.5YR 6/8), with black inclusions, traces of lime mortar on the sides.
21. SV. 734. Inv. no. 9123. 7.5 cm thick and 21 cm radius. Reddish orange paste (Munsell 1992, 5YR 6/8), coarse and with black inclusions.
22. SV. 357. Inv. no. 9823. Complete example 45 cm long, 32 cm wide and 3.5 cm thick. Reddish orange paste (Munsell 1992, 5YR 7/8), coarse and with many mica and with black inclusions less than 1 mm size. Upper side polished and with two finger points. Gravels on the under side.
23. SV. 639. Inv. no. 9824, reddish orange paste (Munsell 1992, 5YR 6/6). Example analysed.
24. SV. 643. Inv. no. 9118, orange paste (Munsell 1992, 7.5YR 7/8) and mica.
25. SV. 644. Inv. no. 9825, orange paste (Munsell 1992, 7.5YR 7/6). 6.5 cm. thick
26. SV. 681. Inv. no. 9115. 22.5 cm long, 20 cm wide and 3 cm thick. Orange paste (Munsell 1992, 7.5YR 7/6), with black inclusions.
27. SV. 715. Inv. no. 9119. 10 cm long, 6 cm wide and 3 cm thick. Coarse orange paste (Munsell 1992, 7.5YR 7/8), with black inclusions and mica.
28. SV. 736. Inv. no. 9120. 11 cm long, 5.7 cm wide and 2.5 cm thick. Coarse orange paste (Munsell 1992, 7.5YR 7/8), with black inclusions and mica.
29. SV. 737. Inv. no. 9121. 11 cm long, 5.7 cm wide and 2.3 cm thick. Coarse orange paste (Munsell 1992, 7.5YR 7/8), with black inclusions and mica.
30. No. SV. 691. Inv. no. 9104. Found in the factory. 14 cm long, 14 cm wide and 6 cm thick. Coarse orange paste (Munsell 1992, 7.5YR 7/6), with black inclusions greater than 2 mm size and mica. It is covered with a beige wash.
31. No. SV. 714. Inv. no. 9102. Found in room A of the factory. 8 cm long, 6 cm wide and 6 cm thick. Coarse orange paste (Munsell 1992, 7.5YR 7/8), with black inclusions and mica. Complete.
32. SV. 646. Inv. no. 9103. 12 cm long, 7.8 cm wide and 7.5 cm thick. Coarse reddish orange paste (Munsell 1992, 5YR 6/6), with black inclusions greater than 2 mm size. It is covered with a beige wash.
33. SV. 663. Inv. no. 9826. 12 cm long, 7.8 cm wide and 7.5 cm thick. Coarse reddish orange paste (Munsell 1992, 5YR 6/6), with black inclusions greater than 2 mm size. Example analysed. It is covered with a beige wash.

REFERENCES

ADAM J.-P. 1994, *L'arte di costruire presso i Romani. Materiali e tecniche,* Biblioteca di archeologia 10, Milano.

AMARI S. (in press), "A late Roman amphorae production in eastern Sicily", *in SOMA '05, Symposium on Mediterranean archaeology 2005. Università degli Studi di Chieti 24-26 febbraio 2005,* British Archaeological Reports.

BRANCIFORTI M.G. 1999, "Siti e insediamenti nella regione etnea", *in Origini e incontri di culture nell'antichità. Magna Grecia e Sicilia, stato degli studi e prospettive di ricerca*, Messina, p. 241-248.

BRANCIFORTI M.G. 2002, *Santa Venera al Pozzo. 25 anni di Beni Culturali in Sicilia*, Regione siciliana. Assessorato dei BB. CC. AA. e P.I. Area Soprintendenza per i Beni Culturali ed Ambientali, Servizio per i Beni Archeologici, Catania.

BRANCIFORTI M.G., PAGNANO G., AMARI S. *et al.* 2006, *L'area archeologica di Santa Venera al Pozzo-Acium. Antiquarium*, Regione siciliana. Assessorato dei BB.CC.AA. e P.I. Area Soprintendenza per i Beni Culturali ed Ambientali di Catania. Palermo.

CALZOLARI M., GIORDANI N. (eds.) 2001, *L'insediamento preistorico e romano di Corte Vanina (Località Fossa di Concordia): nuove ricerche archeologiche nella Bassa Modenese*, Gruppo Studi Bassa Modenese, Modena.

CHAVANE M.J. 1975, *Les petits objets. Salamine de Chypre VI*, Paris.

CUNTZ O. 1990, *Itineraria Romana*, Stutgardiae.

DE SALVO L. 2002, "La Sicilia e le province occidentali in età imperiale e tardoantica", *in L'Africa romana 14, Sassari 2000*, Roma, p. 1601-1616.

FALLICO A.M. 1976-1977, "Naxos: fornaci tardo-romane", *Kokalos* 22-23, II, 1, p. 632-633.

GIANFROTTA P.A., POMEY P. 1981, *Archeologia subacquea. Storia, tecniche, scoperte e relitti*, Milano.

GIULIANI C.F. 1990, *L'edilizia nell'antichità*, Nuova Italia Scientifica, Roma.

GREEN T. 1979, "Techniques for studying comb signature distributions", *in* A. McWhirr (ed.), *Roman brick and tile: studies in manufacture, distribution and use in the Western Empire,* BAR International Series 68, Oxford, p. 363-373.

HOUEL J. 1782-1787, *Voyage pittoresque des îles de Sicile, de Malte et de Lipari*, Paris.

LA ROSA V. 1972, "Un rudere in via dell'Asilo ad Aciplatani", *in Accademia di Scienze, Lettere e Belle Arti degli Zelanti e dei Dafnici*, Memorie e Rendiconti, Acireale, p. 377-386.

LIBERTINI G. 1922, "Scoperte a Casalotto-Acireale", *in Notizie degli Scavi*, Roma, p. 491-499.

LUGLI, G. 1957, *La tecnica edilizia romana, con particolare riguardo a Roma e Lazio*, Roma.

MUNSELL A.H. 1992, *Munsell Soil Color Charts* (revised ed.), Newburgh-New York.

PAGNANO G. 2001, *Le antichità del Regno di Sicilia 1779. I plani di Biscari e Torremuzza per la Regia Custodia*, Palermo, Regione siciliana.

PAPPALARDO L., ROMANO F.P., GARRAFFO S., DE SANOIT J., MARCHETTA C., PAPPALARDO G. 2003, "The improved LNS PIXE-alpha portable system: archaeometric applications", *Archaeometry* 45 (2), p. 333-339.

PARKER A.J. 1992, *Ancient shipwrecks of the Mediterranean & the Roman provinces*, BAR International Series 580, Oxford.

PATERNÒ CASTELLO PRINCIPE DI BISCARI I. 1781, *Viaggio per tutte le antichità della Sicilia*, Napoli.

PELAGATTI P. 1976-1977, "L'attività della Soprintendenza alle Antichità della Sicilia Orientale. Parte I", *Kokalos* 22-23, II, 1, p. 519-550.

Settefinestre 1985, *Settefinestre: una villa schiavistica nell'Etruria romana*. Modena.

SPIGO U. 1980-1981, "Ricerche a Monte S. Mauro, Francavilla di Sicilia, Acireale, Adrano, Lentini, Solarino", *Kokalos* 26-27, II, 1, p. 771-795.

SPIGO U. 1984-1985, "Ricerche e rinvenimenti a Brucoli (cda. Gisira), Valsavoia (Lentini), nel territorio di Caltagirone, ad Adrano e Francavilla di Sicilia", *Kokalos* 30-31, II, 1, p. 863-904.

TOMBER, R. 1987, "Evidence for long-distance commerce: imported bricks and tiles at Carthage", *Rei Cretariae Romanae fautorum acta* 25-26, p. 161-174.

TORTORICI E. 2003, "Contributi per una carta archeologica subacquea della costa di Catania", *in Archeologia subacquea. Studi, ricerche e documenti, III*, Libreria dello Stato, Roma, p. 275-333.

VOLPE G. 2002, "Relitti e rotte commerciali nel Mediterraneo occidentale tardoantico", *in L'Africa romana 14, Sassari 2000*, Roma, p. 239-250.

WILSON R.J.A. 1979, "Brick and tiles in Roman Sicily", *in* A. McWhirr (ed.), *Roman brick and tile: studies in manufacture, distribution and use in the Western Empire*, BAR International Series 68, Oxford, p. 11-43.

WILSON R.J.A. 1990, *Sicily under the Roman Empire: the archaeology of a Roman province, 36 BC-AD 535*, Teddington House, Warminster, Wiltshire, England.

THE FIRST BYZANTINE "GLAZED WHITE WARES" IN THE EARLY MEDIEVAL TECHNOLOGICAL CONTEXT

S.Y. WAKSMAN[1], A. BOUQUILLON[2], N. CANTIN[1], I. KATONA[2]

[1] Laboratoire de céramologie, CNRS UMR 5138, Maison de l'Orient et de la Méditerranée, 7 rue Raulin, 69365 Lyon cedex 7, France, waksman@mom.fr, nadia.cantin@mom.fr

[2] C2RMF, CNRS UMR 171, 14 quai François Mitterand, 75001 Paris, France, anne.bouquillon@culture.fr, ildiko.katona@culture.fr

Byzantine "Glazed White Wares" appear in Constantinople in the 7th century AD (Hayes 1992). They represent a significant break with the Late Roman typological repertoire and correspond to the beginning of the medieval development of glazes in Byzantium. Due to the chronological hiatus of several centuries between the Hellenistic and Roman glazed productions of Asia minor and the first Byzantine Glazed White Ware, the question of the (re)introduction of the glazing technique has been a longstanding question. Glazed production seems to have been continuously present in the Parthian, Sasanian, then Islamic territories, and an Eastern introduction had been conjectured. However, the alternative hypothesis of a Western input has more recently been favoured (Hayes 1992, p. 15), as archaeological excavations gave evidence for the presence and production of glazed wares in northern Italy and in the Balkans in the 5th–7th centuries, that is, in the period just preceding the appearance of Byzantine Glazed White Wares (Paroli 1992; Spieser 1991, p. 250). It was the purpose of the present pilot project to investigate in the laboratory this question of circulation of techniques with the help of an appropriate sampling, which included for the first time examples of the earliest Byzantine Glazed White Wares. Paste and glaze analyses were carried out in order to shed some light on the technological relationships between the latter and both "Eastern" and "Western" productions.

SAMPLING

The first requirement of our selection was that samples should come from well defined archaeological contexts, with dates as close as possible to the 7th century AD. Even though the precise origin of the samples was not identified, they were considered representative of well identified technological environments. The sampling consists of four batches, which were given general names referring to this environment for the sake of clarity. It should however be clear that the samples represent the more restricted contexts introduced below (Fig. 1):

— The "Byzance" batch (samples BYZ718-733) is representative of the first Byzantine Glazed White Wares (GWWI) as defined by J.W. Hayes (Hayes 1968; Hayes 1992).[1] It comes from the excavations at Saraçhane, *i.e.* the church of Saint Polyeuktos, in Istanbul. The samples are dated to the early 7th century AD and are thought to have been manufactured in Istanbul or its region (Hayes 1992, p. 12);

— The "Middle East" batch (LEV402; LEV404-414) comes from the basilica and pilgrimage complex of Saint Symeon, in Northern Syria. These samples, dated to the second half of the 8th–9th century, correspond to the first glazed wares appearing on the site (Orssaud 2001). Typological parallels can be found in Northern Syria, Iraq and Iran. Although later than GWWI, they were thought of interest as being representative of an early Islamic production, to be complemented by other data on earlier glazed wares of the region (see *infra*);

— The "Balkans" batch (BYZ623-625; BYZ627-628) is from Caričin Grad in Serbia, a city probably founded by the emperor Justinian. It had a short existence, from 530 to 615 according to texts, which makes it of particular interest for the present study. The samples are thought to be of local origin (Bjelajac 1990);

— The "Italy" batch comes from Saint-Blaise (LIS 62-63; NMA147) and Hyères (NMA146) in south-eastern France. The samples were found in contexts dated to the 5th–7th centuries and are thought to have come from northern Italy (CATHMA 1992).

All samples are sherds with glazes directly applied to the paste, without intermediary slip, unlike later Byzantine table wares which appear from the 11th century onwards (Papanikola-Bakirtzis 1999).

1 – This name does not imply that the pastes are white. The ware is "light-coloured, mostly light or pale brown or orange, sometimes reddish, on occasions fired partly grey, virtually never plain white" (Hayes 1992, p. 15).

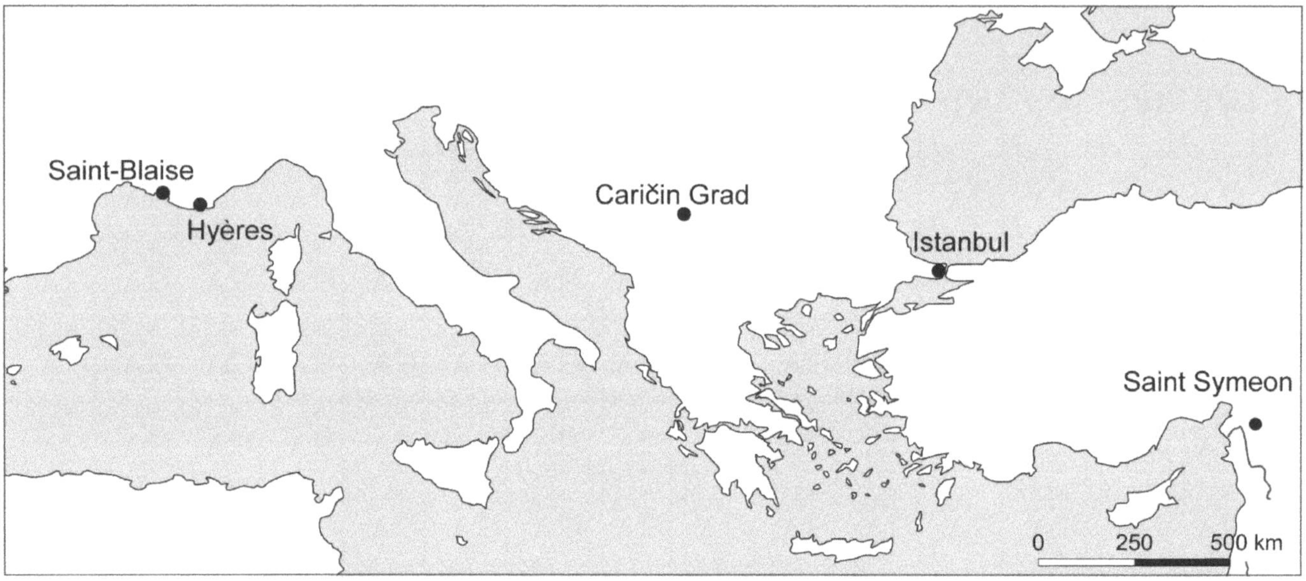

Fig. 1 – Location of the main sites mentioned.

PASTE AND GLAZE ANALYSIS

Pastes and glazes were investigated by elemental analysis. Paste analysis was carried out by WD-XRF at the "Laboratoire de céramologie" in Lyon, and glaze analysis by EDX coupled to a SEM at the "C2RMF" in Paris, thus allowing an examination of the microstructures as well. Special attention was devoted to accurate quantitative analysis in the case of heavy matrices, such as high lead glazes.[2] A larger number of samples was usually analyzed for their paste and a sub-sample selected for glaze analysis.

With the exception of the "Italy" batch, which is further split into (at least) two sub-groups closely related to typology,[3] the four batches of samples are moderately to very homogeneous and are well differentiated by paste analysis (Table 1). Three of them, the "Byzance", "Balkans" and "Italy" batches, have in common high contents of silicon and aluminium, close to 90% SiO_2 + Al_2O_3, which indicate that kaolinitic clays are likely to be main components of these pastes. The "Byzance" batch is further distinguished by lower contents of Ca, Mg and Mn. Batch "Balkans" shows distinctive features in its high contents of Mn, Sr, Ba. Other elements such as Ti and Zr allow further differentiation between the batches of kaolinitic pastes.

The "Middle East" batch is very different, due to high calcium and related high Mn and Sr contents. It constitutes a very homogeneous compositional group. Calcareous pastes, as well as synthetic pastes, are typical for fine table wares of Islamic tradition (*e.g.* Pérez-Arantegui *et al.* 1995; Daszkiewicz and Raabe 1999). Calcareous pastes are also widespread in Mesopotamia owing to the geology of the region (Schneider 1996, p. 131; Le Mière and Picon 2003, p. 179). In most samples high contents of lead in the paste, and in some cases of copper,[4] reveal the nature of the glaze, from which these elements would have diffused.

Glaze analyses show that all the glazes are high lead ones,[5] the alkali content being consistently low (below 1% with few exceptions, Table 2). They are coloured with copper for green glazes, and iron for orange-brown or greenish glazes when applied on oxidized or reduced pastes respectively.[6] Antimony is associated with copper in two samples in the "Middle East" batch, which may have been used to obtain a more vivid colour.

Samples in batch "Byzance" have a quite homogeneous high lead glaze over their kaolinitic pastes, showing little alteration and a fairly high aluminium content of circa 7 to 9% Al_2O_3. Such consistently high Al concentration, at some distance from the paste-glaze interface, may indicate that clay was used as a component of the glaze mixture. However, it is also possible that a lead compound was applied directly onto the paste, as the composition of the glaze after substraction of PbO and recasting to 100% roughly corresponds to the composition of the clay (Tite *et al.* 1998). Another indication of the manufacturing process is given by the thickness of the interface (Fig. 2, left), which suggests a single firing (Tite *et al.* 1998). The compositions of the crystals

2 – The φρz correction procedure, available on Isis-Oxford software, was applied.

3 – NMA146-147 correspond to mortars: CATHMA 1992, p. 67, fig. 2 no. 1; LIS 62-63 to closed forms decorated with applied petals: p. 67, fig. 2 nos. 2 and 3.

4 – Lead and copper had not been determined in old analyses of samples LIS 62-63 and NMA146-147.

5 – Definitions of high lead, lead-alkali and alkali glazes were taken from Tite *et al.* 1998 and Lauffenburger *et al.* 2001.

6 – The later case applies to ceramics either glazed on both sides, or glazed on one side with partial re-oxidation during cooling of the unglazed side.

id.	Na_2O	MgO	Al_2O_3	SiO_2	P_2O_5	K_2O	CaO	TiO_2	MnO	Fe_2O_3	V	Cr	Ni	Cu	Zn	Rb	Sr	Zr	Ba	La	Ce	Pb
Istanbul (batch "Byzance")																						
BYZ718	0.43	0.66	22.37	67.04	0.10	1.19	0.87	1.135	0.0230	6.03	130	84	33	26	48	59	41	274	308	56	106	369
BYZ719	0.30	0.61	17.12	73.84	0.06	1.55	0.27	1.184	0.0207	4.90	90	71	27	33	48	73	42	347	298	43	102	220
BYZ720	0.33	0.47	18.72	71.86	0.10	1.26	0.24	1.081	0.0275	5.79	79	69	31	34	40	54	28	325	272	47	95	87
BYZ721	0.27	0.52	23.51	65.57	0.16	1.43	0.35	1.179	0.0278	6.82	115	92	39	33	49	75	35	277	348	39	94	307
BYZ722	0.26	0.57	21.05	68.55	0.10	1.22	2.39	1.078	0.0271	4.61	94	79	29	33	46	58	55	297	283	53	114	240
BYZ723	0.25	0.55	19.84	69.71	0.08	1.69	0.31	1.239	0.0172	6.11	131	112	31	36	47	109	33	314	260	35	58	651
BYZ724	0.21	0.45	19.93	69.72	0.09	1.01	0.37	1.033	0.0185	6.99	142	88	29	29	34	51	34	282	230	35	67	696
BYZ725	1.46	0.81	16.82	70.99	0.36	1.98	1.46	0.856	0.0349	5.07	120	80	31	20	54	71	*202	242	508	35	69	114
BYZ726	0.21	0.56	19.08	71.20	0.08	1.39	0.37	1.088	0.0156	5.88	110	87	25	19	47	80	33	272	225	45	97	211
BYZ727	0.36	0.62	20.22	69.29	0.11	1.72	0.46	1.187	0.0161	5.86	105	83	26	37	46	80	63	305	382	59	103	181
BYZ728	0.24	0.54	21.63	66.97	0.10	1.53	0.90	1.143	0.0204	6.74	134	103	40	31	39	84	40	282	347	39	82	469
BYZ729	0.20	0.46	18.12	73.14	0.08	1.25	0.38	1.068	0.0174	5.15	88	74	25	31	36	65	32	316	268	49	82	265
BYZ730	0.37	0.72	22.74	66.39	0.13	1.79	0.49	1.211	0.0230	5.94	143	98	49	39	45	86	43	274	371	55	102	699
BYZ731	0.18	0.57	22.34	67.98	0.09	1.31	1.82	1.116	0.0216	4.42	93	88	31	33	45	63	48	285	405	50	103	265
BYZ732	0.17	0.61	23.12	67.75	0.08	1.21	0.45	1.214	0.0225	5.22	126	92	33	25	47	66	40	275	331	46	103	232
BYZ733	1.91	1.09	19.38	67.48	0.14	1.57	1.37	0.857	0.0326	5.88	90	76	33	66	49	60	*223	191	436	32	72	1621
m	**0.45**	**0.61**	**20.37**	**69.22**	**0.12**	**1.44**	**0.78**	**1.104**	**0.0229**	**5.71**	**112**	**86**	**32**	**33**	**45**	**71**	**41**	**285**	**330**	**45**	**91**	**414**
σ	**0.50**	**0.16**	**2.11**	**2.44**	**0.07**	**0.26**	**0.65**	**0.114**	**0.0057**	**0.77**	**21**	**12**	**6**	**10**	**5**	**15**	**10**	**36**	**78**	**8**	**17**	**377**
Caricin Grad (batch "Balkans")																						
BYZ623	1.20	0.98	23.64	62.72	0.22	1.65	1.96	0.745	0.0448	6.66	157	139	56	21	66	82	366	154	570	40	68	135
BYZ624	1.62	1.03	19.89	66.69	0.14	1.94	1.40	0.773	0.0891	6.22	140	101	49	29	63	91	371	183	735	44	77	172
BYZ625	1.63	1.26	20.36	65.47	0.28	1.92	1.79	0.773	0.0607	6.11	134	105	44	23	73	85	365	174	861	40	67	1415
BYZ627	1.77	1.23	19.85	65.20	0.38	2.15	2.38	0.767	0.0814	5.87	133	82	38	29	65	96	422	183	1090	38	71	898
BYZ628	1.59	2.06	18.39	65.23	0.25	2.29	2.06	0.875	0.0702	6.57	161	106	65	38	91	110	226	182	814	45	80	4368
m	**1.56**	**1.31**	**20.43**	**65.06**	**0.25**	**1.99**	**1.92**	**0.787**	**0.0692**	**6.29**	**145**	**107**	**50**	**28**	**72**	**93**	**350**	**175**	**814**	**41**	**73**	**1398**
σ	**0.21**	**0.44**	**1.94**	**1.44**	**0.09**	**0.24**	**0.36**	**0.051**	**0.0174**	**0.33**	**13**	**21**	**11**	**7**	**11**	**11**	**73**	**12**	**190**	**3**	**6**	**1744**
Saint-Blaise and Hyères (batch "Italy")																						
LIS 62	2.06	1.36	19.18	66.09	0.06	1.96	0.71	0.951	0.0250	7.49	122	152	63	nd	59	86	64	295	359	57	80	nd
LIS 63	1.37	1.03	20.99	64.53	0.09	1.73	0.32	1.076	0.0185	8.71	160	127	53	nd	54	88	62	310	423	59	70	nd
NMA146	1.04	1.03	18.24	69.22	0.08	3.06	0.96	0.625	0.0178	5.54	82	65	39	nd	92	176	59	292	1046	42	84	nd
NMA147	1.04	1.26	15.95	70.59	0.10	4.47	1.25	0.573	0.0385	4.58	94	64	39	nd	67	164	83	297	460	54	105	nd
Saint Symeon (batch "Middle East")																						
LEV402	0.88	6.48	13.06	48.50	0.25	3.05	19.07	0.749	0.1198	7.58	154	210	195	47	61	83	293	125	270	19	71	1052
LEV404	1.15	7.24	13.26	47.00	0.41	2.44	19.59	0.720	0.1320	7.70	156	223	218	67	81	78	319	121	265	22	70	1875
LEV405	0.73	7.02	12.72	46.08	0.31	2.87	21.52	0.709	0.1155	7.56	152	209	201	67	73	79	320	120	285	20	62	1896
LEV406	1.16	6.59	13.11	48.13	0.39	2.91	18.88	0.717	0.1323	7.47	148	219	217	146	94	58	302	121	309	23	66	3443
LEV407	1.09	6.41	13.03	47.10	0.43	2.85	20.33	0.717	0.1328	7.51	147	212	214	84	93	63	318	123	323	22	64	2231
LEV408	1.10	7.19	13.31	47.32	0.33	2.62	19.09	0.729	0.1276	7.71	168	224	216	66	76	88	311	121	266	23	69	2966
LEV409	0.83	6.81	12.91	45.67	0.60	2.42	21.39	0.707	0.1213	7.34	146	217	182	775	84	*32	319	112	267	24	66	9845
LEV410	1.13	7.27	13.39	47.46	0.32	2.40	18.96	0.730	0.1242	7.77	163	226	220	89	73	93	303	120	289	25	69	2675
LEV411	1.18	7.23	13.28	47.52	0.33	2.55	18.84	0.726	0.1230	7.71	152	224	218	139	75	84	298	120	282	27	65	3326
LEV412	0.70	6.64	12.88	46.74	0.29	3.09	20.89	0.722	0.1192	7.54	150	207	201	83	84	70	377	120	299	19	64	2140
LEV413	1.08	6.50	13.15	47.02	0.42	2.94	19.80	0.715	0.1315	7.68	157	210	219	156	87	63	325	119	257	19	67	3815
LEV414	1.16	6.49	13.26	47.46	0.29	2.92	19.62	0.726	0.1329	7.62	158	221	217	78	94	69	323	124	295	26	64	1668
m	**1.02**	**6.82**	**13.11**	**47.17**	**0.36**	**2.76**	**19.83**	**0.722**	**0.1260**	**7.60**	**154**	**217**	**210**	**150**	**81**	**75**	**317**	**121**	**284**	**22**	**66**	**3078**
σ	**0.18**	**0.34**	**0.20**	**0.78**	**0.09**	**0.25**	**0.98**	**0.011**	**0.0063**	**0.12**	**7**	**7**	**12**	**200**	**10**	**11**	**22**	**3**	**20**	**3**	**3**	**2281**

Table 1 – Paste analysis (WD XRF): elemental compositions of the samples in the four batches considered. Major and minor elements Na to Fe in pourcents of oxides, trace elements V to Pb in ppm; the means m and standard deviations σ of homogeneous batches are indicated; nd: not determined; data with an asterisk were not taken into account in the calculation of m and σ.

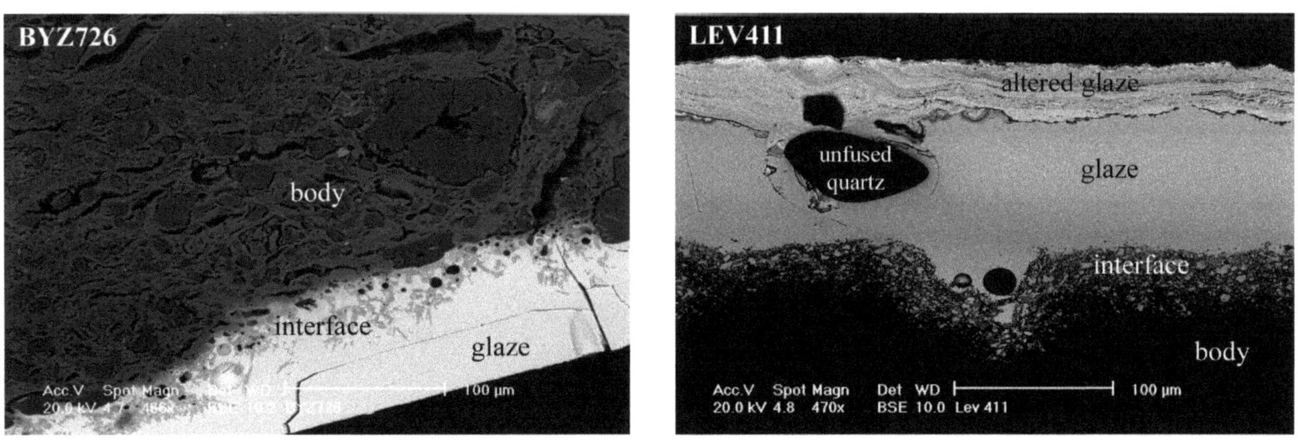

Fig. 2 – SEM backscattered electron images of samples BYZ726 (left, batch "Byzance") and LEV411 (right, batch "Middle East"). In addition to different compositions of bodies and glazes, the structure of the interface and of the localized alteration features are quite dissimilar. © C2RMF.

id.	Na_2O	MgO	Al_2O_3	SiO_2	K_2O	CaO	TiO_2	Fe_2O_3	CuO	PbO	others
Istanbul (batch "Byzance")											
BYZ724	<dl	<dl	8.3	29.8	<dl	0.2	0.2	3.7	<dl	57.7	
BYZ726	<dl	0.2	8.7	31.3	0.5	0.2	0.6	3.6	<dl	54.8	
BYZ727	<dl	0.2	6.9	28.7	<dl	0.7	0.6	2.4	<dl	60.5	
BYZ728	<dl	0.3	8.3	27.4	0.9	2.6	0.0	3.3	<dl	57.1	
BYZ731	<dl	0.1	9.0	26.4	<dl	0.2	0.4	2.2	<dl	61.3	0.3 ZrO_2
Caricin Grad (batch "Balkans")											
BYZ623 (side 1)	0.5	0.3	8.3	25.9	0.4	1.0	0.3	2.8	<dl	60.6	
BYZ623 (side 2)	0.2	0.3	7.6	24.1	0.3	0.7	0.2	3.2	<dl	63.4	
BYZ624 (side 1)	0.5	0.3	10.9	32.5	0.7	1.1	0.5	3.5	<dl	50.0	
BYZ624 (side 2)	0.3	0.2	7.5	26.5	0.4	0.9	0.3	3.0	0.1	60.9	
BYZ625	0.5	0.4	6.4	22.4	0.6	1.0	0.2	2.3	<dl	66.2	
BYZ627 (side 1)	0.3	0.4	6.7	23.0	0.2	1.0	0.2	2.3	<dl	65.9	
BYZ627 (side 2)	0.3	0.3	6.3	22.6	0.5	1.0	0.1	1.8	<dl	66.9	
BYZ628	0.1	0.3	5.9	17.1	0.6	0.9	<dl	2.6	<dl	72.5	
Saint-Blaise and Hyères (batch "Italy")											
LIS 62	0.4	0.7	6.9	26.3	0.3	1.0	0.3	2.8	<dl	60.4	0.6 SnO_2
LIS 63	0.3	0.3	4.9	16.8	<dl	0.2	0.3	2.6	<dl	74.6	
NMA146	<dl	1.0	6.5	30.1	<dl	0.2	0.6	3.8	<dl	57.9	
NMA147	<dl	0.7	5.5	30.4	0.5	0.9	0.0	2.0	<dl	59.9	
Saint Symeon (batch "Middle East")											
LEV405	0.5	0.3	0.2	30.7	0.0	1.2	<dl	0.2	3.6	63.2	
LEV406	0.5	0.5	0.3	26.2	<dl	0.6	<dl	0.1	3.4	67.4	1.1 Sb_2O_3
LEV407 (side 1)	0.5	0.5	0.2	26.2	<dl	0.5	<dl	0.0	3.1	68.2	0.9 Sb_2O_3
LEV407 (side 2)	0.4	0.5	0.1	25.7	<dl	0.7	<dl	0.3	2.3	69.5	0.6 Sb_2O_3
LEV410	0.3	0.4	0.3	31.2	<dl	1.1	<dl	0.5	1.5	64.7	
LEV411	0.4	0.1	0.5	28.5	0.2	0.9	<dl	0.1	4.1	65.2	

Table 2 – Glaze analysis (SEM-EDX): elemental compositions of the samples in the four batches considered. Major and minor elements in pourcents of oxides; two results are given when glaze is present on both sides of the sample; data correspond to the mean of 3 analyses or more; dl: detection limits; SnO_2 is present in LIS 62 as small cristals of cassiterite.

concentrating at the interface, as estimated by EDS spot analyses, seem compatible with the felspars containing lead already described notably by Molera *et al.* (1993). Sample BYZ727 presents atypical features within the "Byzance" batch, as it includes many small crystals of tin oxides[7] unevenly dispersed throughout the glaze.[8] The role of these crystals, which do not act as opacifiers, and the reason for their presence, still need to be clarified.

The "Balkans" batch globally has features similar to the "Byzance" batch, with high lead, aluminium-rich glazes. Interfaces show heterogeneous micro-structures including lead felspars which may be present throughout the thickness of the glaze.[9] Batch "Italy" also shows high-lead, aluminium-rich glazes, with the additional features of cassiterite crystals in sample LIS 62 and of localized zones of iron-rich clay under the glaze in samples LIS 62 and LIS 63. The latter are possibly related to the plastic decoration of these samples. For both batches, though to a lesser extent for the "Italy" one, the composition of the glazes after subtraction of PbO and renormalisation to 100% matches the composition of the body.

The "Middle East" batch shows remarkably pure high lead glazes over calcareous bodies. In this case, it is clear from the compositions that lead was not applied by itself, but in combination with silica. Furthermore, raw materials low in impurities must have been carefully chosen. The interface has a complex structure and includes Si-Ca-Pb crystals and As-rich components. Patches of alteration are frequent at the surface (Fig. 2, right), with typical layered structures being observed which are probably responsible for the opaque appearance and for a change in colour from green to turquoise in these zones.

The main compositional features of the glazes are shown in a ternary diagram $PbO-SiO_2-Al_2O_3$ (Fig. 3, top left). Two groups of high lead glazes are well distinguished according to their aluminium content, the "Middle East" batch constituting one of the groups, being singled out by its low concentrations of aluminium. We may note that several of the Al-rich glazes plot close to the eutectic point of the $PbO-SiO_2-Al_2O_3$ system. The main characteristics of pastes, glazes, interfaces and micro-structures are summarized in Table 3. From this synthesis it is clear that our samples of Byzantine Glazed White Wares share the same general features with samples from Caričin Grad, Saint-Blaise and Hyères, whereas samples from Saint Symeon are very different.

7 – Typical dimensions are below 1 μm.

8 – The average glaze composition of sample BYZ727 is very similar to those of other samples within batch "Byzance".

9 – They are responsible for the opaque aspect of these glazes.

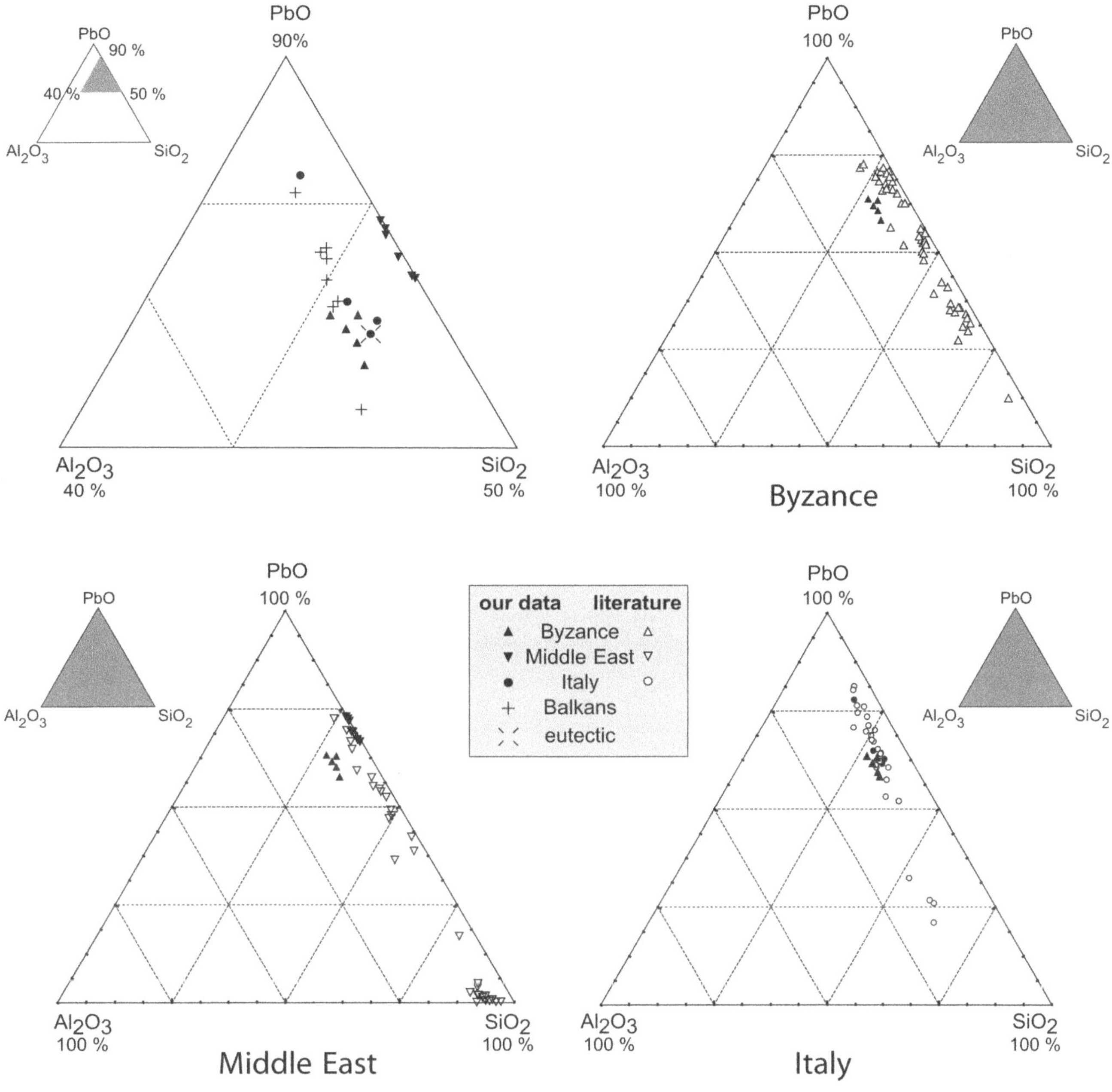

Fig. 3 – Ternary diagrams PbO-SiO_2-Al_2O_3, showing our sampling with the eutectic of the system (top left) and comparisons with data from the literature for Byzantine (top right), Middle Eastern (bottom left) and Italian productions (bottom right). Points next to the SiO_2 pole mostly correspond to alkali glazes.

	batch "Byzance" Istanbul	batch "Balkans" Caričin Grad	batch "Italy" St-Blaise, Hyères	batch "Middle East" St Symeon
body	kaolinitic paste	kaolinitic paste	kaolinitic paste	marly paste
interface	lead-felspars	lead-felspars	lead-felspars	Ca-Mg silicates
glaze	high lead glaze Al-rich	high lead glaze Al-rich	high lead glaze Al-rich	very pure high lead glaze
crystals in glaze	no	sometimes lead felspars	sometimes cassiterite	lead antimoniates, cassiterite, As-rich microcrystals

Table 3 – Summary of the data concerning bodies, glazes, interfaces and microstructures.

COMPARATIVE DATA

In order to evaluate how representative each batch could be for its respective "technological context", we compared our data, obtained on a very limited sampling, with previous research. However this notion of technological context should be seen in a very broad sense, as the data taken into consideration cover a large period, roughly from the 3rd to the 12th century in the case of Italian and Middle Eastern productions.[10] Actually, none of them strictly corresponds to the period of interest. Yet it was thought that such a comparison could on one hand give indications to support or question our own results, and on the other hand enable them to be placed in a chronological perspective.

Fig. 3 (top right) shows the comparison in the PbO-SiO_2-Al_2O_3 system of GWWI with later Byzantine Glazed White Wares (Vogt and Bouquillon 1996; Armstrong *et al.* 1997; Lauffenburger *et al.* 2001).[11] The latter are either high-lead or alkali-lead glazes. Samples of GWWI are not particularly well integrated in this group, as they have some of the highest aluminium contents. It is noticeable at this point that, at a period which cannot be precisely defined as yet,[12] GWW evolved towards more alkali-rich glazes. In the next ternary diagram (Fig. 3, bottom left), GWWI are compared to glazes of Parthian, Sasanian and Middle Eastern early Islamic wares (McCarthy *et al.* 1995; Pérez-Arantegui *et al.* 1996; Tite *et al.* 1998; Daszkiewicz and Raabe 1999; Hill *et al.* 2004).[13] With few exceptions, all of later date,[14] GWWI may be distinguished as the only high lead samples bearing significant aluminium concentrations. This comparative set of data includes high lead, alkali-lead and low to lead-free alkali glazes.[15] Samples from Saint Symeon are included in the former group, and remain among the purest lead-silicon mixtures. The last ternary diagram (Fig. 3, bottom right) shows the comparison between GWWI and Italian "vetrina pesante" (Aurisicchio *et al.* 1993; Walton and Tite 2005).[16] The diagram shows a very good correspondence between GWWI and the majority of the "vetrina pesante" considered, with the exception of a few samples having higher SiO_2 contents. This similarity is however less apparent on raw data, as the average aluminium content in "vetrina pesante" glazes is circa 5% Al_2O_3, and, furthermore, several pastes are moderately calcareous (circa 10% CaO). It would be necessary to examine a well defined corpus of "vetrina pesante" from late Roman production sites to further support what may appear to be a confirmation of our results.

CONCLUSIONS AND PERSPECTIVES

Both pastes and glazes of a sampling of the earliest Byzantine Glazed White Ware present more resemblance to late Roman production from the Balkans and northern Italy than to Middle Eastern production. The former associate kaolinitic pastes with high-lead, Al-rich glazes whereas the latter present high lead Al-poor glazes over calcareous pastes. The manufacturing process also appears to be different. In the first case the glaze may result from the direct application of a lead compound to the paste and from a single firing. In the second case a lead-silica compound was used.

These results support the hypothesis of a re-introduction of the glazing technique in Byzantium via regions which were then the Western provinces of the Byzantine empire, rather than due to an Eastern input. However, lead glazes with Al contents close to the PbO-SiO_2-Al_2O_3 eutectic may have existed in China at the same period (Wood 1999, p. 192) and the role that these productions could have played still needs to be investigated. The connection between the first Byzantine Glazed White Wares and late Roman "Western" productions needs to be confirmed on the basis of a larger sampling coming from well attested production sites.

Although Middle Eastern ceramics do not seem to have been the main influence in the re-introduction of the glazing technique in Byzantium, their role could have been decisive later on, in the evolution of subsequent production of Byzantine Glazed White Ware towards alkali-lead glazes. However further research in this direction would depend critically on the availability of well dated material.

ACKNOWLEDGEMENTS

We would like to thank J. Hayes, J.-P. Sodini, D. Orssaud, B. Bavant, L. Vallauri, J.-C. Treglia and the Archaeological Museum of Istanbul for their help and for providing the samples. We are also very grateful to the Departments of Antiquities of Turkey, Syria and Serbia for permission to sample and export the material studied.

10 – As far as we know, no comparative data are available for balkanic productions.

11 – We excluded of this comparative corpus what had been considered an incidental and short-lasting appearance of alkali glazes in Byzantine Jordan in the 6th century: Freestone *et al.* 2001.

12 – The comparative samples come from museums and not from excavation contexts.

13 – From Hill *et al.* 2004, only means of compositional groups were available. Mean Al_2O_3 of "Moderate Pb" group was taken as 3.29%, Table 3, p. 600. High lead tin-opacified glazes were not included in the comparative corpus.

14 – Sample 15/6 from Raqqa, 9th c.: 4.83% Al_2O_3, Daszkiewicz and Raabe 1999, p. 135; sample SJN 12.4 from Iran, 10-11th c.: 4.1% Al_2O_3, Tite *et al.* 1998, p. 243.

15 – The validity of such a diagram in the latter cases is questionable, but was adopted to show Al contents and alkali and lead-alkali glazes simultaneously, which plot in this case close to the SiO_2 pole.

16 – We did not consider here earlier Italic lead glazed wares, which will be taken into account in further work.

REFERENCES

ARMSTRONG P., HATCHER H., TITE M. 1997, "Changes in Byzantine glazing technology from the ninth to the thirteenth centuries", *in La céramique médiévale en Méditerranée. Actes du 6e congrès*, Aix-en-Provence, p. 225-229.

AURISICCHIO C., LAZZARINI L., MARIOTTTINI M. 1993, "Composizione, tecnologia di fabricazione e provenienza della ceramica a vetrina pesante", *in* O. Mazzucato, *Tipologie e tecniche della ceramica a vetrina pesante. IX-X secolo*, Rome, p. 161-220.

BJELAJAC L. 1990, "La céramique et les lampes", *in* B. Bavant, V. Kondic and J.-M. Spieser (eds.), *Caričin Grad. II, Le quartier sud-ouest de la Ville Haute*, Collection de l'École française de Rome 75/2, Belgrade-Rome, p. 161-190.

CATHMA 1992, "Céramiques glaçurées de l'Antiquité tardive et du haut Moyen Age en France méridionale", *in* L. Paroli (ed.), *La ceramica invetriata tardoantica e altomedievale in Italia*, Firenze, p. 65-74.

DASZKIEWICZ M., RAABE J. 1999, "Untersuchungen der glasierten Keramik", *in* P.A. Miglus, *Ar-Raqqa. I, Die Frühislamische Keramik von Tall Aswad*, Mainz, p. 127-137.

FREESTONE I., POLITIS K.D., STAPLETON C.P. 2001, "The Byzantine glazed pottery from Deir 'Ain 'Abata, Jordan", *in La céramique byzantine et proto-islamique en Syrie-Jordanie (IVe-VIIIe siècles apr. J.-C.)*, Bibliothèque archéologique et historique 159, Beyrouth, p. 197-205.

HAYES J.W. 1968, "A seventh-century pottery group", *in* R.M. Harrison and N. Fıratlı, "Excavations at Saraçhane in Istanbul: fifth preliminary report", *Dumbarton Oaks papers* 22, p. 203-216.

HAYES J.W. 1992, *Excavations at Saraçhane in Istanbul. II, The Pottery*, Princeton University Press.

HILL D.V., SPEAKMAN R.J., GLASCOCK M.D. 2004, "Chemical and mineralogical characterization of Sasanian and early Islamic glazed ceramics from the Deh Luran plain, southwestern Iran", *Archaeometry* 46, p. 585-605.

LAUFFENBURGER J., VOGT C., BOUQUILLON A. 2001, "Technical insights into the working practices of the Byzantine tile maker", *in* S.E.J. Gerstel and J.A. Lauffenburger (eds.), *A lost art rediscovered: the architectural ceramics of Byzantium*, Baltimore, p. 67-87.

LE MIÈRE M., PICON M. 2003, "Appearance and development of cooking and "non-cooking" ware concepts in the Near East", *in Ceramics in the Society, Proceedings of the 6th European Meeting on Ancient Ceramics*, Fribourg, p. 175-188.

MCCARTHY B., VANDIVER P.B., GIBSON M. 1995, "Innovation and continuity in the technology of southwest asian monochrome blue glazes", *in* P. Vincenzini (ed.), *The ceramics cultural heritage*, Faenza, p. 207-218.

MOLERA J., PRADELL T., MARTINEZ-MANENT S., VENDRELL-SAZ M., 1993, "The growth of sanidine crystals in the lead of glazes of Hispano-Moresque pottery", *Applied clay science* 7, p. 483-491.

ORSSAUD D. 2001, "Les céramiques à glaçure monochrome de Qalat Sem'an (VIIIe-IXe siècles)", *in La céramique byzantine et proto-islamique en Syrie-Jordanie (IVe-VIIIe siècles apr. J.-C.)*, Bibliothèque archéologique et historique 159, Beyrouth, p. 215-220.

PAPANIKOLA-BAKIRTZI D. (ed.) 1999, *Byzantine glazed ceramics: the art of sgraffito*, Athens.

PAROLI L. (ed.) 1992, *La ceramica invetriata tardoantica e altomedievale in Italia*, Firenze.

PÉREZ-ARANTEGUI J., QUERRÉ G., KACZMARCZYCK A., BERNUS-TAYLOR M. 1995, "Chemical characterization of Islamic glazed ceramics from northern Syria by Particle Induced X-ray Emission (PIXE)", *in* P. Vincenzini (ed.), *The ceramics cultural heritage*, Faenza, p. 475-482.

PÉREZ-ARANTEGUI J., QUERRÉ G., EVENO M., KACZMARCZYCK A. 1996, "Chemical, SEM and petrographic study of early Islamic glazed ceramics from several specific sites in Syria, Iraq and Iran", *in Archaeometry 94, The Proceedings of the 29th International symposium on archaeometry*, Ankara, p. 219-226.

SCHNEIDER G. 1996, "Chemische und Mineralogische Untersuchungen von Keramik der Hellenistischen bis Frühislamischen Zeit in Nordost-Syrien", *in* K. Bartl and S.R. Hauser (eds.), *Continuity and change in northern Mesopotamia from the hellenistic to the early islamic period*, Berliner Beiträge zum Vorderen Orient 17, Berlin, p. 127-136.

SPIESER J.-M. 1991, "La céramique byzantine médiévale", *in Hommes et richesse dans l'Empire byzantin. Tome II, VIIIe-XVe siècle*, Réalités byzantines 3, Paris, p. 249-260.

TITE M., FREESTONE I., MASON R., MOLERA J., VENDRELL-SAZ M., WOOD N. 1998, "Lead glazes in antiquity: methods of production and reasons for use", *Archaeometry* 40, p. 241-260.

VOGT C., BOUQUILLON A. 1996, "Technologie des plaques murales décorées de Preslav et de Constantinople (IXe-XIe siècles)", *Cahiers archéologiques* 44, p. 105-116.

WALTON M., TITE M. 2005, "Le analisi sulla ceramica invetriata di Carlino", *in* C. Magrini and F. Sbarra, *Le Ceramiche invetriate di Carlino*, Ricerche di archeologia altomedievale e medievale 30, p. 67-68.

WOOD N. 1999, *Chinese Glazes*, London.

THE "POLISHED YELLOW" CERAMICS OF THE CAROLINGIAN PERIOD (9th CENTURY AD): SAMPLES FROM ZALAVÁR, SOUTH-WEST HUNGARY

HAJNALKA HEROLD

Department of Prehistoric and Mediaeval Archaeology and Vienna Institute of Archaeological Science, University of Vienna, Franz Klein Gasse 1, A-1190 Vienna, Austria

hajnalka.herold@univie.ac.at

1. ARCHAEOLOGICAL BACKGROUND

The ceramics analysed and presented here belong to the very distinct group of the so called "polished yellow ceramics", which are the best-quality ceramics of the Carolingian Period (9th century AD) found in Zalavár (South-West Hungary, Fig. 1). Their reddish-yellow colour and fine-grained fabrics are very different from the rest of the mainly brownish-black, coarse-grained ceramics found at the site. The "polished yellow" ceramic vessels are mostly flasks (Fig. 2), but there are also some amphora-like vessels with two handles (Fig. 3) and a variety of other special forms (lids, small pots, bowls) in very small numbers. Similar "yellow ceramics" are known from the same period from different parts of Central- and Eastern-Europe, in a region ranging from today's Austria to Bulgaria.

The primary goal of the investigations presented here is to find and characterise the groups of this special ceramic type in order to obtain information about the standard and the framework of production and their changes through time and to establish the possibility for a later archaeometric (chemical, mineralogical, microstructural) comparison with similar ceramics from different sites and regions. The

Fig. 2 – Polished yellow flask from Zalavár.

Fig. 1 – Location of the site of Zalavár.

Fig. 3 – Polished yellow "amphora" from Zalavár.

investigation methods applied include thin section analysis, XRF, XRD and scanning electron microscopy.

The present project is a part of a series of projects for the investigation of this special ceramic type. In the framework of this project series there have hitherto been samples from Zalavár (Hungary), Mikulčice, Břeclav-Pohansko, Uherské-Hradiště (Czech Republic: Herold in press), Gars-Thunau (Austria) and Pliska (Bulgaria: Herold 2004) analysed. A final publication of the results of the whole project series is in preparation. In the present paper the results concerning the samples from Zalavár are presented.

2. INVESTIGATED SAMPLES

50 samples of archaeological ceramics from Zalavár have been investigated by thin section analysis, X-ray fluorescence analysis and X-ray diffraction analysis. 12 of these samples have also been investigated by scanning electron microscopy. In addition to the samples of archaeological ceramics 7 clay samples from the site have also been analysed by thin section analysis, X-ray fluorescence analysis, X-ray diffraction analysis and scanning electron microscopy (only XRF of clay samples are presented here).

3. CERAMIC TYPES BY PETROGRAPHY

On the basis of the petrographic analysis by polarising microscopy 4 different groups of the archaeological ceramics have been found. Groups 1-3 have practically the same mineralogical composition (quartz, alkaline feldspars, low grade metamorphic rock fragments, small amounts of plagioclase, muscovite mica, grains of opaque phases, practically no carbonates; epidote, tourmaline, rutile, zircon and garnet as accessory minerals), they can only be differentiated by their texture (group 1: fine grained, well sorted; group 2: coarse grained, poorly sorted; group 3: mid-coarse grained, poorly sorted; see also Fig. 4).

The mineralogical composition of group 4 is differentiated from that of groups 1-3 by a relatively large amount of carbonates. Group 4 also has a texture different from all of the other three groups (very fine matrix with a small number of large non-plastic inclusions).

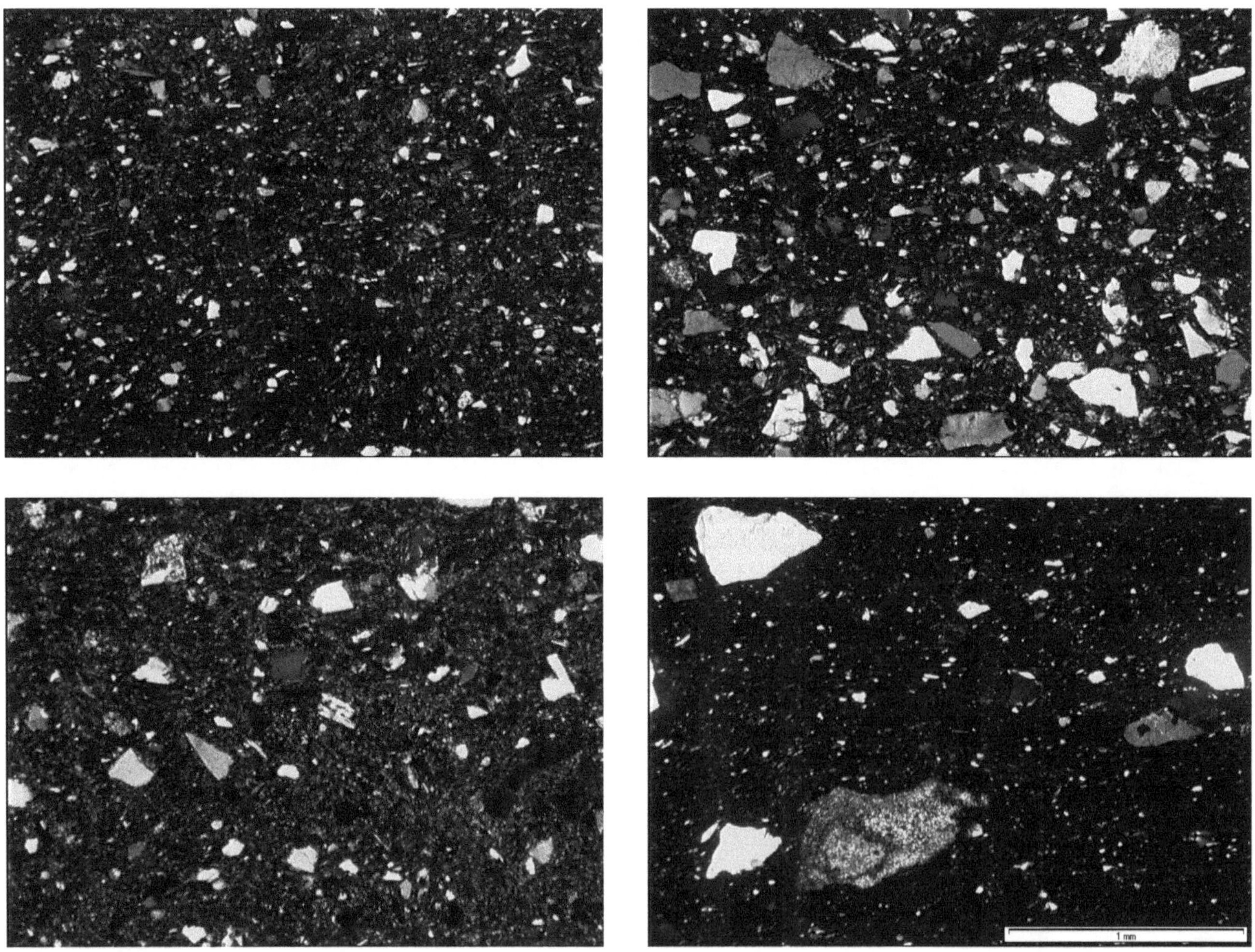

Fig. 4 – The four groups of the "polished yellow ceramics" from Zalavár in the polarising microscope; under crossed polars, areas shown equal 2.6x1.8 mm. Groups can mainly be differentiated on the basis of micromorphological criteria.

group	sample	SiO_2 wt%	TiO_2 wt%	Al_2O_3 wt%	Fe_2O_3 wt%	MnO wt%	MgO wt%	CaO wt%	Na_2O wt%	K_2O wt%	P_2O_5 wt%	SUM wt%	Ba ppm	Cr ppm	Cu ppm	Nb ppm	Ni ppm	Pb ppm	Rb ppm	Sr ppm	Y ppm	Zn ppm	Zr ppm	H_2O- wt%	LOI wt%
1	Z03	65,48	0,83	17,99	7,02	0,07	1,94	1,46	0,56	3,08	0,89	99,50	667	131	31	15	64	32	157	194	27	111	193	1,26	2,07
1	Z04	64,90	0,82	17,73	7,41	0,09	1,92	1,82	0,54	2,91	1,25	99,57	750	121	26	16	72	42	155	223	24	116	194	2,07	3,05
1	Z06	64,80	0,78	17,39	7,45	0,07	1,98	1,99	0,53	2,75	1,32	99,24	733	119	34	15	75	29	151	195	28	117	186	3,04	3,95
1	Z07	65,49	0,83	18,57	6,98	0,07	2,10	1,03	0,93	3,07	0,24	99,47	585	125	42	17	88	29	166	168	27	111	191	0,41	0,81
1	Z08	65,00	0,82	18,27	7,09	0,08	2,05	1,36	0,53	3,09	0,65	99,11	649	125	29	18	73	29	166	190	27	117	184	0,86	1,69
1	Z09	65,38	0,83	18,39	6,97	0,07	2,06	1,15	0,48	3,12	0,43	99,03	592	132	29	16	70	32	164	174	28	112	188	0,50	1,15
1	Z10	64,98	0,79	17,54	7,43	0,07	1,97	1,87	1,03	2,81	1,25	99,92	720	120	37	15	70	29	148	200	31	111	189	2,49	3,59
1	Z11	65,24	0,82	18,34	7,09	0,07	2,04	1,35	0,51	3,00	0,58	99,23	637	121	70	18	80	30	164	188	27	113	183	0,81	1,40
1	Z12	64,84	0,79	17,38	7,35	0,07	1,93	1,87	0,57	2,97	1,38	99,34	729	119	31	16	71	30	150	190	28	122	188	2,66	3,68
1	Z20	64,29	0,78	17,28	7,52	0,08	1,97	2,10	0,52	2,87	1,54	99,11	704	127	27	16	77	26	150	217	29	116	187	3,09	4,49
1	Z25	65,44	0,85	18,73	6,73	0,05	1,96	1,29	0,50	2,94	0,69	99,33	607	127	27	17	76	32	152	152	26	98	190	2,20	3,11
2	Z01	72,12	0,69	14,87	5,49	0,08	1,60	0,81	0,83	2,61	0,14	99,38	462	100	17	14	85	31	132	105	27	92	181	0,32	0,70
2	Z13	71,54	0,67	14,06	5,33	0,16	1,30	1,75	0,84	2,61	1,17	99,59	659	92	19	13	54	23	121	210	22	94	180	1,92	2,95
2	Z15	69,49	0,77	15,87	5,80	0,12	1,66	1,30	0,59	2,87	0,59	99,23	599	104	25	17	56	24	142	175	28	95	214	0,90	1,76
2	Z19	69,06	0,75	15,81	5,68	0,10	1,73	1,74	0,83	2,78	0,38	99,03	582	103	15	16	58	23	148	170	30	99	199	0,95	2,07
2	Z21	69,22	0,75	15,77	5,55	0,11	1,70	2,01	0,76	2,72	0,56	99,31	631	102	17	16	56	23	139	181	32	100	193	1,41	3,03
2	Z22	67,70	0,74	15,46	6,73	0,08	1,78	2,29	0,75	2,64	0,92	99,26	642	100	21	15	67	24	139	169	29	105	201	2,81	4,28
2	Z26	69,29	0,78	15,51	6,17	0,09	1,61	1,23	0,73	2,79	0,72	99,08	591	104	18	17	64	27	134	137	31	84	205	1,64	2,32
2	Z27	69,36	0,79	15,69	6,22	0,10	1,64	1,23	0,69	2,80	0,66	99,33	591	107	25	16	69	27	143	139	30	90	203	1,56	2,05
2	Z28	69,49	0,78	15,66	6,05	0,09	1,64	1,19	2,05	2,85	0,62	100,57	582	112	22	16	71	27	140	130	29	90	209	1,67	2,31
2	Z29	70,41	0,72	14,93	5,93	0,06	1,58	1,33	1,97	2,64	0,50	100,21	550	94	22	15	98	23	124	128	28	74	194	1,86	2,73
2	Z30	69,54	0,79	15,65	6,26	0,11	1,75	1,05	0,91	2,79	0,48	99,47	507	102	38	17	63	22	136	126	32	92	206	0,85	1,26
2	Z33	71,25	0,70	15,23	5,35	0,08	1,59	1,36	1,85	2,71	0,59	100,87	516	102	23	15	66	22	118	130	17	74	188	1,97	2,86
2	Z45	70,86	0,65	14,55	5,50	0,06	1,65	1,55	1,57	2,58	0,74	99,90	792	97	260	14	76	21	127	164	23	105	187	2,45	3,27
2	Z46	70,10	0,79	15,11	4,77	0,07	1,83	2,56	1,52	2,51	0,38	99,82	636	100	50	16	53	31	133	189	19	97	213	0,68	1,26
2	Z48	69,92	0,72	16,13	6,14	0,12	1,95	1,26	1,67	2,60	0,29	100,96	581	113	26	14	63	29	134	136	22	91	176	0,35	0,68
2	Z49	71,05	0,68	14,54	5,29	0,10	1,49	1,59	1,39	2,51	1,23	100,09	1081	106	97	15	60	29	120	216	16	112	185	1,49	2,34
2	Z50	73,83	0,62	13,87	5,04	0,06	1,36	0,89	1,51	2,23	0,14	99,70	641	90	19	12	54	25	117	111	12	82	155	0,30	0,85
3	Z02	68,06	0,87	17,30	5,64	0,12	1,95	1,40	0,66	3,13	0,26	99,55	587	121	24	16	60	36	166	140	29	114	197	0,35	0,94
3	Z05	66,73	0,83	16,74	5,79	0,06	1,97	3,09	0,75	2,51	0,42	99,04	592	105	16	15	54	24	134	190	31	100	191	0,50	1,87
3	Z14	69,23	0,76	15,94	5,82	0,10	1,65	1,14	1,20	2,87	0,41	99,29	653	109	13	18	68	22	145	158	30	99	201	0,44	0,90
3	Z16	65,42	0,85	18,14	6,12	0,06	2,14	1,88	0,39	3,17	1,12	99,46	661	123	29	17	87	24	178	199	27	137	169	2,95	3,99
3	Z17	67,83	0,83	15,71	5,50	0,12	1,78	2,53	0,68	2,87	1,18	99,22	720	99	34	17	63	23	144	208	26	122	198	2,24	3,95
3	Z18	66,24	0,77	17,16	6,17	0,09	1,98	1,81	1,15	3,21	1,35	100,11	758	123	31	15	63	25	157	197	19	118	158	2,30	3,29
3	Z23	67,58	0,85	15,86	5,52	0,10	1,78	2,70	0,69	2,82	1,20	99,27	743	101	16	17	63	25	148	212	26	121	202	2,20	3,90
3	Z24	68,01	0,83	15,54	5,55	0,09	1,76	2,51	0,71	2,73	1,14	99,03	729	104	18	16	61	23	146	208	26	115	203	2,28	3,82
3	Z31	67,81	0,85	16,03	6,40	0,07	1,82	2,12	0,79	2,71	0,36	99,12	516	108	33	17	119	34	129	117	26	95	217	2,27	3,92
3	Z32	67,05	0,85	16,15	6,25	0,10	1,77	2,07	1,36	2,80	0,71	99,26	609	106	91	17	82	32	131	136	28	93	222	2,32	3,64
3	Z41	62,13	0,88	18,82	7,35	0,10	2,32	3,44	1,31	2,98	0,35	99,87	736	122	29	16	71	32	156	172	26	107	188	0,60	1,65
3	Z43	63,22	0,89	19,13	6,97	0,09	2,32	3,34	1,39	3,02	0,32	100,88	788	117	29	17	66	30	153	172	26	111	183	0,59	3,39
3	Z47	65,97	0,91	17,76	6,15	0,08	2,03	2,55	1,65	2,82	0,70	100,81	756	140	52	18	67	29	138	192	25	104	219	0,91	1,86
4	Z34	58,98	0,85	19,08	7,38	0,10	2,33	5,25	0,54	2,88	1,41	99,01	872	112	26	18	68	40	151	309	30	120	160	2,06	4,53
4	Z35	58,83	0,79	18,49	7,46	0,10	2,24	5,87	1,95	2,85	1,26	100,04	884	124	27	17	72	27	149	293	36	117	164	1,88	5,39
4	Z36	56,16	0,82	17,74	9,15	0,10	2,33	7,09	0,53	2,73	2,10	98,99	1007	121	23	18	80	24	146	348	30	134	166	3,44	7,84
4	Z37	58,89	0,81	18,37	7,56	0,10	2,44	6,33	0,46	2,90	1,07	99,15	882	119	121	16	68	30	152	283	31	134	144	1,26	3,99
4	Z38	59,48	0,81	18,73	7,25	0,11	2,60	5,57	1,39	3,04	0,86	100,05	908	126	41	17	71	35	160	250	23	126	138	0,95	3,05
4	Z39	56,93	0,79	18,52	7,33	0,12	2,94	7,15	1,23	2,98	0,96	99,19	1118	130	56	15	72	35	153	269	24	127	131	1,16	4,28
4	Z40	59,03	0,80	18,76	7,40	0,13	2,68	5,82	1,15	2,99	1,11	100,10	1145	129	51	17	68	34	159	265	22	121	134	1,15	3,34
4	Z42	57,61	0,82	18,49	7,08	0,14	2,66	7,77	1,36	3,03	1,60	100,83	1200	103	43	18	71	34	150	353	22	122	149	2,13	5,79
4	Z44	57,27	0,80	17,97	6,95	0,12	2,91	8,06	1,58	2,89	1,12	99,89	972	125	50	16	70	33	147	295	22	117	148	1,38	5,95
clay sample	ZT01	71,74	0,93	13,96	5,31	0,11	1,90	2,25	1,75	2,45	0,14	100,70	625	111	21	17	45	23	118	106	27	76	326	0,56	2,13
clay sample	ZT02	85,05	0,49	7,47	2,07	0,06	0,65	1,16	1,55	1,51	0,25	100,32	311	54	9	11	18	12	67	89	9	32	163	0,16	0,62
clay sample	ZT03	79,29	0,43	10,60	3,96	0,04	1,16	1,65	1,47	1,86	0,18	100,72	313	64	16	8	43	24	85	91	19	52	109	0,46	1,24
clay sample	ZT04	73,58	0,52	7,91	2,38	0,05	2,54	9,94	1,55	1,53	0,21	100,31	341	51	14	11	20	19	61	156	8	39	190	0,32	7,05
clay sample	ZT05	83,53	0,53	8,22	2,77	0,04	0,77	1,42	1,56	1,62	0,32	100,89	532	44	13	9	21	14	70	92	9	51	152	0,25	0,73
clay sample	ZT06	79,93	0,46	7,93	3,38	0,06	0,91	3,65	1,46	1,56	0,53	99,95	299	50	7	9	22	11	65	140	9	35	134	0,32	2,06
clay sample	ZT07	79,21	0,54	8,98	3,65	0,07	1,04	3,09	1,68	1,71	0,59	100,68	465	60	55	11	32	36	76	131	8	99	161	0,42	1,37

Table 1 – Numeric data from the X-ray fluorescence measurements of the samples from Zalavár. Groups 1-3 of petrography have a similar chemical composition and thus cannot be distinguished by XRF. Group 4 can well be differentiated from groups 1-3 chemically, first of all on the basis of its higher CaO content.

4. CERAMIC TYPES BY X-RAY FLUORESCENCE ANALYSIS

Groups 1-3 of petrography cannot be distinguished by X-ray fluorescence analysis (Table 1; total iron measured as Fe^{3+}; see also the plot of archaeological samples in Fig. 8), which supports the conclusion from the petrographic analysis that these three groups have a very similar mineralogical composition and are only different in their texture. Group 4 differs from groups 1-3 first of all in its CaO and SiO_2 content and also in some trace elements. Thus the results of the XRF analysis strengthen the results of the polarising microscopy, but do not provide a basis for the quantitative separation of groups 1-3.[1]

5. CERAMIC TYPES BY SCANNING ELECTRON MICROSCOPY, MICROMORPHOLOGY

The petrographic groups 1-3 cannot be distinguished by X-ray fluorescence analysis; they can only be separated on the basis of their microstructure. In order to separate them on a firm basis, we need quantified data for their microstructure. The problems of the quantification of microstructural data in polarising microscopy are well known: under crossed polars, grains of the same mineral can have different colours and grains of different minerals can show the same colour. This makes a digital quantitative micromorphological analysis of polarising microscope pictures almost impossible.

To avoid the above described optical problems when quantifying micromorphological data from the polarising microscope, scanning electron microscopy has been used.[2] On a SEM image all particles are shown in different shades of grey according to the atomic weight of their constituents. Moreover it is possible to produce mappings of the distribution of chemical elements within the sample. The images obtained in the SEM are thus best suited to digital quantitative analysis and allow the recording of different features of grain size and shape. Therefore a quantitative differentiation between groups of samples on a micromorphological basis is possible.

In the present project mappings of the chemical element Si have been selected for the micromorphological analysis (Fig. 5). Each of the 12 samples was measured at two different places; the measured area was 2.6x1.8 mm in all cases. The subsequent image analysis was performed using the free UTHSCSA ImageTool program (developed at the University of Texas Health Science Center at San Antonio, Texas and available from the Internet from http://ddsdx.uthscsa.edu/dig/download.html). Various measures of quartz grains have been recorded (area, perimeter, major axis length, minor axis length etc.). The plot of the perimeter of quartz grains shown here (mean value against standard deviation in each sample, Fig. 6) makes a separation of groups 1-3 possible on the basis of micromorphological criteria. This novel methodology for quantitative micromorphological analysis developed in the present project not only solves the research question set to the micromorphological analysis in the Zalavár project, but also provides a perspective for answering further micromorphological questions in future projects.

6. FIRING TEMPERATURE AND VITRIFICATION BY X-RAY DIFFRACTION ANALYSIS AND SCANNING ELECTRON MICROSCOPY

On the basis of the phase composition of the samples of archaeological ceramics detected by XRD, the firing temperature has been estimated (Table 2; temperature estimation based on data from Maggetti *et al.* 1984, p. 175, fig. 12 and Noll 1991, p. 99, fig. 21). Groups 1-3 of the archaeological ceramics are somewhat heterogeneous with regard to the firing temperature estimated; group 4 shows the most homogeneous picture in this respect. This means that the potters producing the vessels belonging to group 4 are likely to have had better control over the firing process than the potters of the groups 1-3.[3]

The samples of group 4 show a higher degree of vitrification in the scanning electron microscope than the samples of the groups 1-3 (samples Z02 and Z39 shown in Fig. 7). This is probably due to the higher Ca-content of group 4 (shown also by X-ray fluorescence analysis), with Ca acting as a flux in the firing process.

7. PROVENANCE OF THE ARCHAEOLOGICAL CERAMICS BY X-RAY FLUORESCENCE ANALYSIS

In order to investigate a possible local provenance for the archaeological ceramics, six clay samples were taken from the site of Zalavár and one clay sample (ZT1) from a distance of 2 km from the site. Since the clay samples are much coarser (higher Si-content) than the archaeological ceramics, the data from the X-ray fluorescence analysis is plotted below as ratios of elements in order to eliminate

1 – The measurements were performed on a Philips PW 2400 machine equipped with a Rh tube. Major elements (Si, Ti, Al, Fe, Mn, Mg, Ca, Na, K, P) and trace elements (Ba, Cr, Cu, Nb, Ni, Pb, Rb, Sr, Y, Zn, Zr) of 57 samples (50 samples of archaeological ceramics and 7 fired clay samples) were analysed. Analyses were performed on glass tablets for both major and trace elements. Glass tablets were prepared by starting with about 3 g of powdered sample dried for 6-8 hours at 110 °C and calcined at 1000 °C for one hour. Samples were weighed after each step in order to calculate the loss on ignition (LOI). 0.700 g of calcined powder was then mixed with 0.350 g of Li fluoride and 6.650 g of Li tetraborate, put into a Pt crucible and melted at 1150 °C for 10 min using a Philips Perl'X-2 machine. Calibration of major and trace elements was made on 50 standards. Analytical reproducibility and detection limits for major and trace elements on analyses of standards are reported in DiPierro 2002.

2 – A scanning electron microscope from FEI (Focused Electrons and Ions) XL 30 Sirion FEG was used with various settings.

3 – The Powder XRD measurements were performed with a Philips PW1800 X-ray diffractometer (40mA/40kV, Cu-Kα radiation, 2 Theta 2-70°, step size 0.020° 2 Theta, time per step 1 sec, PC-APD diffraction software). The samples were ground in a tungsten-carbide mill. For the identification of the peaks the freeware MacDiff was used (Version 4.2.5, Rainer Petschick, Johann Wolfgang Goethe University, Frankfurt am Main, Germany).

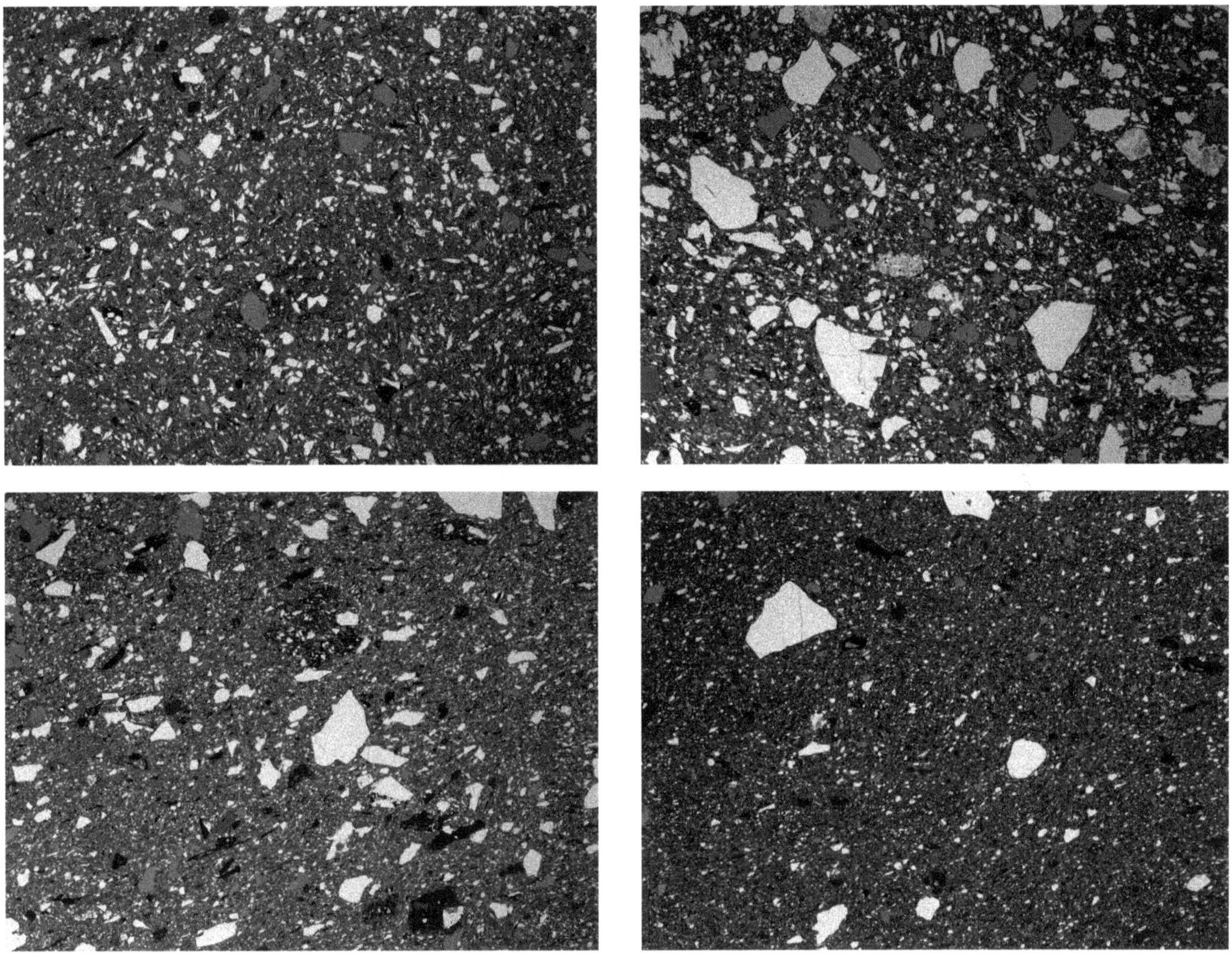

Fig. 5 – The four groups of the "polished yellow ceramics" from Zalavár by scanning electron microscopy; Si mappings from SEM EDX (the whiter the particles are, the more Si they contain: the light grey particles are quartz, the mid-grey particles are feldspars); between 800-1200 quartz particles could be detected in the areas of 2.6x1.8 mm shown here.

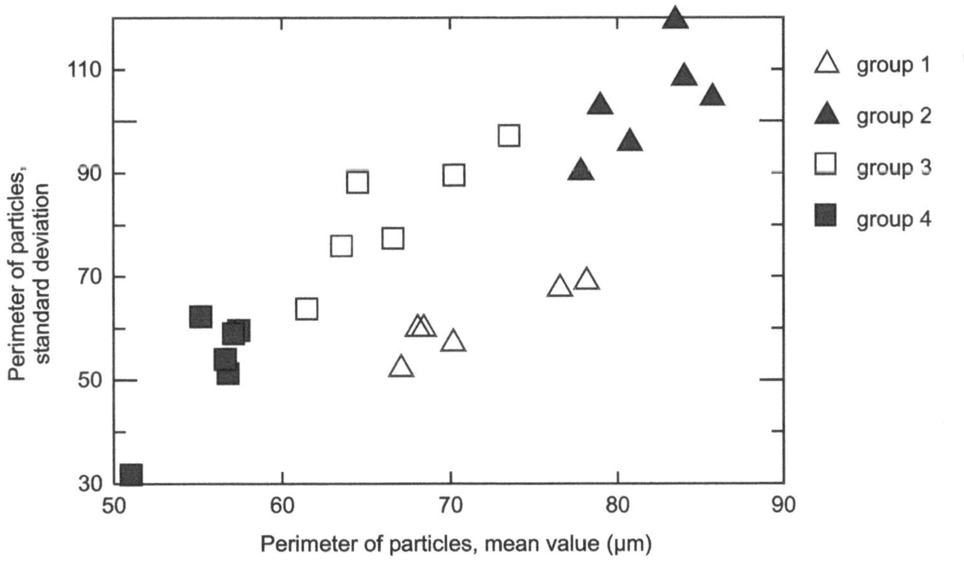

Fig. 6 – The four groups of the "polished yellow ceramics" from Zalavár: Bivariate plot of the perimeter of the measured quartz grains in the archaeological samples; mean value of the perimeter of each sample against the standard deviation of the perimeter of each sample; perimeter measured in µm. The plot makes a separation of groups 1-3 possible on the basis of micromorphological criteria; these groups cannot be separated clearly on the basis of chemical data.

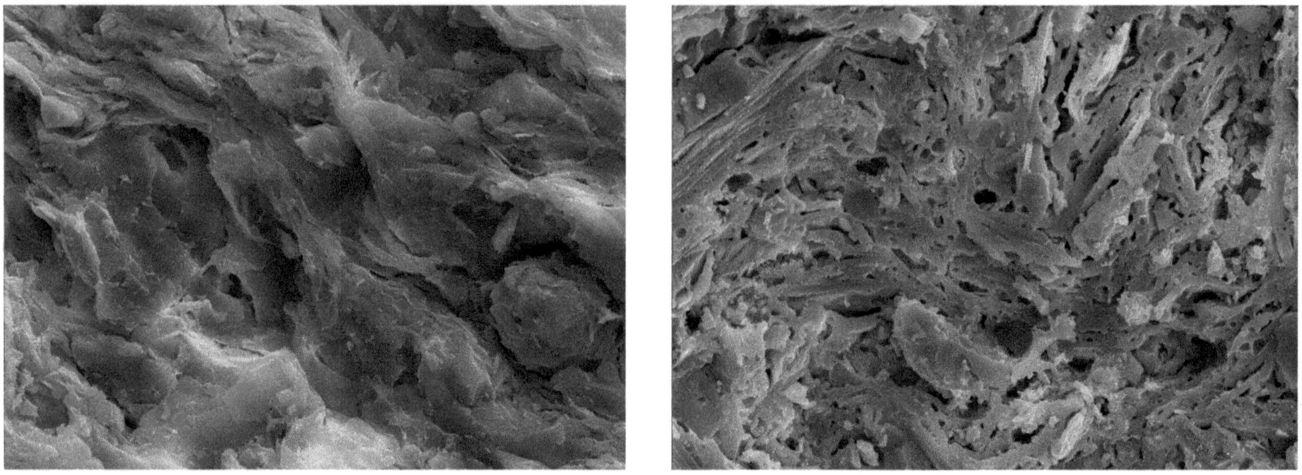

Fig. 7 – Scanning electron microscopy images of the internal structure of samples Z02 (group 3) and Z39 (group 4); longer side of images represents 65 µm. Sample Z39 of group 4 shows a higher degree of vitrification in the SEM than sample Z02 of group 3. This is probably due to the higher Ca content of group 4.

	1a	1b	2a	2b	2c	2d	3a	3b	3c	3d	4a	4b
Chlorite	■		■				■	■			■	
Illite+Muscovite	■	■	■	■	■		■	■	■	■	■	■
Feldspar	■	■	■	■	■	■	■	■	■	■	■	■
Quartz	■	■	■	■	■	■	■	■	■	■	■	■
Calcite			■	■				■	■		■	■
estimated T ºC	< 650	650-850	< 650	650-750	650-850	~850-900	< 650	< 650	650-750	650-850	< 650	650-750
Samples	Z06, Z10, Z12, Z20, Z25	Z03, Z04, Z07, Z08, Z09, Z11	Z22, Z29	Z19, Z21	Z01, Z13, Z15, Z26, Z27, Z28, Z30, Z33, Z45, Z48, Z49, Z50	Z46	Z16, Z18	Z24, Z31, Z32	Z02, Z05, Z17, Z23, Z47	Z14, Z41, Z43	Z36	Z34, Z35, Z37, Z38, Z39, Z40, Z42, Z44

Table 2 – Phase composition and estimated firing temperature of the four groups of the "polished yellow ceramics" from Zalavár. Groups 1-3 are somewhat heterogeneous with regard to the estimated firing temperature on the basis of XRD analyses; group 4 shows the most homogeneous picture in this respect.

the dominance of Si. (For the numeric results of the XRF analysis see Table 1, for details of the XRF analysis see footnote 1).

Both main and trace elements show that raw material with a chemical composition similar to the four groups of archaeological ceramics is available at and in the vicinity of the site (Fig. 8). Clay samples ZT1-3 and ZT5 are well compatible with groups 1-3 of the archaeological ceramics; clay samples ZT6-7 are well compatible with group 4 of the archaeological ceramics. Clay samples ZT2-7 were collected directly at the site and thus allow for a provenance of the archaeological ceramics local to the site itself; however clay sample ZT1 was collected 2 km east of the site, and is still compatible with the archaeological ceramic composition. The extent of the area where sediments of a similar composition occur can only be defined by analysing more clay samples from the vicinity of the site.

8. RESULTS

The analyses (thin-section analysis, XRF, XRD, scanning electron microscopy) show that the 50 investigated samples of archaeological ceramics can be divided into four groups.

Of these four groups, groups 1-3 have a similar mineralogical and elemental composition and differ mainly in textural parameters. These differences were best shown by means of digital micromorphological analyses of element mappings in the SEM. However, on the basis of the investigated samples it seems that these three groups are part of a "continuous textural set", i.e. it is difficult to draw exact borders between the groups. This can either be interpreted as a raw material source with slightly different types of raw materials being exploited or as somewhat different clay preparation techniques being used.

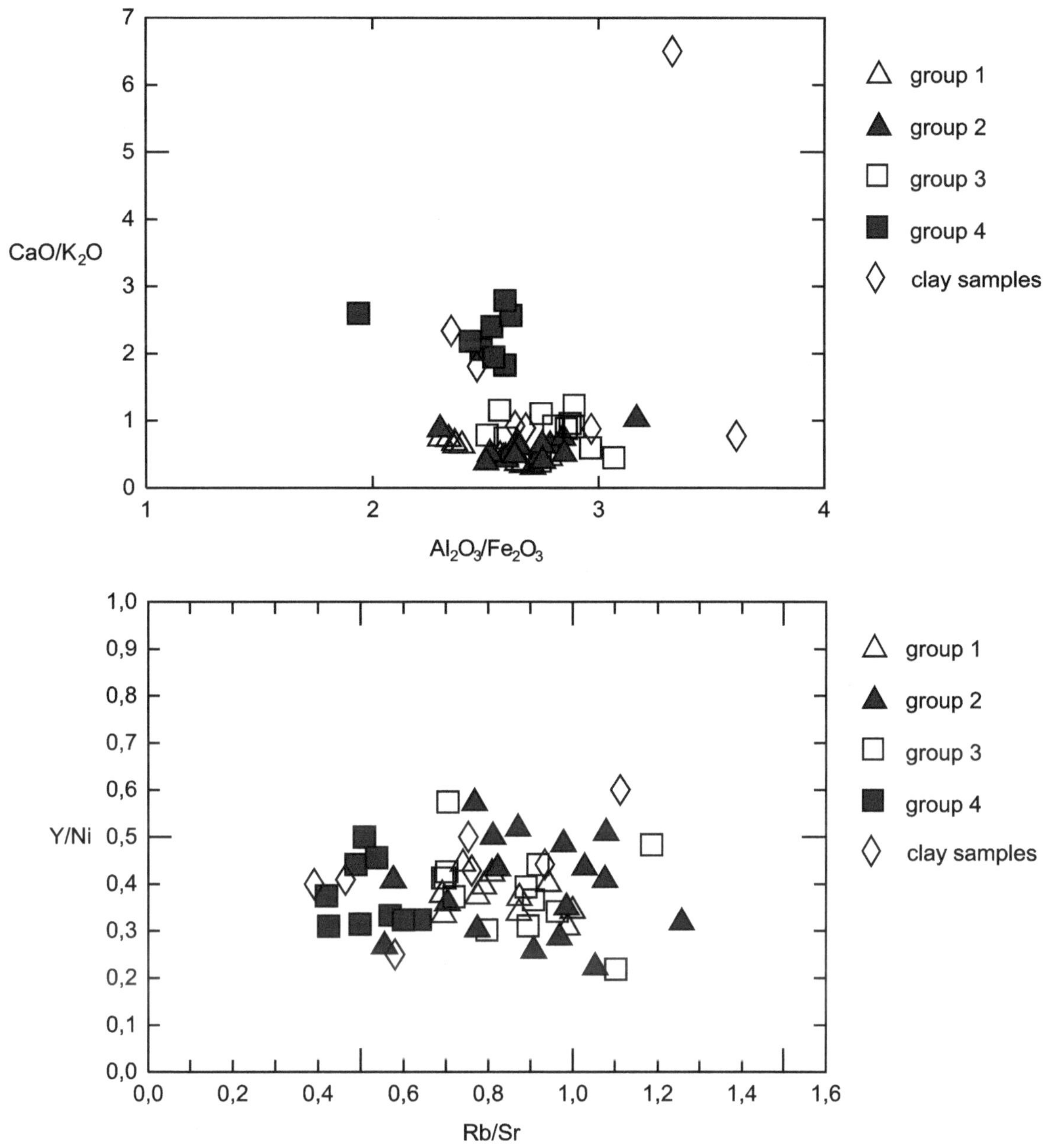

Fig. 8 – The four groups of "polished yellow ceramics" and the clay samples from Zalavár by X-ray fluorescence analysis, plots of Al_2O_3/Fe_2O_3 against CaO/K_2O and Y/Ni against Rb/Sr. Both main and trace elements show that raw material with a chemical composition similar to the four groups of archaeological ceramics is available at the site.

For each of the three groups a set of macroscopic criteria can be defined, which makes it possible to assign potsherds with no archaeometric analysis with confidence to the groups defined by archaeometric methods. A comparison with the full archaeological record is thus made possible and could clarify whether the groups 1-3 are more likely to have been produced at the same time or whether they are more likely to have been phases of a continuously changing production.

Group 4 diverges from the above three groups in mineralogical and elemental as well as in textural parameters. Again, a comparison with the full archaeological record could clarify whether group 4 is more likely to have been produced at the same time as groups 1-3 using a different raw material source or whether group 4 represents a different chronological phase from groups 1-3.

The investigated clay samples (mainly from the results of XRF analysis) show that a local provenance is possible for all four groups of the archaeological ceramics.

From a methodological point of view it can be concluded that the digital micromorphological analysis carried out on the basis of element mappings in the SEM does add an extra dimension to the investigations and allows the analysis of features which are not accessible by other means. However, the opportunities offered by this type of microstructural analysis and possible ways for its further application have yet to be investigated in detail in order to be able to utilise the full potential offered by this investigation method.

ACKNOWLEDGEMENTS

I would like to thank Prof. Marino Maggetti and Prof. Bernard Grobéty for making possible the use of the equipment (XRD, SEM, etc.) at the Department of Geosciences, University of Fribourg (CH). I am also indebted to them and to the members of the Archaeometry Working Group in Fribourg for their advice and support. I especially thank Prof. Vincent Serneels for conducting the XRF measurements, Dr. Gisela Thierrin-Michael for advice on questions of optical microscopy and Christoph Neururer for advice concerning the SEM.

The samples were provided by Dr. Béla Miklós Szőke (Budapest), I am also grateful for his support. The thin sections were prepared by Jean-Paul Bourqui (polished thin sections; Fribourg) and Andreas Wagner (covered thin sections; Eggenburg, Austria).

For the duration of the project the author was supported by the Foundation "Aktion Österreich-Ungarn", by the Federal Commission for Scholarships of the Swiss Confederation and by the Austrian Science Fund (Project P16468-G02).

REFERENCES

DIPIERRO S. 2002, *Domestic production versus pottery exchange during the Final Neolithic: characterization of the Auvernier-cordé ceramics of the Portalban and Saint Blaise settlements, Western Switzerland*, Ph.D. Thesis. University of Fribourg, Switzerland, 2002.

HEROLD H. 2004, *Dünnschliffanalysen frühmittelalterlicher Keramik aus Pliska, Bulgarien*, unpublished project report, Vienna 2004; to be published as an appendix to the Ph.D. thesis of V.P. Vasileva at the "Graduiertenkolleg Archäologische Analytik", Johann Wolfgang Goethe University, Frankfurt am Main, Germany.

HEROLD H. (in press), "Frühmittelalterliche Prunkkeramik aus Mikulčice, Mähren: Archäometrische Analysen und ihre Interpretation", *in* L. Poláček (ed.), *Das wirtschaftliche Hinterland der frühmittelalterlichen Zentren,* Internationale Tagungen in Mikulčice VI, Spisy Archeologického Ústavu AV ČR Brno.

MAGGETTI M., WESTLEY H., OLIN J. 1984, "Provenance and Technical Studies of Mexican Majolica Using Elemental and Phase Analysis", *in* J.B. Lambert (ed.), *ACS Advances in chemistry,* Series, No. 205, Archaeological chemistry III, American chemical society, p. 151-191.

NOLL W. 1991, *Alte Keramiken und ihre Pigmente*, Studien zu Material und Technologie, Stuttgart 1991.

LEAD-GLAZED SLIPWARE OF 10th -11th CENTURY AKHSIKET, UZBEKISTAN

C. HENSHAW[1], TH. REHREN[1], O. PAPACHRISTOU[2], A.A. ANARBAEV[3]

[1] Institute of Archaeology, 31-34 Gordon Square, London WC1H 0PY

[2] Haras 12. Neo Iraklio. 14121 Athens, Greece

[3] Institute of Archaeology, Samarkand, Uzbekistan

INTRODUCTION

Akhsiket was a city in the Ferghana Valley of north eastern Uzbekistan, a centre for trade and industry in the region. Earliest habitation layers so far discovered date to the 2nd/1st century BC with the foundation of the city itself in the 6th/7th century AD (Anarbaev 1988, p. 171). By the early Islamic period, it was one of the largest cities in the Ferghana Valley—with 40 ha enclosed by fortification walls. During the campaign by Kutayba ibn Muslim in the 8th century AD, Akhsiket fell under the influence of the growing Arabian caliphate. Historical references to pre- and early Islamic Akhsiket are few and controversial—primary evidence for the cultural character and economic structure of the city consequently resting almost entirely on archaeological evidence and numismatics. Regular excavations by local archaeologists since the 1950s have revealed the nature of the habitation, fortification and industrial aspects of the inner city (*shakhristan*) and parts of the suburbs (*rabad*).

Collection and microanalysis of samples of pottery is currently being undertaken as a doctoral research project, the first of its kind for wares from the Ferghana Valley.

POTTERY AT AKHSIKET

Lead-glazed slipware forms the bulk of Akhsiket's glazed pottery assemblage, used from the 10th century until the time of Akhsiket's abandonment in the early 13th century, when the Mongols arrived. Lead-glazed slipware is a technique whereby fine earthenware vessels were decorated by an all-over coating of white slip with a high concentration of quartz (engobe), which was then decorated with a variety of coloured slips and covered with a lead glaze. In this paper, we look at samples from the 10th and 11th century, when this technology was at its height. The development of this technology is attributed to the production centres of Samarkand and Nishapur (Mason 2004, p. 122).

It was used to decorate a limited range of vessels, particularly bowls, plates and lamps. By far the most common vessel type is the sectional or hemispherical bowl with a simple rim and concave base. Cono-segmental bowls, ring-foot bases, and more complicated rims are occasionally seen. Decorative styles incorporate a limited palette, and a series of standard motifs, many of which are related to common themes seen throughout the caliphate (Fig. 1).

Calligraphic styles based on Koranic inscription (kufic) is particularly common in Central Asia, and is historically called "Samanid" ware because it was developed to a high degree during the rule of the Central Asian Samanid dynasty. The original design was legible and repeated quotes from the Koran, phrases from the epic Shah Nemah, or blessings. In Central Asia, the letters shortly became highly stylised—having no readable content. This latter type is what is generally found at Akhsiket. These wares show a direct link to the artistic traditions of the wider Islamic world (Lane 1957, p. 6). Other designs include linear bands of black, red and olive on a white ground and motifs such as "pseudo-rope", "eye", "peacock eye", and "knot". Zoomorphic and human representations are extremely rare, although not unheard of.

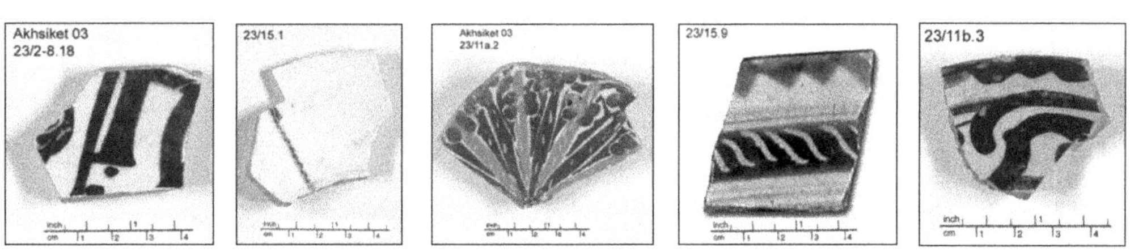

Fig. 1 – Examples of lead-glazed decorative motifs from 10th-11th century AD Akhsiket.

Alongside the lead-glazed slipware at Akhsiket, there are also alkaline glazed wares (the earliest glazed type), and other types of lead glazes such as monochrome green, black, brown and yellow that are applied directly to the body.

Unglazed pottery includes a diverse range of domestic wares, stoneware (highly vitrified) spheroconical vessels, and industrial kaolin-based crucibles. Provenancing any of these wares is difficult, as Akhsiket has so far produced no definite evidence of the manufacturing process—as has no other site in the Ferghana Valley.

Although there is still a long way to go before finishing analysis of the range of pottery types found at the site, the data are already providing clues to the nature of the technology of glazed pottery, and the range of variation to be found at Akhsiket.

THE LEAD-SILICA GLAZE

The data used in this paper consist of normalised concentrations of major and minor elements detected by EDX coupled to a scanning electron microscope (SEM). These represent averages of several area scans for each sample, obtained at magnifications of 100-800x depending on the material. This method is accurate to within a few percent for the major and minor elements in glazes, and avoidance of spot analysis minimizes loss of soda.

The microstructure of the glazes shows a relatively homogenous, smooth, blemish-free surface and cross-section, with occasional relict quartz grains, depending on the state of preservation. As lead glaze is a fluid, low-firing glaze, with a similar thermal expansion coefficient to the ceramic bodies used, crazing and flaking are seen much less often than with the alkali glazes.

Lead glazes can be classified according to the amount of lead present in the glaze. The Akhsiket sherds fall into the "lead-silica" classification, with appreciable amounts of silica, often of equal concentration with lead oxide. The element concentrations given here were taken from the mid-point to near the surface of the glaze, where leaching from the body is minimised.

The chart (Fig. 2) shows that Akhsiket's lead glazes form a linear progression with a trend towards slightly higher concentrations of lead than silica. They form a relatively tight group, with the exception of one outlier with a high lead concentration. The silica concentration of all glazes are affected by the taking up of silica from the body of the vessel during firing, and this is apparent in many glazes where silica is a few percent higher near the body than the surface of the vessel—even accounting for a few percent error in the analysis, a consistently higher concentration near the body is evidence of some diffusion. It is not possible to determine the exact nature of the glaze recipe, due to this, and it could be that a nearly pure lead oxide slurry was applied to the surface of the vessel, and all the silica present comes from the body and/or slips underneath. Further work on the application and firing of glazes is needed to approach the question of glaze recipes.

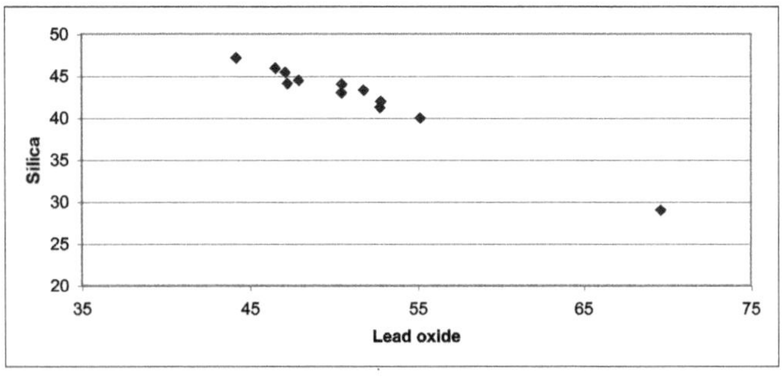

Fig. 2 – Chart showing lead oxide v. silica in weight % for colourless lead-silica glazes on twelve Akhsiket sherds.

GLAZE COLORANTS

In the majority of cases where there is black or brown decoration on these slipwares, the colour is not produced using a slip, but with a dark-coloured glaze. It is unclear how the coloured glazes were produced, although bulk SEM-EDX analysis shows that the colorants are Fe_2O_3 and (usually) MnO, and the base glaze is very similar in chemical composition to the colourless glazes (Fig. 3). It is possible that the colorants were added as a pigment either under the glaze, or over the glaze, which, during firing, diffused more or less evenly through the glaze thickness.

Crystals present in the glaze can provide clues to the manufacturing process. These crystals are due to the interaction of clay minerals with the glaze during firing, and are present along the interface between glaze and body (Fig. 4). In the cases studied so far, it seems that the *colourless* glazes have very limited and occasionally no crystalline interaction at the interface point, whereas *coloured* glazes have many alumina-potash rich crystals. This can indicate a glaze was applied over a biscuit-fired body, or that the coloured glazes or pigments were brushed on (which can facilitate such an interaction layer) (Tite *et al.* 1998, p. 250). Initial analyses of body fabric under polarised light microscopy indicate that primary organic residue (such as plant fibers and seeds) is present—if this is the case, then it is unlikely the vessels were fired twice. The presence of these well-formed, angular crystals can also indicate a long firing period (Vandiver and Kingery 1985, p. 187), although again in our case, that would seem to be disproved by the existence of organic residue.

In the coloured glazes, there is occasionally a second type of crystalline phase found both along this interface area and further into the glaze (Fig. 5). This latter type is high in iron and manganese oxide, and may be unfused remnants of the colouring minerals used in the glaze.

In a few cases, a very fine-grained, thick black slip is used for black decoration, although the glaze covering this is also coloured—either by the diffused MnO and Fe_2O_3 from the slip into the glaze, or by deliberately applied coloured glaze or pigment in addition to the dark coloured slip.

Black glaze	Na_2O	Al_2O_3	SiO_2	K_2O	CaO	MnO	Fe_2O_3	PbO
23/2-8.3	0.0	4.0	41	1.3	0.6	2.1	3.6	48
23/2-8.17	0.7	7.4	37	2.3	0.5	1.8	8.2	43
23/12.15	0.0	3.5	37	0.8	0.7	0.0	3.9	54
23/13.1	0.0	2.5	40	1.1	0.0	0.7	3.1	53
23/14.3	1.0	2.9	36	2.4	1.6	2.7	5.3	48
23/15.1	0.5	3.6	44	1.7	0.5	0.6	2.2	47
Brown glaze								
23/12.2	0.0	2.8	37	1.3	1.2	1.3	3.5	53
23/16.3	0.4	3.1	42	1.5	0.5	0.9	2.5	49
Glaze over black slip								
23/11a.2	0.0	2.6	42	1.6	0.7	0.0	1.5	51
23/13.3	0.6	2.8	42	1.6	0.4	1.3	1.5	50
23/14.15	1.0	3.8	42	0.6	0.0	1.4	3.2	48

Fig. 3 – Table of SEM-EDX results showing major and minor elements in the coloured lead glazes, in weight %.

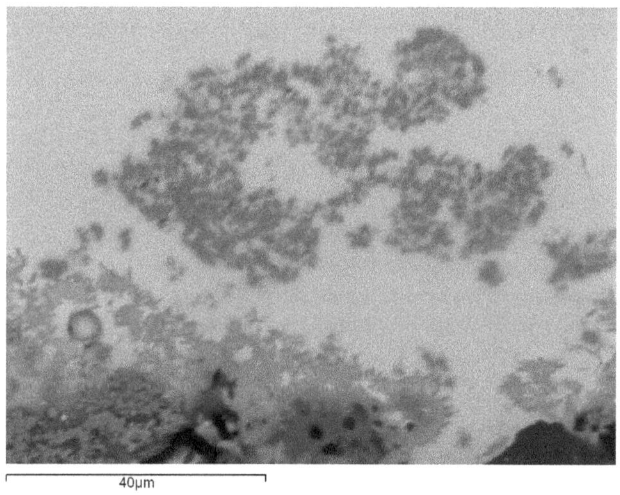

Fig. 4 (left) & 5 (right) – BSE images showing crystalline formation in coloured lead-silica glazes.

THE SLIPS

The lead-glazed slipware described here employs a layering technique. A clay-based slip engobe, high in microscopic quartz grains, is applied to the fineware body forming a smooth, white surface. This is most often on the inside of bowls (and the outside of lamps) and just over the rim, although sometimes it is seen covering the outer surface of bowls and under the base as well. In a small minority of samples, a red engobe is applied instead.

The most common colours used for painting designs are red, olive, black and brown. Yellow, light green and orange are also seen, although more often on 12th century wares. The coloured slips are usually between 50 and 100 microns thick. Their compositions, like the engobe, are similar to the body fabric (Fig. 6, Fig. 7), although with a lower CaO content, due to the higher levels of quartz in relation to clay. It is likely that the slips were made with the same clay as the bodies, and that the quartz-rich slip base is the same for all the coloured slips. The concentration of colorants varies depending on the shade.

Black shows the strongest colouration, with the highest concentration of iron and manganese oxides (while brown is similar, with slightly reduced amounts). Red uses the same elements, in far less quantities, and the olive colour is due to chromium oxide. The lead oxide amounts for all but the red could either have diffused through during firing from the glaze, or have been part of the original slip composition. In the case of red slips, it would seem to have been added deliberately, with double the amount, on average, as the other slips.

SLIP	Na_2O	MgO	Al_2O_3	SiO_2	K_2O	CaO	TiO_2	MnO	Fe_2O_3	PbO	Cr_2O_3
Black	**0.9**	**0.8**	**11.0**	**43.0**	**3.1**	**3.9**	**0.9**	**5.1**	**22.0**	**9.2**	
St.d.	0.9	0.5	1.9	5.2	0.8	5.0	0.3	1.5	5.5	1.1	
Red	**0.5**	**1.1**	**16.0**	**50.0**	**3.9**	**2.6**	**0.7**	**1.3**	**4.4**	**17.0**	
St.d.	0.6	2.0	8.2	16.5	3.9	4.3	1.3	5.9	1.3	5.0	
Olive	**0.5**	**2.1**	**10.0**	**62.0**	**3.4**	**1.8**	**0.3**		**3.2**	**9.1**	**7.4**
St.d.	0.5	0.9	1.6	3.2	1.1	1.6	0.5		1.4	4.3	4.1
White	**0.8**	**1.7**	**16.0**	**62.0**	**4.0**	**4.9**	**1.2**		**0.8**	**9.1**	
St.d.	0.7	1.3	6.2	13.1	2.2	5.2	3.0		1.1	8.7	

Fig. 6 – Table of average compositions in weight % for slips (black: n=3, red: n=5, olive: n=3, white: n=12).

BODY	Na_2O	MgO	Al_2O_3	SiO_2	K_2O	CaO	TiO_2	Fe_2O_3
Average	**1.1**	**3.5**	**12.0**	**56.0**	**3.2**	**16.0**	**0.7**	**6.0**
St.d.	0.7	1.6	2.8	5.4	1.8	5.1	0.3	1.5

Fig. 7 – Table of average concentrations in weight % of major and minor elements in the bodies of twelve Akhsiket sherds.

CONCLUSION

From the preliminary analysis it can be seen that the lead-glazed slipware at Akhsiket has a number of characteristics that can lead to better understanding of the glazed pottery technology in the region.

The concentrations of lead and silica, and the presence of other oxides puts Akhsiket's lead-glazed ware firmly in the lead-silica category, with a single example of high-lead glaze. Further work is needed to determine whether there are any subgroups, or whether there is simply a gradual continuum within a single group. The other elements (alumina, soda, potash, lime) may have been taken up from the slip underlying the glaze during firing or the result of a clay slurry being used as a base for applying the lead oxide to the surface, and it is unknown how much silica was in the original glaze recipe.

Other than colourless, transparent glaze, there is also a coloured brown or black glaze. In some of these, globule-like crystals high in iron and manganese oxides may be unfused remnants of the colouring oxides used in the glaze or pigment. The differing levels of interaction at the glaze/body interface may show that, in the case of black or brown decorated vessels (and taking into account evidence of organic residue in the bodies), the coloured glaze or pigment was brushed onto an unfired vessel, then dipped in a glaze slurry, and fired at a relatively low temperature.

The white engobe and coloured slips appear to be finely applied versions of the same clay as the body, with an added quantity of crushed quartz, and the coloured slips having added colouring oxides of iron, manganese (or a mixture of the two) or chromium. Some of the designs produced using this technology show the closest resemblance, and thus cultural connections, to the wider Islamic world of any of the pottery types found at Akhsiket from the 10th–11th century.

Further study is needed on a wider sample of lead-silica glazed wares, particularly 12th century examples from Akhsiket and contemporary pottery from nearby sites such as Kuva and Tashkent. Such comparison will show how typical the technological characteristics of Akhsiket's pottery are. Work is currently underway on all of the above, as well as the collection of compositional and petrographical data on the bodies of these sherds. The aim is to gain a better understanding of how Akhsiket's pottery technology fits into a wider chronological and geographical setting, as well as its own cultural and economic background.

REFERENCES

Anarbaev A.A. 1988, "Ancient and medieval Akhsiket (research results and perspectives)", *Sovetskaya archaeologiya* 1, p. 171-187.

Lane A. 1957, *Early Islamic pottery*, London.

Mason R. 2004, *Shine like the sun*, Costa Mesa, Ca.

Tite M.S., Freestone I., Mason R., Molera J., Vendrell-Saz M., Wood, N. 1998, "Lead glazes in Antiquity: methods of production and reasons for use", *Archaeometry* 40, p. 241-60.

Vandiver P., Kingery W.D. 1985, "Variations in the micro-structure and microcomposition of pre-Song, Song, and Yuan Dynasty ceramics", *in* W.D. Kingery (ed.), *Ancient technology to modern science. 1*, p. 181-227.

ARCHAEOMETRIC INVESTIGATION ON 13th CENTURY GLAZED AND SLIPPED POTTERY FOUND IN LIGURIA AND PROVENCE

C. CAPELLI[1], R. CABELLA[1], S.Y. WAKSMAN[2]

[1] Dipartimento per lo Studio del Territorio e delle sue Risorse (DIP.TE.RIS.), Università degli Studi di Genova, Corso Europa 26, 16132 Genova (Italy)

[2] Laboratoire de céramologie, CNRS-UMR 5138, Maison de l'Orient et de la Méditerranée, 7 rue Raulin, 693657 Lyon cedex 7 (France)

1. INTRODUCTION

A great number of Mediterranean pottery types were imported from the 10th c. onwards to Provence and Liguria because of the strategic position of the ports of Marseille, Savona and Genoa in the medieval trade routes. They came from both Islamic and Byzantine areas and their coatings typically consist of tin-opacified glaze and white slip under transparent glaze respectively (Benente 2001; Berti and Gelichi 1999; Vallauri *et al.* 2003).

In the late 12th c., a local production of glazed and slipped ware—the first in the Western Mediterranean—suddenly appeared in Savona, apparently without any period of experimentation. As early as the beginning of the 13th c., Savona products were widespread in Liguria, Provence and the Tyrrhenian area (Varaldo 1997a).

Various types of glazed and slipped dishes and bowls with similar shapes, but different quality levels, were produced in Savona (Mannoni 1975): the "Graffita arcaica tirrenica" ("Savona archaic sgraffito", with green and brown-yellow painting), the "Ingobbiata policroma" (very similar to the previous ware, but without incised decoration) and the "Ingobbiata monocroma" (with plain colourless, green, or yellow monochrome glaze). Thousands of kiln wasters have been found in the Priamar Fortress excavations (Lavagna and Varaldo 1986).

The best acknowledged archaeological model (Berti *et al.* 1997) points to the Middle East as the probable source of the high technical knowledge introduced in Savona, for several reasons: the strong commercial links between the Middle East and Liguria, already existing before the Crusades; the presence of the white slip and the sgraffito techniques, that are strictly related to the Eastern Mediterranean; the discovery of later (14th c.) documents quoting local potters with Jewish names. However, many uncertainties still exist.

In particular, little is known about the relationships between the "Savona archaic sgraffito" (SAS) and two other 13th c. wares sharing similar typological features: the "Port Saint Symeon ware" (PSSW) and the "Ligurian Protomajolica" (LPM).

PSSW is known as a Middle Eastern production of sgraffito with yellow-brown and green decorations (Lane 1937). At present, only five sherds have been identified in the north Tyrrhenian area (Genoa and Marseille), partly with the help of archaeometric analyses (Capelli *et al.* 2005). However, the scarcity of the finds could be due to the difficulty of distinguishing PSSW from SAS with the naked eye, especially in the case of small and altered fragments.

Following different hypotheses based on archaeological data (Riavez 2001), the relationships between SAS and PSSW could be either direct, with movement of potters from the Middle East to Savona (or vice-versa: the chronological evidence is still scarce) or indirect, through other Mediterranean productions.

LPM is also generally similar in shape and decoration to SAS (Gardini 1997; Varaldo 1997b), but it can be easily distinguished by the presence of a tin-opacified glaze. Besides, in LPM paintings are in green and black and there are no sgraffito decorations. Its distribution is very limited: only a few tens of sherds have been recognised in military or religious contexts of Liguria and, occasionally, in Provence, Tuscany and Sardinia (Lunardon 2001). No kiln wasters have been found.

The archaeological hypotheses about the origin of LPM (Berti *et al.* 1997; Gardini 1997; Varaldo 1997b), mainly based on distribution and typological data, are the following: a) LPM was produced in Savona, after the start of SAS, in the same workshops and with the same raw materials as SAS; b) LPM was a short-lived experiment in a luxury production; c) the presence of slip under the enamel in LPM is due both to the traditional use of slip in Savona and to the aim of saving tin.

However, recent archaeometric studies (Capelli 1999b; Capelli and Mannoni 2003) do not support most of these hypotheses, evidencing strong incompatibilities between SAS and LPM. On the contrary, thin section analyses of both ceramic bodies and coatings (Capelli *et al.* 2001; 2002) showed close relationships between LPM and other types different from SAS: rare protomajolicas with decorations not matching SAS and the "typical"

LPM, rare polychrome or monochrome sgraffitos, and abundant ceramics with monochrome (green or colourless) transparent glaze. In some cases, they were considered by Ligurian archaeologists as imports from the Eastern Mediterranean. Furthermore, some of them seem to be more ancient than the first production of SAS (Capelli *et al.* 2001).

In this paper we compared SAS, PSSW, LPM and related types from an archaeometric point of view. In order to integrate the published data, new analyses of ceramic bodies, white slips and glazes were carried out by optical microscopy, SEM-EDS and WD-XRF.

This study aimed at better characterising the different wares and productions, determining the compositional and technical relationships between them, verifying the archaeological hypotheses and, more generally, contributing to improved knowledge of the routes of trade and technology in the Mediterranean.

2. ANALYTICAL RESULTS

2.1. Ceramic bodies

2.1.1. "Savona archaic sgraffito" and other Savona productions

The thin section study of a few hundred samples—including several kiln wasters—of SAS and other Savona medieval productions (Capelli and Mannoni 2001; Capelli *et al.* 1999; 2003) evidenced quite homogenous compositions and production techniques.

The clay matrix of Savona table wares is generally carbonate-rich, with variable percentages of Fe-oxides, and not vitrified (the body is not hard). The aplastic inclusions are rather frequent and medium- or coarse-grained. They are mainly composed of metamorphic rock fragments (paragneisses, metagranites and rare amphibolites, reaching 1 mm in size) and related mineral grains (micas, feldspars, quartz, amphiboles and several heavy minerals), together with calcareous and siliceous microfossils. The raw materials used for SAS are thought to be local fossiliferous Pliocene marine sediments sometimes mixed with Fe-rich alluvial clays, both of them including natural temper derived from the alteration of the Palaeozoic metamorphic basement outcropping in the Savona area.

XRF analyses carried out on SAS and other medieval types show that Savona productions are globally well differentiated from PSSW and LPM (Table 1, Fig. 1), even though their chemical compositions are not very homogeneous and further sub-grouping may be distinguished. The moderately high Mg and the high Cr and Ni contents (Fig. 1B-C) generally observed derive from the mafic mineral inclusions, while the fairly large compositional range of some other elements, such as Ca (Fig. 1A), is related not only to the variability of the marly clays but also to the contribution of the alluvial Fe-rich clays mixed with them in different amounts. It must be noted that SAS is generally more Ca-rich and Fe-poor than other later productions of Savona, such as the "Maiolica arcaica" (Capelli 1999a) and some related glazed monochrome wares (Capelli and Mannoni 2001).

2.1.2. "Port Saint Symeon" ware

Recent archaeometric analyses carried out on PSSW samples from Genoa and Marseille supported the typological attributions, evidencing close resemblance with similar wares found in Beirut excavations (Capelli *et al.* 2005).

Under the polarising microscope, the ceramic bodies of the studied assemblage show a carbonate-rich matrix sometimes partially vitrified (the body is relatively hard) with moderately abundant and fine-grained inclusions, consisting of calcareous microfossils, fragments of limestone, volcanic and ophiolitic rocks, and more generic mineral grains.

The analysed samples show overall distinctive chemical features, such as high Ca, Mg, Cr, and Ni contents (Table 1, Fig. 1). However, several productions of PSSW may be represented, although the sampling is too small to enable any conclusion at this stage.

Both petrographic and chemical features of these samples are consistent with the use of marine carbonate-rich clays with detrital supply from ophiolitic and carbonate rocks. They are in good agreement with the geological characteristics of the region of Antioch and the gulf of Iskenderun, considered by archaeologists as the main production area of PSSW (Lane 1937; Hild and Hellenkemper 1990; Redford *et al.* 2001).

2.1.3. "Ligurian Protomajolica" group

The fabrics of the LPM group are more variable in mineralogical composition and in texture than the previous ones. The matrix is carbonate- to Fe-oxide-rich (sometimes a mixing of two different clays cannot be excluded) and often vitrified (the body is hard). The aplastic inclusions are usually relatively scarce and fine-grained, but in rare cases a medium-grained, well-sorted temper is observed. Mono- and polycrystalline metamorphic quartz fragments (up to some hundred microns in size) predominate on acid metamorphic, sedimentary (limestone, sandstone), volcanic (ophiolitic?) rock fragments and feldspar, mica, clinopyroxene, and other occasional fine-grained mineral grains. Rare and small calcareous and siliceous microfossils are also present, as well as *chamotte* and/or argillaceous rock fragments, sometimes rather abundant and reaching millimetric dimensions (Capelli 1999b; Capelli and Mannoni 2003).

Chemical analyses (Table 1, Fig. 1) confirm on one hand the similarities, already evidenced by the thin section study, between LPM and the other related types with monochrome transparent or polychrome opacified glaze (Capelli *et al.* 2001; Capelli and Mannoni 2002; 2003). On the other hand, the low values in Mg, Cr and Ni allow us to discriminate LPM from both SAS and PSSW samples.

	CaO	Fe_2O_3	TiO_2	K_2O	SiO_2	Al_2O_3	MgO	MnO	Na_2O	P_2O_5	Zr	Sr	Rb	Zn	Cr	Ni	La	Ba	V	Ce
productions of Savona: "Savona archaic sgraffito" and other wares (n=29)																				
m	5,82	6,70	0,764	2,57	63,23	15,17	3,78	0,0721	1,41	0,22	193	228	133	106	282	246	44	403	105	85
σ	3,00	0,30	0,042	0,17	2,76	0,83	0,50	0,0127	0,11	0,12	12	82	9	9	19	42	9	54	10	12
"Ligurian Protomajolica" group: "Ligurian Protomajolica" and related monochrome glazed wares (n=24)																				
m	10,34	5,79	0,667	2,22	62,50	14,55	2,55	0,0571	0,67	0,29	144	453	111	90	158	89	36	325	117	79
σ	1,99	0,33	0,052	0,21	2,05	0,95	0,31	0,0107	0,13	0,14	12	79	13	10	25	13	4	122	16	5
"Port Saint Symeon Ware" (n=11)																				
m	18,94	6,99	0,717	2,19	50,91	12,92	5,20	0,1476	1,33	0,39	134	420	62	86	359	192	23	509	110	59
σ	2,52	0,53	0,058	0,11	1,01	0,70	0,58	0,0273	0,19	0,11	13	57	4	7	31	18	3	108	18	6

Table 1 – Chemical composition of ceramic bodies (WD-XRF analyses): means (m), standard deviations (σ) and numbers of samples (n) of the main compositional groups. Major and minor elements in weight percent of oxides, trace elements in ppm.

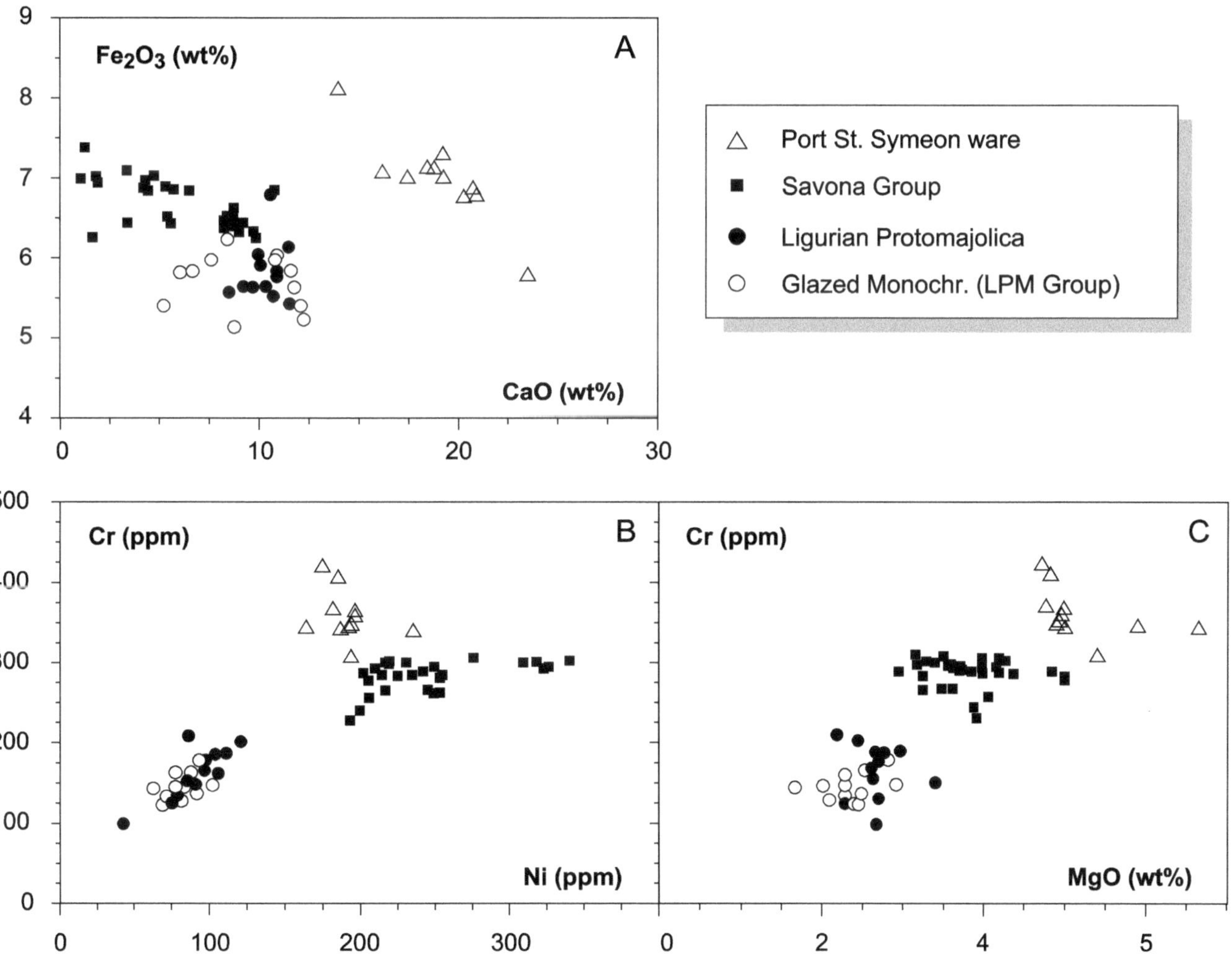

Fig. 1 – Representative binary plots of the chemical composition of ceramic bodies (by WD-XRF analyses). The Savona group includes SAS and glazed monochrome wares together with other later types.

2.2. White slips

Thin section and SEM analyses showed that the three groups discussed above can also be differentiated from each other by the white slip characteristics (see also Capelli and Cabella in these Proceedings).

SAS and the other Savona wares are generally distinguished by a low-quality slip, composed of: i) a clay matrix with lamellar texture, ii) relatively abundant and coarse-grained metamorphic inclusions (mica schist fragments and quartz, feldspar, and mica grains), and iii) scattered Fe-oxide micro-aggregates which often give the slip a macroscopic light orange colour (Capelli and Marescotti 2000).

In contrast, the majority of the LPM group samples show good-quality slips, formed by a compact clay matrix with rather scarce and fine-grained quartz inclusions (Capelli 1999b).

Finally, the PSSW slip is mainly composed of abundant angular inclusions of quartz and subordinate K-feldspar, while the clay matrix is scanty (Capelli *et al.* 2005).

According to Capelli and Cabella (in these Proceedings), the white slips described above can be attributed to the following categories: "Clay-rich types: Coarse-grained slips" (SAS), "Clay-rich types: Fine-grained slips" (LPM) and "Clay-poor, inclusion-rich types: Slips with prevailing quartz inclusions" (PSSW).

2.3. Glazes

The thickness of the analysed glazes is generally low (less than 100-200 microns). All of them are transparent, with the exception of the LPM glaze, which is tin-opacified. In this case, the cassiterite micro-aggregates are abundant and homogeneously scattered; SEM-EDS analyses point to average contents in SnO_2 > 10 wt%. These data, in contrast to the hypothesis of a tin saving based on macroscopic observations, demonstrate that the LPM glaze, even if thin, is completely opaque. Hence, the white slip would not be necessary to obtain a white background and the simultaneous presence of the two coatings is unexplained.

EDS analyses, carried out avoiding altered areas, evidenced that all the glazes are "high-lead, low-alkali glazes" (Tite *et al.* 1998), with no important or systematic differences in Si, Pb and alkali (Fig. 2A). In the same sample, concentration trends from the surface to the base are often observed for some elements, such as Si, Al and K (increasing), and Pb (decreasing), due to diffusion processes from and to the slip during the firing (Fig. 2B).

However, the glazes of the Savona group are distinguished from the other two groups by lower Ca contents ($0.2 \leq$ CaO wt% ≤ 0.9 and $1.8 \leq$ CaO wt% ≤ 4.4 respectively; Fig. 2C). This distinction can be attributed to different recipes in glaze production. In fact, the possibility of a significant Ca mobilisation from the underlying coating can be excluded in this case, because the white slips are Ca-poor (Capelli and Cabella in these Proceedings).

The colouring agents are always Fe for yellow-brown (in SAS and PSSW only), Mn for black (in LPM only), and Cu for green.

3. DISCUSSION AND CONCLUDING REMARKS

Our study showed the clear compositional and technical distinctions existing between SAS, PSSW and LPM, necessitating a revision of previous archaeological models mainly based on morphological and stylistic similarities.

With regard to the SAS-PSSW relationships, the results of the analysis of the coatings do not support the hypothesis of a direct technical transfer between the Middle East (or, at least, the area of origin of PSSW) and Savona. Together with the differences in the Ca contents of the glazes, the contrasting textural and compositional characteristics of the white slips reflect quite incompatible production techniques: Savona slips are clay-rich and derived from clayey raw materials, while PSSW slips are clay-poor and possibly made by mixing ground quartz-rich raw materials with small quantities of clay (in a way which is similar to "stonepaste" ceramic bodies of Middle-Eastern Islamic productions: see Capelli and Cabella in these Proceedings).

For LPM and SAS, the archaeological hypothesis indicating a common origin could be excluded by the analytical results:

— the body compositions are completely different; moreover, several petrographic components of LPM seem poorly compatible with the rocks outcropping in the territory of Savona and possibly of Liguria;

— LPM is generally characterised by good-quality, fine-grained white slips, differing from the coarse-grained, low-quality slips of SAS;

— the Ca contents of glazes are low in SAS, while they are quite high in LPM; besides, in the latter the glaze is always tin-opacified and, for paintings, Mn was used together with Cu (in SAS, only Fe and Cu were used);

— the degree of hardness and some textural characteristics of the bodies suggest that LPM was generally fired at higher temperatures than SAS.

Moreover, there are close compositional (and morphological) similarities between LPM and other "non-Ligurian" types. Therefore, LPM—very rare, even in Liguria—could represent a minor part of an important and complex commercial flow of imported wares, mainly composed of glazed monochrome pottery.

The localization of the manufacturing centre—or region—where this flow originated is unknown at present. In fact, LPM fabrics are different from those of every Ligurian pottery production known until now, as well as from the main medieval Mediterranean ones belonging to our thin section and chemical databases. In addition, the scarcity of diagnostic petrographic or chemical features in the LPM

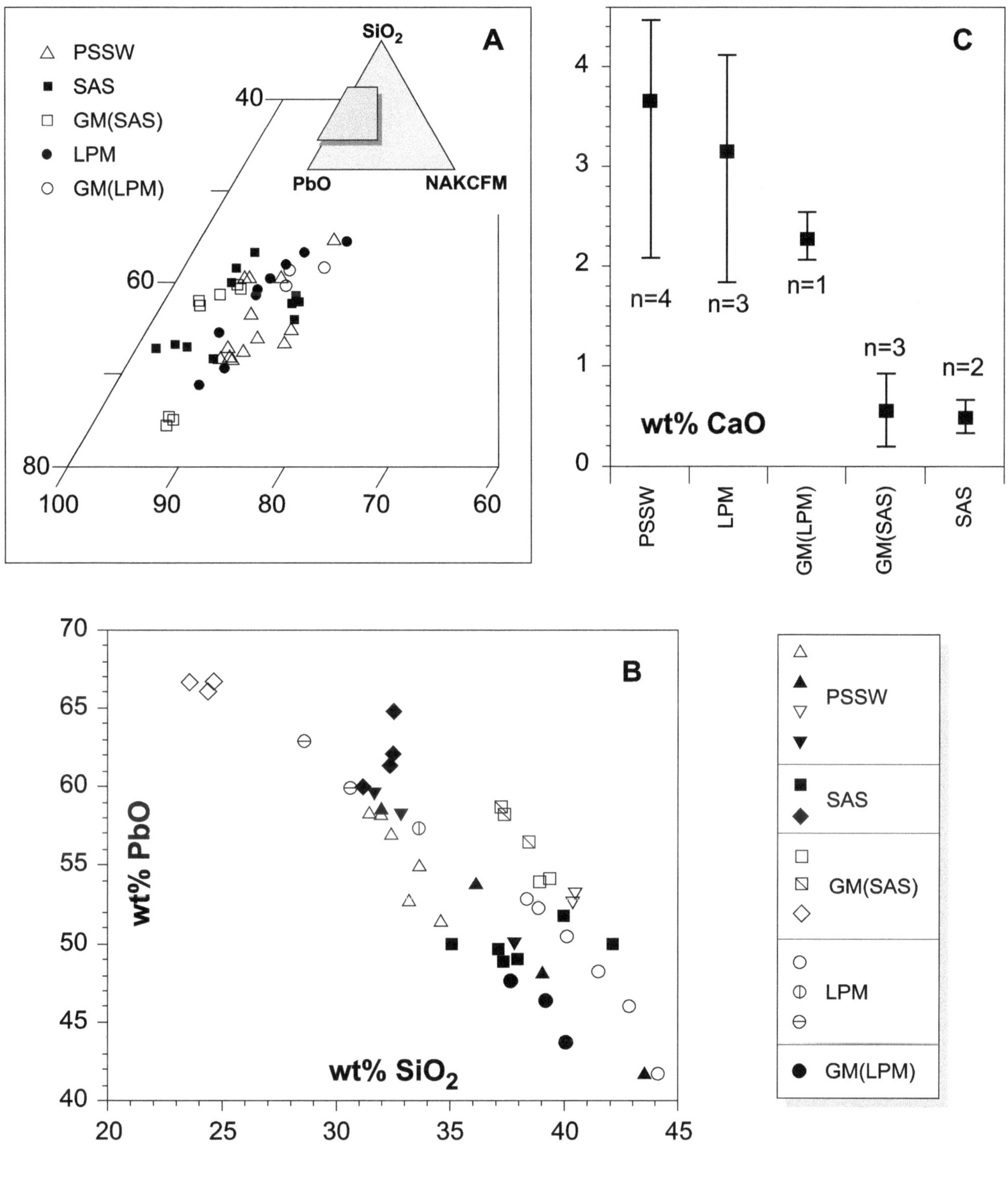

Fig. 2 – Representative diagrams of the chemical composition of glazes (by SEM-EDS analyses).
2A: Ternary diagram PbO - SiO_2 -Na_2O+Al_2O_3+K_2O+CaO+FeO+MgO (from Tite et al. *1998, modified).*
2B: Binary diagram SiO_2% - PbO%; each point corresponds to an analysis across the glaze profile, each symbol represents a single glaze sample.
2C: CaO% variability in the glazes of the different ceramic types (squares: mean values; n: number of analysed samples).
PSSW: Port Saint Symeon ware; SAS: Savona archaic sgraffito; GM(SAS): glazed monochrome wares related to SAS; LPM: Ligurian protomajolica; GM(LPM): glazed monochrome wares related to LPM.

body prevents us from suggesting precise provenance hypotheses without any archaeological constraints.

We should probably look for a place where the technique of the tin-opacified glaze—known at that time only in the Islamic areas—met the white slip, sgraffito, and lead glaze techniques, related to the Byzantine world but also present in the Islamic territories of the Eastern Mediterranean. The Middle East could well be that region—the petrographic data do not exclude this hypothesis—, but neither LPM, nor the associated types have been recognised there yet.

REFERENCES

Benente F. 2001, "Ceramica di importazione islamica e bizantina", *in* C. Varaldo (ed.), *Archeologia urbana a Savona: scavi e ricerche nel complesso monumentale del Priamàr. II, 2, Palazzo della Loggia*, Bordighera-Savona, p. 131-155.

Berti G., Gelichi S. 1999, "Trasmissione di tecnologie nel medioevo: tendenze e linee di ricerche attuali", *Albisola* XXXII, p. 23-42.

Berti G., Gelichi S., Mannoni T. 1997, "Trasformazioni tecnologiche nelle prime produzioni italiane con rivestimenti vetrificati (secc. XII-XIII)", *in Actes du VIe colloque sur la céramique médiévale en Méditerranée, Aix-en-Provence 1995*, p. 383-402.

Capelli C. 1999a, "Il contributo delle analisi minero-petrografiche per la caratterizzazione delle produzioni savonesi e pisane: considerazioni preliminari sulle maioliche arcaiche", *Albisola* XXXII, p. 43-48.

Capelli C. 1999b, "Indagini archeometriche sulla protomaiolica ligure", *Albisola* XXXII, p. 73-86.

Capelli C., Cabella R., "The archaeometric study of white slips: a contribution to the characterisation of the Medieval Mediterranean productions", in these Proceedings.

Capelli C., Mannoni T. 2001, "Ricerche archeometriche per una caratterizzazione delle terre savonesi: le produzioni basso-medievali di Graffita arcaica tirrenica e Ingobbiata monocroma", *in* C. Varaldo (ed.), *Archeologia urbana a Savona: scavi e ricerche nel complesso monumentale del Priamàr. II, 2, Palazzo della Loggia*, Savona, p. 533-542.

Capelli C., Mannoni T. 2003, "Caratteristiche tipologiche ed archeometriche di un'area produttiva del XIII secolo da scoprire", *in Actes du VIIème colloque sur la céramique médiévale en Méditerranée, Thessaloniki 1999*, p. 115-124.

Capelli C., Marescotti P. 2000, "Caratterizzazione mineralogico-petrografica degli ingobbi delle ceramiche basso-medievali savonesi", *in Atti del Primo Congresso Nazionale dell'Associazione Italiana di Archeometria (AIAr)*, Verona 1999, p. 389-400.

Capelli C., Ramagli P., Ventura D. 1999, "Attrezzature fittili da infornamento dagli scavi di Savona e Albisola: treppiedi o distanziatori a zampa di gallo", *Albisola* XXXII, p. 57-72.

Capelli C., Gavagnin S., Gardini A., Mannoni T. 2001, "Ingobbiate monocrome di produzione locale e d'importazione a Genova tra XI e XIII secolo. Problemi tipologici e archeometrici", *Albisola* XXXIV, p. 25-35.

Capelli C., Gavagnin S., Vallauri L. 2002, "Nuovi ritrovamenti di ceramiche medievali con smalto su ingobbio a Genova e in Provenza", *Albisola* XXXV, p. 217-222.

Capelli C., Gardini A., Ramagli P. 2003, "Importazione e produzione locale di piastrelle con rivestimento vetrificato in Liguria tra XIV e XVI secolo: dati archeologici e archeometrici", *in Atti del III Congresso di archeologia medievale, Salerno 2003*, p. 649-658.

Capelli C., Cabella R., Riccardi M.P., Waksman Y. 2005, "Caratterizzazione archeometrica di ceramiche graffite medievali (Port Saint Symeon ware) rinvenute a Beirut, Genova e Marsiglia", *in* C. D'Amico (ed.), *Atti del III Congresso nazionale di archeometria (Bressanone 2004)*, p. 193-202.

Gardini A. 1997, "La protomaiolica a Genova e nella Liguria di Levante", *in Quaderni di archeologia medievale II*, Firenze, p. 75-84.

Hild F., Hellenkemper H. 1990. *Kilikien und Isaurien*, Tabula Imperii Byzantini 5, Wien.

Lane A. 1937, "Medieval finds at Al Mina in North Syria", *Archaeologia* 87, p. 19-78.

Lavagna R., Varaldo C. 1986, "La graffita arcaica tirrenica di produzione savonese alla luce degli scarti di fornace dei secoli XII e XIII", *Albisola* XIX, p. 119-130.

Lunardon G. 2001, "Protomaioliche savonesi", *in* C. Varaldo (ed.), *Archeologia urbana a Savona: scavi e ricerche nel complesso monumentale del Priamàr. II, 2, Palazzo della Loggia*, Bordighera-Savona, p. 199-205.

Mannoni T. 1975, "La ceramica medievale a Genova e nella Liguria", *Studi Genuensi VII* (1968-69), Bordighera.

Redford S., Ikram S., Parr E.M., Beach T. 2001, "Excavations at Medieval Kinet, Turkey: a preliminary report", *Ancient Near Eastern studies* 38, p. 58-138.

Riavez P. 2001, " 'Atlit: Ceramica Port St. Symeon 1217-1291, Graffite 'Crociate' del Mediterraneo orientale", *Archeologia medievale* XXVIII, p. 505-532.

Tite M.S., Freestone I., Mason R., Molera J., Vendrell-Saz M., Wood N. 1998, "Lead glazes in antiquity: Methods of production and reasons for use", *Archaeometry* 40 (2), p. 241-260.

Vallauri L., Démians d'Archimbaud G., Parent F., Richarté C. 2003, "La circulation des céramiques byzantines, chypriotes et du Levant chrétien en Provence, Languedoc et Corse du Xe au XIVe siècle", *in Actes du VIIe colloque sur la céramique médiévale en Méditerranée, Thessaloniki 1999*, p. 137-152.

Varaldo C. 1997a, "La graffita arcaica tirrenica", *in Actes du VIe colloque sur la céramique médiévale en Méditerranée, Aix-en -Provence 1995*, p. 439-452.

Varaldo C. 1997b, "La protomaiolica a Savona e nella Liguria di Ponente", *in Quaderni di Archeologia Medievale II*, Firenze, p. 63-74.

THE ARCHAEOMETRIC STUDY OF WHITE SLIPS: A CONTRIBUTION TO THE CHARACTERISATION OF THE MEDIEVAL MEDITERRANEAN PRODUCTIONS

C. CAPELLI, R. CABELLA

Dipartimento per lo Studio del Territorio e delle sue Risorse (DIP.TE.RIS.),
Università degli Studi di Genova, Corso Europa 26, 16132 Genova (Italy)

1. INTRODUCTION

The technique of applying a white slip between the ceramic body and the transparent glaze spread throughout the Mediterranean basin from its eastern parts (Byzantine areas) during the late Middle Ages (Berti *et al.* 1997). In most cases, it is linked to a double firing and represents an alternative to the tin-opacification of the glaze, which originated in the Islamic world, in order to obtain a white background masking the colour of the body and enhancing the effect of decorations (Tite *et al.* 1998).

Most of the archaeometric studies aiming at characterising ancient ceramics and localising the production centres are mainly based on body analysis. Instead, the white slips are generally little considered because of two main reasons: i) the submillimetric thickness, which creates great difficulty in sampling and prevents from an easy analytical approach by routine methods; ii) the relative homogeneity (in colour, grain size, texture, and mineralogical composition), which is thought a limiting factor for typological and provenance distinctions. Besides, XRD analyses are of little use in determining the nature of primary clay minerals, because their stability field is below the maximum temperature range usually reached in a double firing process.

However, a systematic thin section study carried out on several hundred samples of medieval glazed slipwares—archaeologically well characterised and coming from several Mediterranean production and consumption sites (mainly located in Provence, Liguria and Tuscany)—pointed to the existence of important variations even among white slips (Berti *et al.* 2002). Moreover, the integration between typological and analytical data revealed that particular white slip types are directly attributable to specific ceramic typologies, productions or production areas.

In this paper the main types of white slips that we recognised are discussed. In order to integrate the data resulting from optical microscopy, new analyses by electron microscopy (SEM-EDS) were carried out on a few selected samples. The attention was focused in particular on 12th-13th century Mediterranean productions.

2. TEXTURAL, MINERALOGICAL AND PETROGRAPHIC VARIABILITY OF WHITE SLIPS

In order to produce a white-coloured slip after firing, raw materials must contain a very low percentage of Fe-oxides/hydroxides and other chromophore elements. For that reason, it is commonly thought that most white slips are made from kaolinitic clays, even in absence of specific investigation. However, that does not appear to be always true: the clay matrix is generally prevalent, but we observed a range from very pure, inclusion-free to inclusion-rich slips. Clay-poor and even non-clay coatings exist as well. In this case, the common definition of slip as a "clay coating" cannot be used.

2.1. Clay-rich types

2.1.1. Fine-grained slips

This group is represented by coatings of good quality, compact and pure white in colour, characterised by fine homogeneous textures, with a microcrystalline matrix and scarce, fine-grained mineral inclusions (generally smaller than few ten microns). They are probably obtained by refining clay, often kaolinitic, raw materials in water.

A nearly inclusion-free slip is the distinctive feature of the Byzantine "Zeuxippus ware" found in Liguria (Fig. 1A), whose ceramic body is formed from a generic fine-grained fabric lacking in diagnostic inclusions (Capelli *et al.* 2002).

Relatively fine-grained slips are also present in the majority of the "Ligurian Protomajolica" (Fig. 1B) and other related monochrome glazed types (Capelli 2001; Capelli *et al.* in these Proceedings).

Finally, a peculiar fine-grained slip characterises the sgraffito "Aegean ware" (Fig. 1C) and other related monochrome glazed types found in Provence (unpublished data, work in progress in collaboration with L. Vallauri and co-workers of the LAMM-CNRS, Aix-en-Provence). The presence of fairly abundant red-coloured, very fine-grained aggregates, that reach greater dimensions than the non-plastic silicate inclusions, is the peculiarity of this coating.

2.1.2. Coarse-grained slips

The group consists of medium- or low-quality coatings, white to light buff or orange in colour, characterised by the presence of abundant, more or less coarse (up to some hundred microns) inclusions, generally composed of single mineral grains (mainly quartz, feldspar, and mica) and rarer acid (either metamorphic or igneous) rock fragments. Fine-grained Fe-oxide aggregates are often present, which are responsible for the colour on the macroscopic scale. Sometimes, the coarse-grained inclusions of the slip cause obvious irregularities on the external surface of the overlying glaze.

A lamellar texture is frequent, due to the parallel orientation of both non-plastic inclusions (mainly micas, tabular minerals and schistose rock fragments) and phyllosilicate-derived phases in the matrix, which are often birefringent and relatively coarse-grained (several microns in size). Textural and compositional evidences point to the use of poorly refined raw materials, which could also be non-kaolinitic clays.

The features described above could both affect the final look of the glaze and favour its detachment, especially in post-depositional stages.

An example of coarse-grained slip (Fig. 1D) can be found in most of the medieval productions of Savona (the "Graffita arcaica tirrenica", or "Savona archaic sgraffito", and other monochrome or polychrome glazed types; see Capelli *et al.* in these Proceedings). In this case, the non-plastic inclusions (mainly quartz-schist fragments and quartz and mica grains) are of metamorphic origin. The petrographic composition and the presence of the same type of slip in different wares over a large span of time (a few centuries) lead us to hypothesise a local provenance for the raw materials. They could come from limited outcrops of illitic-montmorillonitic sediments deriving from the alteration of Palaeozoic acid metamorphic rocks which outcrop in the Savona-Finale area (Capelli and Marescotti 2000).

A small assemblage of "archaic sgraffito" and monochrome glazed wares, not exactly matching Savona wares in typological features and body compositions, was recently discovered in Finale Ligure (30 km south-west from Savona; Capelli *et al.* 2006). The hypothesis of a local imitation of Savona products was supported by the striking similarities in the characteristics of the slips. For this component, a common source of raw materials should be considered.

2.2. Clay-poor, inclusion-rich types

2.2.1. Slips with polymineralic inclusions

These are coatings of medium or low quality, often light orange in colour because of several Fe-oxide micro-aggregates dispersed in the matrix. Non-plastic inclusions predominate on the clay matrix, even when they are fine-grained. Their mineralogical-petrographic composition is heterogeneous: individual grains of quartz, mica, feldspar, amphibole and other minerals, together with rarer rock fragments, are observed.

A slip of this type (Fig. 1E) characterises the production made up of the so-called "Glazed slip ware with green splashed decorations" and related monochrome types found in Northern Italy, whose provenance from the Aegean-western Anatolian region is assumed on the basis of the distribution data and the presence of both acid metamorphic (dominant) and volcanic (rarer) rock fragments in the ceramic body (Berti and Capelli 2001; Capelli *et al.* 2002). The origin of the raw materials of this slip and their processing are not clear: either superficial layers of weathered acid rocks or artificially crushed rocks (or sediments) could have been used, with a possible sieving step to eliminate the coarser granulometric fraction.

2.2.2. Slips with prevalent quartz inclusions

The coatings of this group are of medium or high quality and generally a pure white colour. They are mainly composed of abundant, angular, coarse-grained fragments of quartz and subordinate K-feldspar with a scarce clay matrix. Textural features point to the use of ground quartz-rich rocks or sands as raw materials, mixed with low percentages of clay. However, the possibility of the application of a "frit" in some cases cannot be completely excluded.

A quartz-rich slip is characteristic of a group of samples found in Liguria, Provence and Beirut attributed to the "Port Saint Symeon ware" (Fig. 1F), whose origin is probably to be located in the region of Antioch (Capelli *et al.* 2005; Capelli *et al.* in these Proceedings). In this pottery, the slip matrix is frequently vitrified (Fig. 1G) and the mineral grains sometimes show rounded rims because of the reaction with the Pb-rich glaze. The peculiar compositional and textural features of the slip considerably increase the brilliance of the glaze.

Quartz-rich slips are also described in the late Byzantine pottery from Sardis (Scott and Kamilli 1981), in 10th-11th century lead-glazed slipwares from Akhsiket, Uzbekistan (Henshaw *et al.* in these Proceedings) and in polychrome sgraffito wares of the Mamelouk period found in Harim, Syria (Gelichi 2003).

Finally, we can find more recent examples in the Iznik production, started in the 15th century and characterised by a "stonepaste" body (Paynter *et al.* 2004; Tite 1989). Quartz-rich slips are present in the early production (Colomban *et al.* 2004) as well as in the modern one. Present-day potters of Kütahya, near Iznik, make the slip by mixing ground quartz with a subordinate quantity of clay (Atasoy and Raby 1989).

3. CHEMICAL COMPOSITIONS OF WHITE SLIPS

Preliminary SEM-EDS analyses were carried out on 14 white slip samples of the types mentioned above. The electron beam was focused on rectangular areas a few tens of microns wide, excluding the parts closest to the glaze and

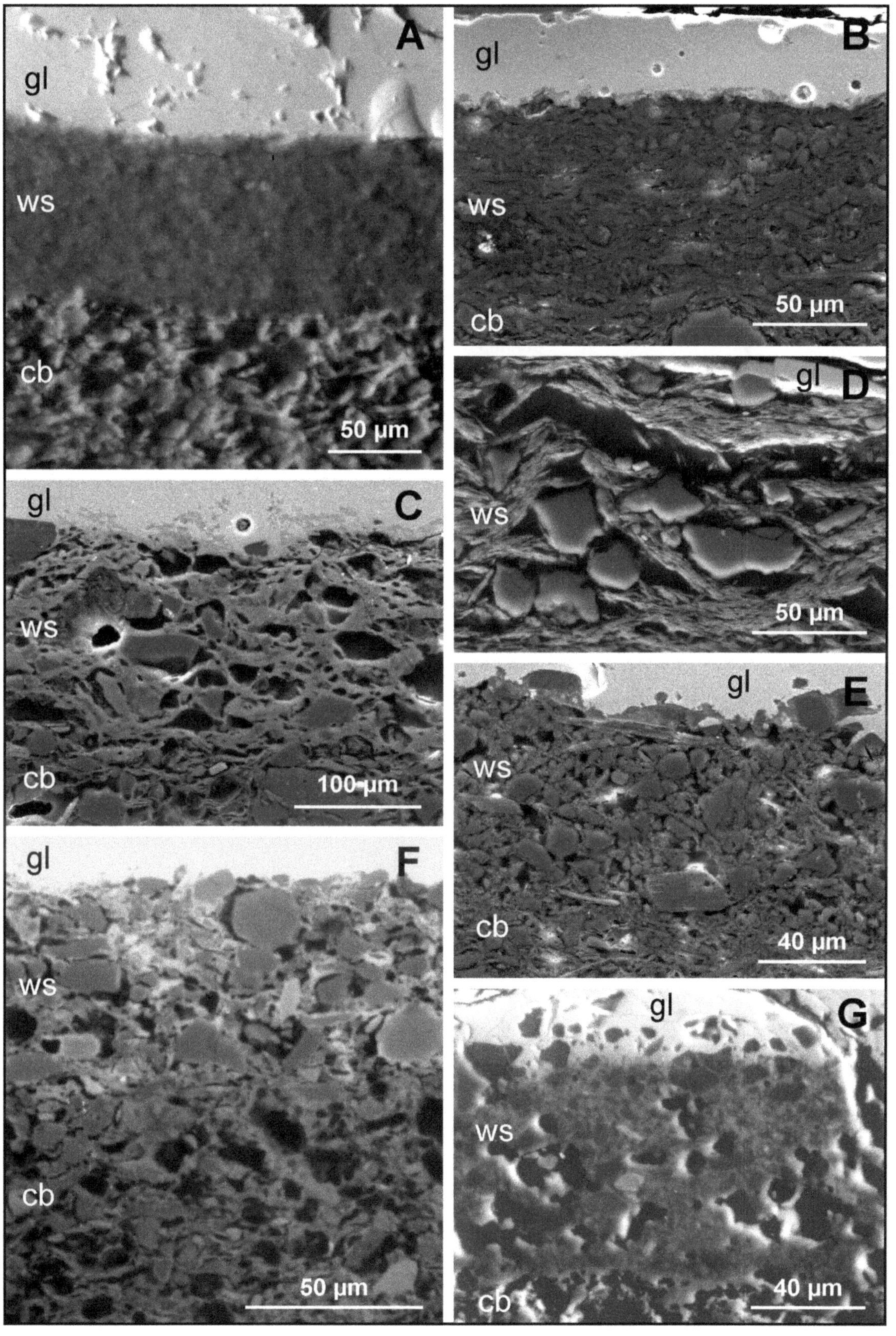

Fig. 1 – SEM images of white slips representative of the types discussed in the text. Fine-grained slips, Clay-rich types: A (Zeuxippus ware), B (Ligurian Protomajolica), C (Aegean ware); Coarse-grained slips, Clay-rich types: D (Savona archaic sgraffito); Slips with polymineralic inclusions, Clay-poor, inclusion-rich types: E (Glazed slip ware with green splashed decorations); Slips with prevailing quartz inclusions, inclusion-rich types: E, F (Port Saint Symeon ware); cb: ceramic body; gl: glaze; ws: white slip.

the ceramic body. Two or three analyses (whenever possible) were carried out for each sample and then averaged. Finally, the resulting data were normalised to 100%.

The possibility of a large imprecision in the results, due to analytical reasons, textural and granulometric heterogeneities and interaction processes with the glazes must be taken into account.

Lead contamination from the glaze is generally evident throughout the whole slip thickness: PbO average ranges between 3 and 9 wt% in most cases, but even higher values were also observed (mainly in Savona wares); it is less than 1 wt% in only two cases ("Zeuxippus ware" and "Ligurian Protomajolica" samples).

In general, the analysed slips are essentially composed of Si, Al and K, agreeing with the mineralogical composition, which is rich in quartz and K-Al-rich silicates. The large variability (SiO_2: 40-80 wt%, Al_2O_3: 5-35 wt%, K_2O: 2-8 wt%) is due to the different relative ratios of the main mineral components. Some chemical sub-groups, in good agreement with the types differenciated under the microscope, can be recognised (Fig. 2A, B).

The samples of "Clay-poor, inclusion-rich types" can be easily distinguished from those of the "Clay-rich types" by their higher Si contents (Fig. 2A) and lower Al contents (Fig. 2B).

The slips of Savona and Finale wares ("Coarse-grained slips, Clay-rich types") are similar in chemical composition as well and form a separate cluster (Fig. 2B). The high Al and K values are due to the numerous K-feldspar and white mica inclusions, but also a primary illitic composition of the clay matrix cannot be excluded. In the same way, the low K and high Al values (Fig. 2B) peculiar to the "Zeuxippus ware" sample ("Fine-grained slips, Clay-rich types") could be due to use of kaolinitic raw materials.

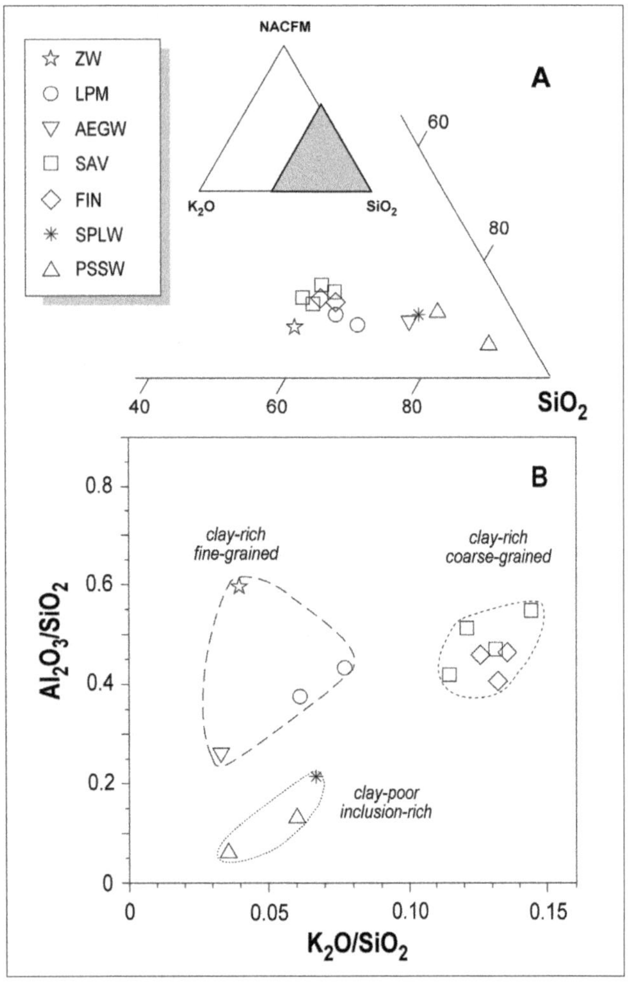

Fig. 2 – Chemical composition of white slips by SEM-EDS analyses. A: ternary diagram K_2O-SiO_2-$Na_2O+Al_2O_3+CaO+FeO+MgO$. B: binary plot K_2O/SiO_2-Al_2O_3/SiO_2. ZW: Zeuxippus ware; LPM: Ligurian Protomajolica; AEGW: Aegean ware; SAV: Savona wares; FIN: Finale wares; SPLW: Glazed slip ware with green splashed decorations; PSSW: Port Saint Symeon ware.

4. GLAZE-WHITE SLIP INTERACTIONS

Since body and white slip are generally fired before the application of the glazing mixture (at least in the cases studied), rather limited reaction processes, which are important to bind the two cover layers together, occur between glaze and slip during the second firing. The diffusion of Pb from the glaze, discussed above, may induce partial vitrification of the slip (Fig. 1G), essentially due to the melting of clay minerals in the matrix and of K-feldspars in the temper. These processes lead to a diffusion of Si, Al and K towards the glaze, as well as to the growth of microcrystalline K-Pb feldspars (Molera *et al.* 2001) at the glaze-slip interface (Fig. 1C). The lack of Ca-rich reaction phases, such as clinopyroxenes, at the glaze base is due to the Ca-poor composition of white slips. The thickness of the glaze-slip interaction layer is variable (up to a few hundred microns) depending on the glaze composition and firing conditions.

5. CONCLUSIONS

The analyses carried out by optical microscopy and SEM-EDS showed the existence of different white slip types in medieval productions. The relatively large variability in chemical, mineralogical and petrographic compositions, as well as in textural and granulometric features, depends on both the nature of the raw materials and the production techniques.

Clay-rich slips derive from the processing of more or less refined (in water or by sieving) clay sediments, which are not necessarily kaolinitic, but poor in chromophore elements or impurities. Instead, most of the clay-poor, quartz-rich slips were obtained from quartz-rich raw materials (rocks, pebbles or sands), which were crushed, sieved and probably mixed with small quantities of clay.

While clay-rich slips are found in Italian as well as in eastern medieval productions, the technological area related to inclusion-rich, clay-poor slips seems to be strictly linked to the Eastern Mediterranean, some evidence pointing to a possible origin of the quartz-rich slips from more eastern areas (see for example: Henshaw *et al.* in these Proceedings). Moreover, the possibility of a relationship between the quartz-rich slip technology and that of the "stonepaste" (Mason and Tite 1994) should be considered.

In conclusion, the first results of our research showed that a careful study of white slips could become a powerful tool to improve knowledge of medieval pottery, in the framework of an integrated approach together with body analysis and archaeological study. The characterisation of slips could strengthen typological distinctions, lead to recognition of the complexity of productions composed of different ceramic types (also monochrome wares, generally disregarded by archaeologists), contribute to the reconstruction of the routes of technical knowledge, and even give important information about provenance.

REFERENCES

ATASOY N., RABY J. 1989, *Iznik: The pottery of Ottoman Turkey*, London.

BERTI G., CAPELLI C. 2001, "Considerazioni e analisi sulle ceramiche del tipo dei bacini bizantini di S. Nicolò di Albisola", *Albisola* XXXIII, p. 281-288.

BERTI G., GELICHI S., MANNONI T. 1997, "Trasformazioni tecnologiche nelle prime produzioni italiane con rivestimenti vetrificati (secc. XII-XIII)", *in Actes du VIème colloque sur la céramique médiévale en Méditerranée, Aix-en-Provence 1995*, p. 383-402.

BERTI G., CAPELLI C., MANNONI T. 2002, "Ingobbio/ingobbi e gli altri rivestimenti nei percorsi delle conoscenze tecniche medievali", *Albisola* XXXIV, p. 9-15.

CAPELLI C. 2001, "Indagini archeometriche sulla protomaiolica ligure", *Albisola* XXXII, p. 73-86.

CAPELLI C., MARESCOTTI P. 2000, "Caratterizzazione mineralogico-petrografica degli ingobbi delle ceramiche basso-medievali savonesi", *in Atti I Congresso nazionale AIAr (Verona 1999)*, p. 389-400.

CAPELLI C., GAVAGNIN S., GARDINI A., MANNONI T. 2002, "Ingobbiate monocrome di produzione locale e d'importazione a Genova tra XI e XIII secolo. Problemi tipologici e archeometrici", *Albisola* XXXIV, p. 25-35.

CAPELLI C., CABELLA R., RICCARDI M.P., WAKSMAN Y. 2005, "Caratterizzazione archeometrica di ceramiche graffite medievali (Port Saint Symeon ware) rinvenute a Beirut, Genova e Marsiglia", *in* C. D'Amico (ed.), *Atti del III Congresso nazionale di archeometria (Bressanone 2004)*, p. 193-202.

CAPELLI C., PALAZZI P., PARODI L., CABELLA R., PIAZZA M., WAKSMAN Y. 2006, "Una nuova produzione di Graffita arcaica tirrenica? Caratterizzazione archeologica e archeometrica di ceramiche ingobbiate da Finale Ligure (SV)", *Albisola* XXXVII, p. 113-127.

CAPELLI C., CABELLA R., WAKSMAN Y., "Archaeometric investigation on 13th century glazed and slipped pottery found in Liguria and Provence", *in* these Proceedings.

COLOMBAN P., MILANDE V., LE BIHAN L. 2004, "On-site Raman analysis of Iznik pottery glazes and pigments", *Journal of Raman spectroscopy* 35 (7), p. 527-535.

GELICHI S. 2003, "Il castello di Harim: un sito fortificato tra musulmani e crociati nella Siria del nord", *Archeologia medievale* XXX, p. 431-452.

HENSHAW C., REHREN T., PAPACHRISTOU O., ANARBAEV A., "Lead-glazed slipware of 10th-11th century Akhsiket, Uzbekistan", *in* these Proceedings.

MASON R.B., TITE M.S. 1994, "The beginnings of Islamic stonepaste technology", *Archaeometry* 36 (1), p. 77-91.

MOLERA J., PRADELL T., SALVADO N., VENDRELL-SAZ M. 2001, "Interactions between clay bodies and lead glazes", *Journal of the American ceramic society* 84, p. 1120-1128.

PAYNTER S., OKYAR F., WOLF S., TITE M.S. 2004, "The production technology of Iznik pottery: a reassessment", *Archaeometry* 46 (3), p. 421-437.

SCOTT J.A., KAMILLI D. 1981, "Late byzantine glazed pottery from Sardis", *in Actes du XVe congrès international des études byzantines (Athens 1976)*, vol. II, p. 679-696.

TITE M.S. 1989, "Iznik pottery, an investigation of the methods of production", *Archaeometry* 31 (2), p. 115-132.

TITE M.S., FREESTONE I., MASON R., MOLERA J., VENDRELL-SAZ M., WOOD N. 1998, "Lead glazes in Antiquity: methods of production and reasons for use", *Archaeometry* 40 (2), p. 241-260.

FROM FURNACE TO CASTING MOULDS: AN EXCEPTIONAL 14th CENTURY COPPER-METALLURGY WORKSHOP STUDIED IN THE LIGHT OF REFRACTORY CERAMIC MATERIALS

ILDIKO KATONA[1], DAVID BOURGARIT[1], NICOLAS THOMAS[2], ANNE BOUQUILLON[1]

[1]Centre de recherche et de restauration des musées de France (C2RMF), CNRS UMR171, 14 quai François Mitterrand, 75001 Paris, France.

[2]Institut national de recherches archéologiques préventives (INRAP); Équipe d'histoire des techniques, Université Paris 1 Panthéon-Sorbonne - CNRS, UMR 8589, LAMOP.

ABSTRACT

A 14th century metallurgy workshop mostly dedicated to the production of small copper alloy jewellery items and metal vessels has recently been excavated in the centre of Paris. This workshop is unique in France firstly because copper metallurgy as a whole, and copper-alloy objects production in particular, are dramatically under-documented for this period. Secondly, an unusually comprehensive corpus of ceramic materials—called "refractories" in archaeological usage—has come to light. The present study aims to discuss the production and workshop organisation from the refractory point of view. Within the framework of a French national research program with support from the Ministry of Culture, petrographic and chemical investigations have been carried out on a large corpus of refractories.

Despite the wide range of states of use encountered, from almost unused ceramics to highly vitrified material, well-defined groups can be identified based upon clay composition and the distribution and nature (organic and mineral) of the temper. This grouping allows us to assign specific refractory types to the different types of artefact. Based on this, the skill of the coppersmith in handling ceramic materials is discussed, as well as the provenance of raw materials and manufactured items. Two main groups can be distinguished. Firstly, some items were manufactured at the copper workshop, which is to say the furnace clay and the moulds, for which high quantities of clay were required, which probably came from the vicinity of the site. Secondly, the large melting crucibles seem to have been manufactured elsewhere, in a specialized workshop. Recent archaeological excavations 60 km north of Paris have brought to light a site of similar date involved in refractory ceramic production: the crucible typology and clay are very similar to that of the Parisian items.

Fig. 1 – The melting furnace (N. THOMAS, © INRAP).

1. INTRODUCTION

In 2003, the "Institut national de recherches archéologiques préventives" excavated the main courtyard of a 17th c. building, the "Hôtel de Mongelas", in the centre of Paris before restoration (Thomas and Bourgarit 2006). The site is located at 62, rue des Archives on the right bank of the River Seine, in an area which is just outside the 13th c. fortifications built by Philippe August. The urban development of this area began at the end of the 13th c., initially due to the Templars, and continued during the entire 14th c. On the excavated site, a metallurgical workshop specialising in the processing of copper alloys appeared before the middle of the 14th c. Despite the small area excavated, the archaeological remains of this industrial activity are varied and abundant. The main structures are the cellar of the workshop house and a melting furnace. The artefacts comprise fragments of refractory ceramics, copper alloy debris (sheet fragments and melting waste), small copper jewellery items, and various other non-metallic materials (such as small glass cabochons). The most noteworthy find is a large melting furnace (70 cm internal diameter), of which the brick structure was quite well preserved, presenting the complete lower working platform, the so-called ash bin, as well as the first layers of the building (Fig. 1). Five types of refractory artefacts can be distinguished, dedicated either to melting, or to casting.

Numerous bricks (1) and lining fragments (2) coming from the upper part of the furnace, as well as refractories attributed to the drilled hearth (sole, 3) were found in the furnace pit. Around 16 kg of crucible fragments (4) were recovered. The crucibles are large—20 cm diameter, 25 cm height—with a typical flattened cone shape (Fig. 2). Some crucible sherds are extremely vitrified, while others seem to be little or not used. Accidents such as thermal shocks are clearly visible on some fragments. The mould fragments (5) are very abundant in all the 14th c. layers (85 kg). They all exhibit a red oxidised outer surface and a black reduced inner face, as is usually observed on clay moulds after use. On approximately 30% of these fragments a negative trace of the cast item is to be seen (Fig. 3). However, the complete form of the casts is not always easy to determine due to the high fragmentation rate .

A French kiln-site, Lacroix-Saint-Ouen (Oise, *circa* 60 km north of Paris) dated to the beginning of the 14th c. has also been excavated (Cartier and Durand 1990; Durand 2004). The majority of the ceramic assemblage is made up of crucible and common-ware ceramic sherds. The forms, sizes and appearance of the clay are very similar to the crucibles found at the "Hôtel de Mongelas". A common ware ceramic, a crucible, and a tuyere from this kiln-site were sampled for comparative analytical investigation.

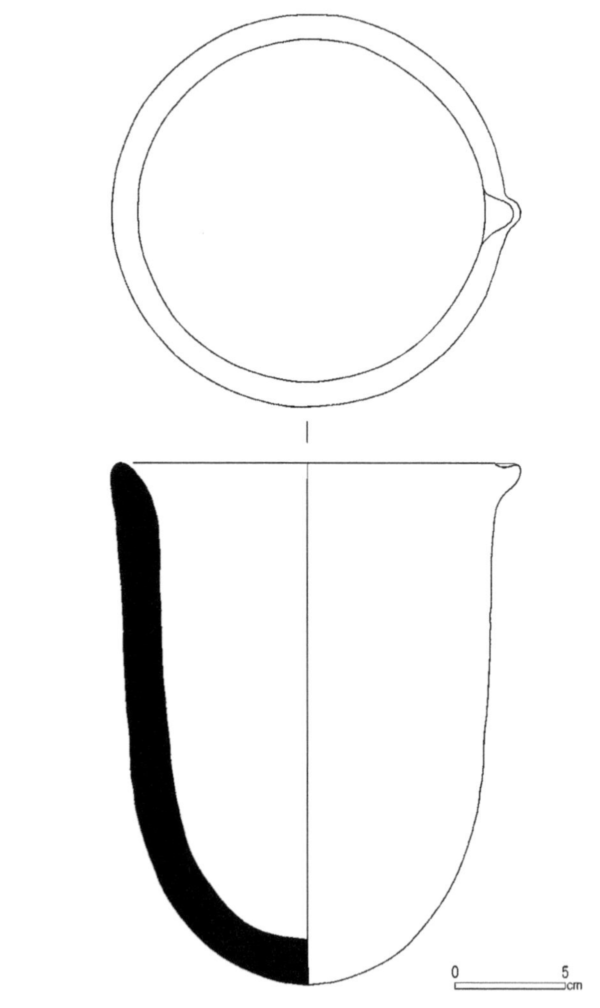

Fig. 2 – Crucible reconstruction (N. THOMAS, © INRAP).

2. METHODOLOGY

27 ceramic sherds representing the different types of refractory materials at the "Hôtel de Mongelas" were sampled for analytical investigation after careful macroscopic observation: 6 furnace lining fragments, 6 pieces of mould debris, 12 crucible sherds, and 3 furnace brick fragments were selected. The three samples from Lacroix-Saint-Ouen were prepared and investigated at the same time. Thick and thin sections using epoxy resin were prepared for petrographic analysis. Bulk chemical composition was determined by Particle Induced X-ray Emission (PIXE) at the AGLAE facility of the C2RMF (Dran *et al.* 2004): the sherds were sampled with a tungsten carbide micro-drill, the resulting 100 mg powder was ground and homogenised in an agate mortar before forming into pellets. The resulting 5 mm diameter pellets were submitted to a 3 MeV proton microbeam, with scanning over a 1 by 1 square mm area for homogenising. Each spectrum recorded was processed with the GUPIX data reduction package (Campbell *et al.* 2000). It was thoroughly checked on samples of standard materials (DR-N from the CRPG, Brill A and Brill B glasses from Corning Museum).

3. RESULTS

3.1. Moulds and furnace lining

Macroscopic examination of the mould sections shows a fine sandy clay, exhibiting both a reddish and a black coloured area due to variation in redox conditions from the internal to the external face. The vegetal impressions of an abundant organic temper which volatilised can be observed, generating a very porous body. Petrographic investigations reveal a fine-grained texture (Fig. 4a) with numerous mineral inclusions (50 vol%), mainly fine quartz grains (50-75 μm) distributed homogeneously. A few K-feldspars, a few grains of mica, as well as iron oxide patches are also present. The non-plastic inclusions are often angular.

The chemical compositions of the 6 analysed moulds show some variation with 80-85 wt% SiO_2 and 8-13 wt% Al_2O_3 (Fig. 5 and Table 1). K_2O contents are quite high (around 2 wt%), due to the K-feldspars and the micas. Traces of Zn and Cu were measured (up to 0.5 wt%), indicating contamination during use or burial.

The 6 furnace-lining samples studied show similar textures and chemical compositions (Fig. 4b and Fig. 5). Only one feature varies when compared to the mould paste: the vegetal temper is clearly coarser for the furnace lining.

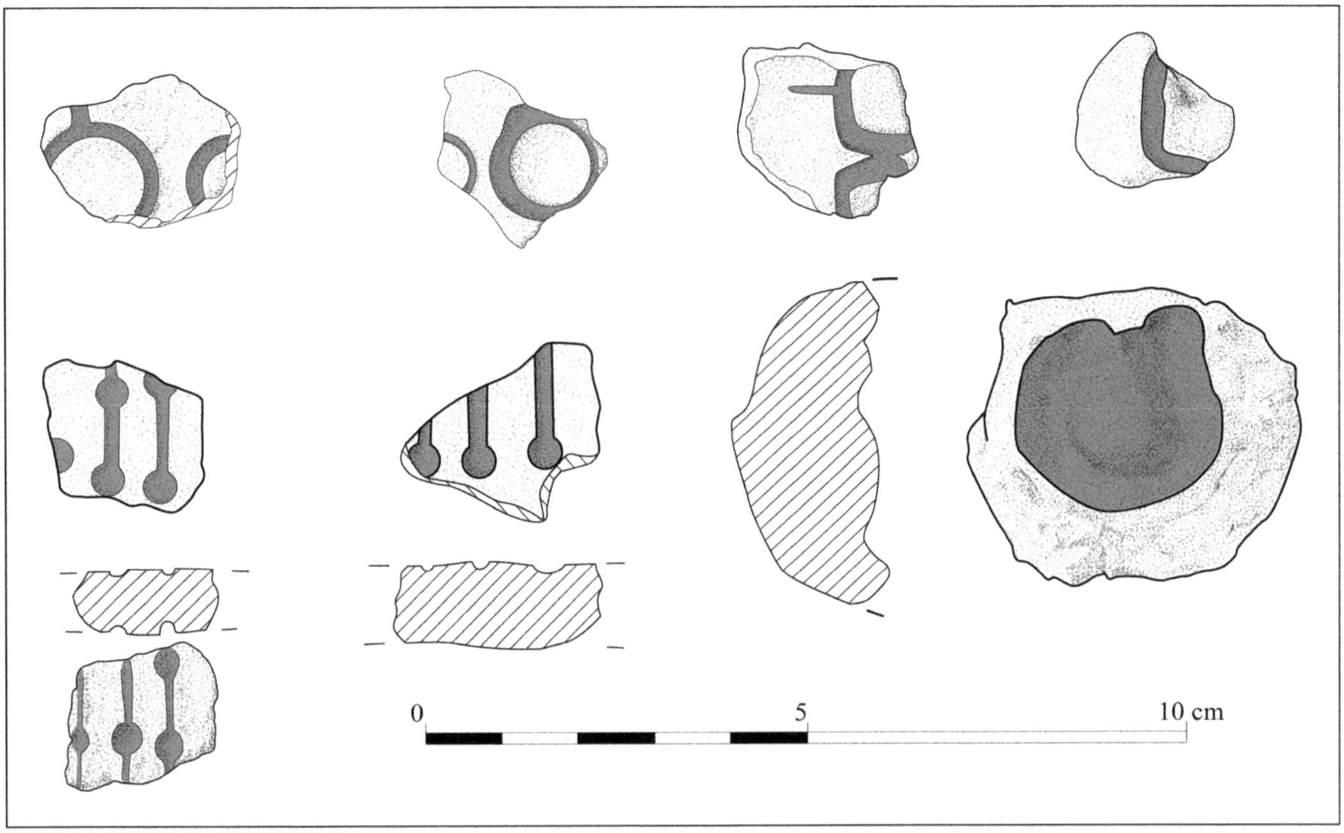

Fig. 3 – Mould fragments (C. Micouin, © INRAP).

	Moulds		Furnace clay		Q-tempered crucibles		Lacroix Saint-Ouen		Grog-tempered crucibles		Bricks	
	m	σ	m	σ	m	σ	m	σ	m	σ	m	σ
SiO_2	80,8	*2,7*	81,5	*3,3*	79,2	*1,2*	77,8	*2,0*	69,4	*1,9*	61,0	*0,7*
TiO_2	0,6	*0,2*	0,5	*0,1*	0,9	*0,1*	1,0	*0,1*	1,3	*0,2*	1,1	*0,3*
Al_2O_3	10,9	*1,8*	10,5	*2,0*	15,4	*1,1*	16,5	*2,8*	17,6	*1,4*	17,2	*2,7*
Fe_2O_3	3,2	*0,4*	2,9	*0,1*	1,9	*0,4*	2,0	*0,4*	2,5	*0,2*	3,2	*0,4*
MnO	< dl		< dl		< dl		< dl		0,1		0,2	*0,1*
MgO	0,9	*0,2*	0,8	*0,3*	0,4	*0,1*	0,3		0,9	*0,2*	1,4	*0,2*
CaO	0,5		0,5	*0,1*	0,8	*0,4*	1,5	*0,5*	1,8	*0,3*	3,4	*0,4*
Na_2O	0,4	*0,3*	0,4	*0,3*	0,3	*0,2*	0,1	*0,1*	1,5	*0,7*	2,1	*0,1*
K_2O	1,9	*0,3*	1,9	*0,4*	0,4	*0,2*	0,4	*0,2*	1,0	*0,1*	1,3	*0,3*
P_2O_5	< dl		< dl		< dl		< dl		< dl		0,2	*0,1*
SO_3	0,1	*0,1*	0,1	*0,1*	0,2	*0,2*	0,1	*0,1*	0,2	*0,1*	1,6	*1,0*
CuO	0,1	*0,1*	0,2	*0,2*	< dl		< dl		0,8	*0,3*	2,4	*0,4*
ZnO	0,1	*0,1*	0,2	*0,1*	0,3	*0,3*	< dl		2,0	*1,2*	2,8	*0,2*
SnO_2	< dl		< dl		< dl		< dl		0,1	*0,1*	0,4	*0,4*
Sb_2O_5	< dl		< dl		< dl		< dl		< dl		< dl	
PbO	0,1	*0,1*	0,1	*0,1*	< dl		< dl		0,1	*0,1*	1,2	*1,4*

Table 1 – Bulk elemental compositions of the studied ceramics measured by PIXE (oxide wt%). m = mean value, σ = standard deviation, dl = detection limits.

3.2. Crucibles

Two groups of crucibles can be distinguished by the nature of the temper, either quartz or grog. In the following discussion, these samples will be called quartz and grog-tempered crucibles respectively. Note that both mineralogical textures and chemical compositions are different.

Nine of the studied crucible samples are tempered with quartz. Two of them are slightly vitrified on the external surfaces and exhibit a grey colour, whereas the seven other samples show very little alteration and are orange in colour. A fine-grained clay matrix moderately tempered (20-30 vol%) with homogeneously scattered rounded quartz grains (125-150 μm) is observed in all crucibles (Fig. 4c). The chemical compositions show very

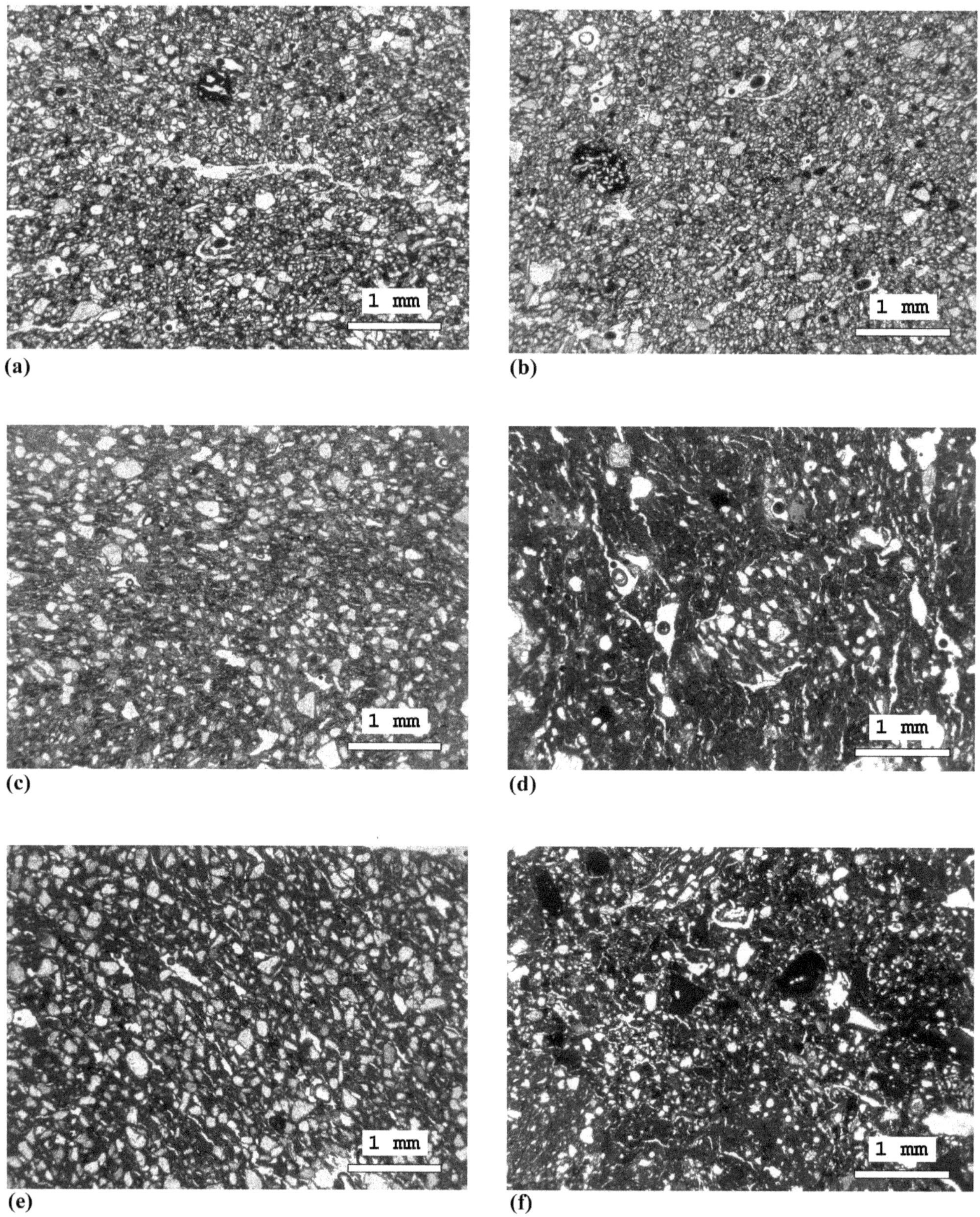

Fig. 4 – Photomicrographs showing the petrographical textures (a) Mould, (b) Furnace lining, (c) Quartz-tempered crucible, (d) Grog-tempered crucible, (e) Crucible from the Lacroix-Saint-Ouen kiln site, (f) Brick (Transmitted plane-polarized light).

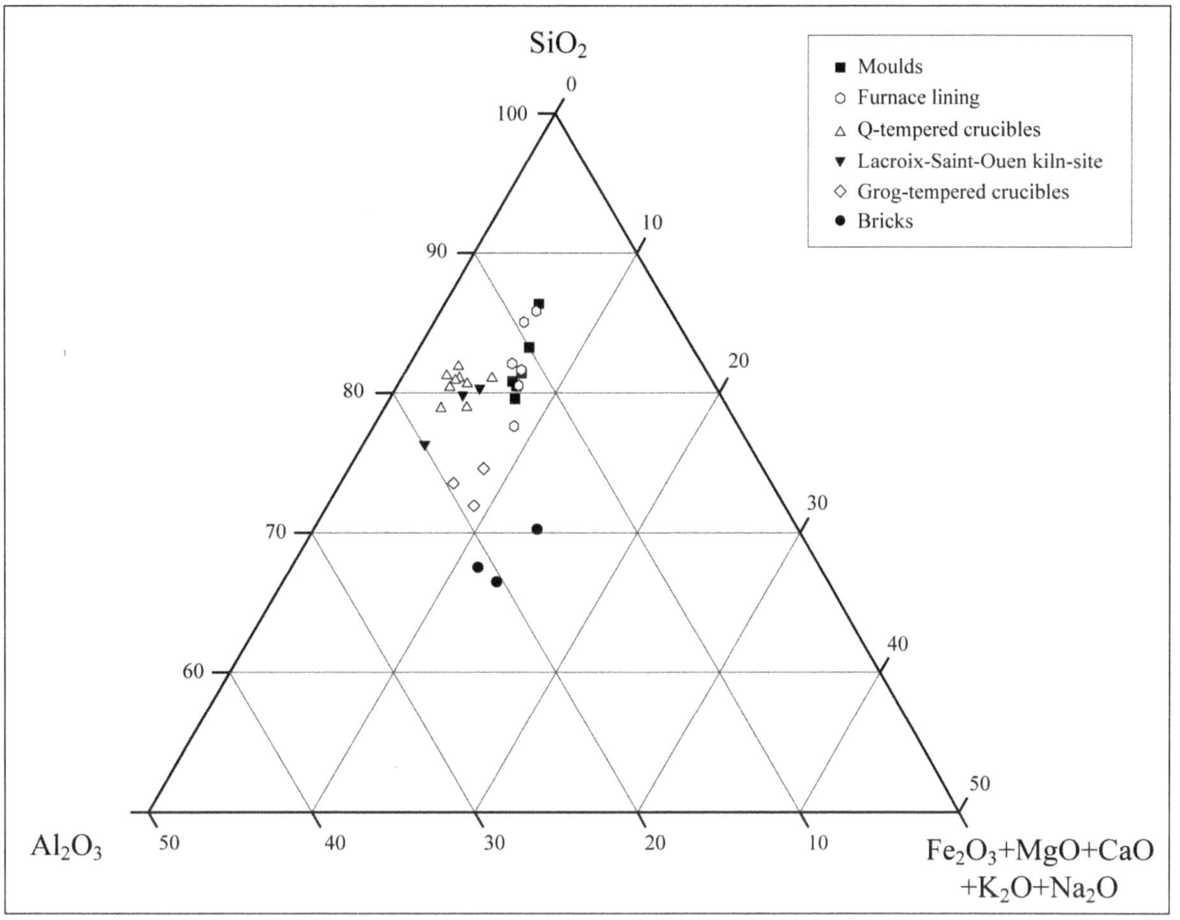

Fig. 5 – Bulk elemental compositions of the studied ceramics measured by PIXE (oxide wt%), plotted on a ternary diagram SiO_2–Al_2O_3–(Fe_2O_3+MgO+CaO+K_2O+Na_2O).

few discordant results (Fig. 5), with high silica contents (80 wt% SiO_2), 15 wt% Al_2O_3, and low calcium and alkali concentrations(CaO+Na_2O+K_2O < 2 wt%). The Zn contents are similar to those measured in the moulds (around 0.3 wt%, Table 1).

The other three crucible samples studied are tempered with grog (Fig. 4d). They are all highly vitrified and exhibit a grey to reddish colour. The main temper used is comprised of large rounded grog inclusions (500–2000 μm) distributed homogeneously (10 vol%). Grog inclusions show mineralogical textures similar to those of the previous quartz-tempered crucibles. A few small quartz grains (100–125 μm and 5 vol%) are also observed as non-plastic inclusions. Silica contents are less than for the quartz-tempered crucibles (*ca.* 70 wt% SiO_2, Fig. 5 and Table 1), while Al_2O_3 concentrations are slightly higher (*ca.* 17 wt% Al_2O_3). Calcium and alkali concentrations (CaO+Na_2O+K_2O = 3-5 wt%) as well as Fe, Mn, and Mg contents are much higher. In addition, remarkably high amounts of Cu and Zn have been measured (up to 3 wt% ZnO).

The pastes of the three samples from the Lacroix-Saint-Ouen site are very similar from both a petrographic and chemical point of view to the paste of the quartz-tempered crucibles: further analyses are necessary to confirm these observations (Fig. 4e and Table 1). In this particular case it can be seen that preparation for the refractory clay by the potter is non-specific.

3.3. BRICKS

The paste of the three brick fragments studied stands out in its general appearance. The mineralogical texture is extremely heterogeneous, with the main temper being coarse grog together with a relatively small amount (15 vol%) of vitrified grog inclusions (2000–3000 μm). Very small quantities (less than 5 vol%) of small-medium quartz grains (*ca.* 75 μm) can also be seen (Fig. 4f). The bulk chemical composition is also very different from the crucible and mould compositions (Fig. 5 and Table 1). Silica contents are much less (*ca.* 60 wt% SiO_2), whereas the proportion of alumina is noticeably higher (*ca.* 20 wt% Al_2O_3). Alkalis are also very abundant, especially Na and Ca (CaO+Na_2O+K_2O > 6 wt%), as well as iron (*ca.* 3 wt% Fe_2O_3), sulphur (*ca.* 2 wt% SO_3) and phosphorous. Note also the high amounts of copper and zinc (*ca.* 5 wt%), as well as tin (up to 1 wt%), partly due to the metallic inclusions which are frequently observed in the grog.

4. DISCUSSION AND CONCLUSION

Most of the clay used at the site is non-refractory by modern criteria, as can be seen by the relatively low alumina contents (around 10–15 wt%). This phenomenon is frequently observed in the ceramics used at high temperatures in ancient times, including the medieval period (Freestone 1989): suitable tempering is required to provide the general and refractory properties needed. In the late-medieval copper-workshop at the "Hôtel de Mongelas", the specificity of the temper for each particular refractory use, as well as the homogeneity of the ceramic compositions for each group, point to well-defined paste recipes. Leaving aside the furnace bricks, two main recipes can be distinguished.

Firstly, both mineral and vegetal tempers are encountered in the moulds, with common compositions whatever the size of the object being cast. The high proportion of fine-quartz-grain (50 vol%), together with the fibrous vegetal temper provide a high level of toughness and thermal shock resistance (Tite *et al.* 2001), capable of withstanding the impact of the melted copper or copper-alloy (around 1100 °C), whatever the pre-heating of the mould. In addition, the vegetal temper generates the high open porosity required for the gas to be released during casting. Note that the high proportion of quartz increases thermal contraction of the body during cooling (Hamer and Hamer 1997) and thus may be deleterious to the dimensional stability. Two layers of clay can sometimes be observed macroscopically. Yet, the clays are both the same in the two layers and consequently do not suggest a special inner mould lining as encountered elsewhere (Freestone 1989): here, the temper and clay matrix are sufficiently fine-grained to ensure the necessary smoothness for a cast surface of good quality.

Secondly, a relatively fine-grained mineral temper distributed in a homogeneous manner (20-30 vol%), enables the crucible paste to withstand relatively big thermo-mechanical stresses, such as those generated during copper and copper alloy melting. With regard to thermal stresses, the crucibles are expected to resist both high temperatures of up to more than 1100 °C over a relatively long period, probably about half an hour according to our initial experiments, and subsequent air quenching at the pouring stage. In relation to mechanical stress, one should bear in mind that these large crucibles were used to hold heavy loads (some 4-5 kg of copper alloy), and that they had to be handled several times, particularly when removing the crucibles from the furnace while still very hot. Incidentally, two different types of mineral temper have been observed in the crucibles, namely quartz grain or grog, depending on the artefact. Such a bimodal recipe remains unexplained. It may refer to a specific use of the crucible: the type of alloy melted, melting/alloying, or brass making processes… Or it may relate to a change in supply.

The supply of refractories at the "Hôtel de Mongelas" site is primarily of two types, depending on the skill required to produce the ceramics. Moulding clearly belongs to the copper-worker's activity and craft: moulds are manufactured at the metallurgical workshop. This is further demonstrated by the similarity of the clays used for the moulds and the furnace lining. Only the size of the vegetal temper varies, which lends support to the assumption that the vegetal temper was added at the workshop. The clay suppliers are probably to be found in the vicinity of the site, as suggested both by the huge quantities consumed and the relatively common appearance of the clay. On-going studies on 16th c. clay do not refute this hypothesis, and may help to locate the clay source more precisely (Ravoire *et al.* 2004). Conversely, the crucibles are of relatively large dimensions and are made by wheel throwing. This requires a specialised skill to be found in the ceramist sphere. Crucibles of this kind were traded therefore. Archaeological excavations have revealed several kiln-sites, dating from the 14th c., which produced crucibles. The kiln-site of Lacroix-Saint-Ouen, contemporaneous with the "Hôtel de Mongelas" site, has already been mentioned for its production of crucibles which are typologically similar to those from the "Hôtel de Mongelas". In Kingston, Surrey, another kiln-site producing melting crucibles has also been excavated recently (Miller and Stephenson 1999). Moreover, the present study has shown that the paste compositions of the quartz-tempered crucibles from the "Hôtel de Mongelas" are similar to the Lacroix-Saint-Ouen production: the quartz-tempered crucibles may have been supplied by the latter site, located 60 km north of Paris. This hypothesis could be confirmed by further research. As far as the grog-tempered crucibles are concerned, the question of supply remains largely unaddressed. That being said, we should point out that high quantities of crushed ceramics, macroscopically similar to the crucibles, have been found in all the sieving residues from the workshop layers. The high mechanical strength of the ceramics clearly proves that this debris came from deliberately crushed crucibles. A sediment study has shown that the average content of crucible fragments in the soil attains 1 to 6% in weight, a considerable amount if we extrapolate to all the layers on the site (more than 100 kg). Further analytical investigation will be required in order to carry out a proper comparison. Nonetheless, the crushed crucibles are macroscopically very similar to the grog in the furnace bricks and the grog-tempered crucibles. Several questions are thus raised, such as the existence of a trade in grog.

ACKNOWLEDGEMENTS

We are most grateful to Marc Durand who provided the samples from Lacroix-Saint-Ouen, and Benoît Mille and Dave Coxall for reading this paper.

REFERENCES

Campbell J.L., Hopman T.L., Maxwell J.A., Nejedly Z. 2000, "The Guelph PIXE software package III: Alternative proton database", *Nuclear instruments and methods in physics research B* 170, p. 193-204.

Cartier J., Durand M. 1990, "Un four de potier médiéval à Lacroix-Saint-Ouen (Oise) à la limite de la forêt de Compiègne", *Revue archéologique de Picardie* 3/4, p. 127-137.

Dran J.C., Salomon J., Calligaro T., Walter P. 2004, "Ion Beam analysis of art works: 14 years of use in the Louvre", *Nuclear instruments and methods in physics research B* 219/220, p. 7-15.

Durand M. 2004, "À propos des cônes en céramique du four de potier de Lacroix-Saint-Ouen (Oise)", *Revue archéologique de Picardie* 3/4, p. 71-74.

Freestone I.C. 1989, "Refractory Materials and their procurement", *in* A. Hauptmann *et al.* (ed.), *Archäometallurgie der Alten Welt = Old World Archaeometallurgy,* Der Anschnitt, Beiheft 7, Deutsches Bergau-Museum, Bochum, p. 155-162.

Hamer F., Hamer J. 1997, *The potter's dictionary of materials and techniques*, 4th edition, Pitman Publishing Ldt, London.

Miller P., Stephenson R. 1999, *A 14th century pottery site in Kingston upon Thames, Surrey*, Museum of London Archaeology Service (MOLAS), London.

Ravoire F., Bouquillon A., Dufournier D. 2004, "La poterie de terre renaissance en Ile de France (XV-XVIe siècle)", *Techne* 20, p. 53-60.

Thomas N., Bourgarit, D. 2006, "Une industrie médiévale du bronze", *La recherche* 403, p. 56-58.

Tite M.S., Kilikoglou V., Vekinis G. 2001, "Strength, toughness and thermal shock resistance of ancient ceramics, and their influence on technological choice", *Archaeometry* 43 (3), p. 301-324.

THE DECORATIVE AND ARCHITECTURAL TERRACOTTAS IN FERRARA

RITA FABBRI[1], SERENA CILIANI[2], MONICA BAGATIN[2], FABIO BEVILACQUA[3]

[1] Teaching Fellow at University of Study in Ferrara-Department of Architecture
[2] Architect
[3] Technician of restoration, C.R.C. Bologna

The decorative and architectural terracottas of Ferrara have been closely connected to the appearance of the town since the 13th century AD. The close relationship between the territory and the resources it provides, such as a variety of clays, determined the preferential use of bricks and tiles and the consequent consolidation of its manufacturing techniques. Terracotta is used as both structural and decorative material; therefore Ferrara represents an example of the building art typical of the Po valley, which is characterised by plastered brick masonry and the very varied use of terracottas for important architectural elements.

The use of stone materials, formerly defined "preda viva" to distinguish them from "predacotta" (brick), was limited to buildings such as the "Palazzo dei Diamanti" (Fig. 1), the "Duomo" (Fig. 2) and its campanile, and further limited to particular architectural elements such as the doorways of a house or palace.

During the Middle Ages and the Renaissance, "cotto" ornamentation was used almost exclusively in Ferrara for cornices, labels and string-courses (Fig. 3, Fig. 4).

The Renaissance forms and methods of production were the primary sources of inspiration for the architectural decorations that flourished in Ferrara at the beginning of the 19th century AD. The cemetery of the "Certosa" is a fine example of how the use of terracotta is mediated by the culture of the time and revisited through history and taste: ornamental terracotta is thereafter dark-red. Terracottas are low value pottery products with coloured paste characterised by a porous texture; their processing differs from that of brick in developing a finer clay material.

Terracotta was moulded to produce decorations by two fundamental processes: sculpture "consistenza cuoio" (Fig. 5), which is obtained after drying or after a brief heating in order to harden only the exterior of the brick; and sculpture "a stampo" (Fig. 6), which is made by manipulating the raw clay into wooden (Fig. 7) or plaster moulds (Fig. 8), letting it dry and putting it into the furnace.

The relationship between decorators, model or form designers, mould producers, moulders and people

Fig. 1 – "Palazzo Diamanti".

Fig. 2 – Ferrara's "Duomo".

responsible for the firing, is not clear yet. Potters or "boccalari" were thought in the past to be the manufacturers of terracottas, while more recent studies attribute the production of the ornamental elements of "cotto" to the "fornaciari"[1] supported by the stonemasons, also known as "taglia pietracotta" or "taglia predacotta".

Most of ornamental cottos were produced with moulds, then refined with the chisel; therefore the mixture had to be very carefully pressed into the mould to fill its cavities completely.

The clay was processed to remove big clasts, which would ruin the decoration, before being put into these occasionally disposable moulds. After being moulded, the pieces were left to dry and then fired in the kiln.

After assembling, the terracottas were treated with different finishes in order to make the general appearance of the decoration homogeneous, transforming them from a group of pieces into a unique architectural element. The finishing layer had an aesthetic and protective function.

The most common finish was the "intonachino" which was a compound of lime, chalk, sometimes sand from the river Po and ruddle pigment; sometimes "chamotte" (powdered cotto) was added to the mixture that was brushed on with a fine paintbrush to make it even. A range of reds from ochre to purple, was very commonly used on artefacts from the Middle Ages until the end of the 15th century AD.[2]

During the 16th century AD the colour finishes simulated the natural colours of calcareous stones. The white-ivory or white-grey colouring was achieved through the use of plaster and lime (Fig. 9), sometimes blended together, and the addition of minimal quantities of black pigments and yellow or red clays; this operation was probably performed in order to lessen the excessive brightness of white.

Occasionally a black colouring was used, whose only purpose was to pick out some elements of the cornices (Fig. 10).

Fig. 3 – An example of "cotto" with "foglia spinosa" ornamentation from the Middle-Ages.

Fig. 4 – The Renaissance doorway of the "Casa Rossetti".

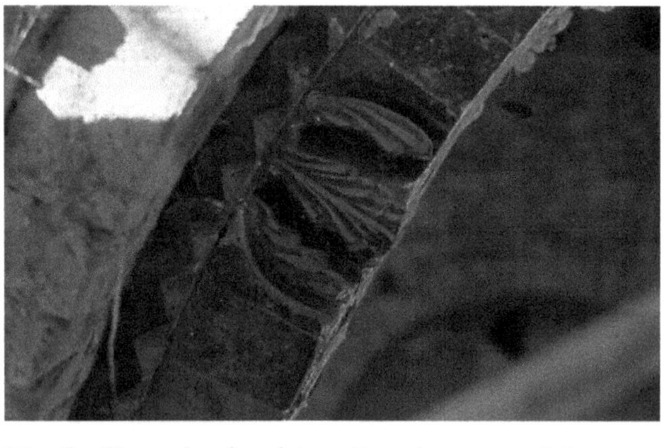

Fig. 5 – Example of sculpture "consistenza cuoio".

Fig. 6 – Example of manufacture "a stampo".

1 – F. Bevilacqua, "La terracotta architettonica: la produzione e la messa in opera", *in* C. Di Francesco (ed.) 1995, p. 41.

2 – C. Di Francesco, "Ferrara e il cotto, spunti di conoscenza, dalla città al museo", *in* C. Di Francesco (ed.) 1997, p. 16-17.

Fig. 7 – An example of wooden mould.

Fig. 8 – A disposable plaster mould.

Fig. 9 – In the cornice of San Francesco's church the colour finish simulated the natural colours of calcareous stones.

Fig. 10 – In this cornice a black colour was used, to pick out some of its elements.

The terracottas in Ferrara are to be considered merely as a support, on which the colour of the finish could vary according to the requirements of the architectural and decorative project of the façade.[3]

THE TERRACOTTAS SURVEY

The study on ornamental and architectural terracottas, so numerous in Ferrara, started from a classification and survey of such elements. The study was extended to include the ways of manufacturing and creating the architectural terracottas and cataloguing of the elements which are present in Ferrara. The survey of decorative terracottas was preceded by research into the surveys previously conducted, in order not to leave out any useful information relevant to the accomplishment of this task.[4] The result is a rich and complete card-index.

The survey was conducted by a careful search throughout the town, during which terracottas were photographed and catalogued (Fig. 11).

After the first part of this survey, 350 buildings including churches and palaces were analyzed.

The comparison between previous surveys, counts and graphic representations revealed that some elements had disappeared because of decay, others had been destroyed to make space for other decorations, others had been substituted by almost perfect copies and others had been treated with finish to avoid ageing, thus compromising their original appearance.

Great attention must be paid to all the elements "in stile": in some cases ornamental elements are perfect copies of the original Late-Gothic and Renaissance ones.

The terracottas were divided into three large categories, namely medieval, Renaissance and "neo-estense". Terracottas which were produced between the end of the 19th and the beginning of the 20th century, either for restoration or for "in stile" constructions, have been precisely dated. This was possible thanks to restoration documents which were kept during the works, or thanks to Ferrariae Decus's[5] documents or research conducted by Eugenio Righini and Giuseppe Agnelli.

After that the sampling of the 350 buildings analyzed was reduced; deteriorated or heavily restored decorations were excluded, while elements which were thought to be typical of every period or better preserved were chosen.

The definitive index cards contain a photograph which shows the site of the building and its surroundings as well as its geographical location (Fig. 12).

3 – C. Di Francesco, F. Bevilacqua, "Il cotto nella decorazione architettonica a Ferrara: tradizione storica e problemi di restauro", *in* Biscontin and D. Mietto (eds.) 1992, p. 475-476.

4 – With reference to some books and atlases such as: Runge 1852, (Fig. 13); Strack 1889 (Fig. 14); Righini 1911-1912; Agnelli 1925; Di Francesco (ed.) 1995 (Fig. 15); Savioli (ed.) 1996.

5 – A society founded by Giuseppe Agnelli in Ferrara in 1906, to protect historical monuments and art.

Ubicazione: *via e numero civico*
Numero scheda: *data con codice strada.01-02*
Tipo di edificio: *chiesa, abitazione*
Prospetto: *est, ovest, facciata principale*
Descrizione: *breve descrizione del prospetto*
Posizione dei cotti: *pt, pp, ps*
Collocazione del pezzo: *cornicione, camino, portale d'ingresso….*
Laterizio: *elemento seriale o multiplio, non intercambiabile*
Numero e tipologie dei pezzi: *se possibile fotografare ogni singolo pezzo e vicino alla descrizione scrivere il numero della foto macro che avrà riferimento in un'altra scheda Es: tre formelle a scacchiera (foto 7 scheda ….)*
Spessore giunti:
Finitura:
Stato di conservazione:
Restauri evidenti:
Rilievo esistente: *si/no*
Riferimenti storici: *Tumiati, Agnelli…*
Riferimenti ad altre schede:
Compilatore: *S/M*
Foto: *digitale/manuale*
Numero rullino
Numero foto: 1,2,3,4,

Fig. 11 – The first index-card.

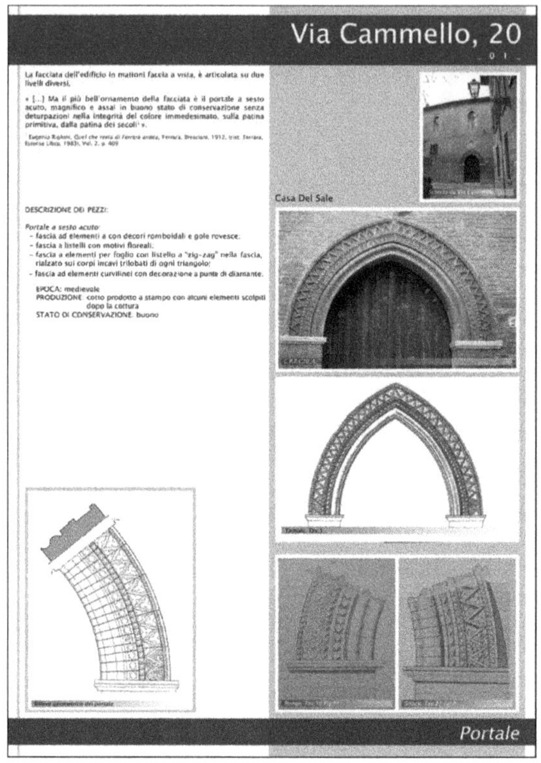

Fig. 12 – The definitive index-card.

Each index card contains information on the decorative elements and the precise date and historical period to which it belongs. In some cases drawings and historical illustrations of the element were added.

The study of these artefacts includes the composition of the raw materials, manufacturing procedures, the embedding and the finishing of architectural terracottas.

THE COLOUR FINISHES AND THE SURFACE TREATMENTS

Specific diagnostic investigations allow identification of the colour finishes and surface treatments which were carried out on the elements of the terracotta after it had been laid and in subsequent restorations. The studies on the colours and finishes were made through specific analysis of the mineralogical composition of the raw materials by chemical analysis of the colouring agents and plasters.

Terracottas can be defined as "artificial stone-like material", resistant to atmospheric agents, though they can vary depending on their composition. The porosity of the pottery causes cracks, splintering and flaking. Terracotta also deteriorates because of the infiltration of soluble solutions. The surfaces can also be attacked by microflora. These are the most frequent alterations; obviously the restoration procedures for such artefacts are the same as those used for stone materials.

The procedures are "Preconsolidamento" (pre-consolidation), "Pulitura" (cleaning), "Incollaggio" (glueing), "Integrazioni" (integration), "Stuccatura" (plastering), "Consolidamento" (consolidation), "Protezione" (protection).

Cleaning is performed with an alkaline poultice (cellulose, sepiolite), and is used to remove polluting substances and/or improperly applied layers such as colouring or paints, or to extract salts. Sometimes cleaning is performed mechanically with an air-abrasive instrument. Application of a biocide to get rid of algae and lichens is another possible treatment.

Polishing has a direct effect upon whether or not the layer, or layers, of finish are preserved. In the cornice of Convento San Giorgio, the grey layer has been recovered, removing the colouring that was applied in the last century.

Hydraulic limes with a low percentage of salts, or with added powdered cotto, sand or powdered limestone are used for plastering or filling.

Consolidation is performed with acrylic, acrylic-silicone and ethyl silicate resins (*e.g.* the northern facade of S.Maria's in Vado church). "Scialbatura" has a protective and aesthetic function and completes the decorative work. In extreme cases, where pieces have deteriorated so badly as to be unable to bear any consolidation treatment or protection with "scialbo", substitution is preferred, to create the original effect.

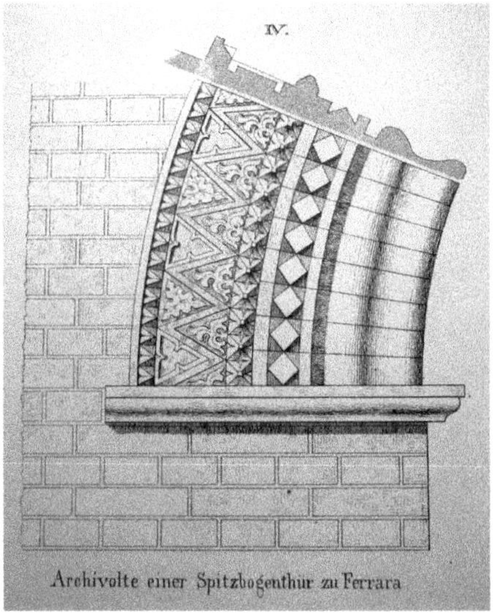

Fig. 13 – A drawing made by L. Runge (Runge 1952).

Fig. 14 – A drawing made by Heinrich Strack (Strack 1889).

Fig. 15 – A drawing made by Domenico Tumiati (Di Francesco [ed.] 1995).

REFERENCES

AGNELLI G. 1909, *Ferrara: Porte di Chiese, di Palazzi, di Case*, Bergamo.

AGNELLI G. 1925, *Ferrara e Pomposa*, Bergamo.

BAGATIN M., CILIANI S. 2004, *Progetto di valorizzazione delle terrecotte architettoniche a Ferrara: studi e proposte per percorsi turistici*, unpublished thesis, University of Ferrara.

BISCONTIN G., MIETTO D. (eds.) 1992, *Le superfici dell'architettura: il cotto. Caratterizzazione e Trattamenti, Proceedings of meeting (Bressanone, 30 giugno-3 luglio 1992)*, Padova.

DI FABRIO M.R. (ed.) 1999, *Architettura e Restauro. Una città: Ferrara*, Ferrara.

DI FRANCESCO C. (ed.) 1995, *Avanzi di antichi cotti che si conservano ancora a Ferrara. Album Domenico Tumiati (1874)*, Ferrara.

DI FRANCESCO C. (ed.) 1997, *Il cotto tra storia e ricerca. Contributi allo Studio, Proceedings of meeting (Ferrara, 28 settembre 1995)*, Firenze.

DI FRANCESCO C., SCARDINO L. (eds.) 1991, *Giuseppe Agnelli. Restauro e arti figurative a Ferrara tra Ottocento e Novecento*, Ferrara.

EMILIANI T. 1971, *La tecnologia della ceramica*, Faenza.

FABBRI R. (ed.) 2005, *Ferrara Architettura 2, Arte di Costruire*, Ferrara.

Raccomandazioni Normal 1/88, Alterazioni Macroscopiche dei Materiali Lapidei, CNR Centri di Studio Milano e Roma sulle Cause di Deperimento e sui Metodi di Conservazione delle Opere d'Arte - ICR Istituto Centrale per il Restauro.

RIGHINI E. 1911-1912, *Quel che resta di Ferrara antica*, 4 volumes, Ferrara.

RUNGE L. 1852, *Beiträge zur Kenntnis der Backstein-Architektur Italiens nach seinen Reiseskizzen herausgegeben von L. Runge*, Ed. Ernst Wasmuth, Berlin. Reprinted with the Italian title *Taccuini di viaggio sulle architetture in cotto*, 1991, Milano.

SAVIOLI G. (ed.) 1996, *I disegni ritrovati di Dora Noyes*, Ferrara.

STRACK H. 1889, *Ziegelbauwerke Des Mittelalters und der Renaissance in Italien nach Original-Aufnahmen herausgegeben von H. Strack*, Berlin.

VIRGILI V. 2002, *Le decorazioni in cotto a Ferrara. Problemi ed esperienze di intervento*, unpublished thesis, University of Ferrara.

ZANONI M. 1994, *Le decorazioni in cotto nell'architettura ferrarese: l'uso della terracotta nel progetto ottocentesco di ampliamento e trasformazione della Certosa in cimitero pubblico ideato da Ferdinando Canonici (1786-1873)*, unpublished thesis, University of Venezia.

ARCHAEOMETRIC CHARACTERIZATION OF MIDDLE AGE AND RENAISSANCE TIN LEAD GLAZED POTTERY FROM BARCELONA

J. GARCIA-IÑAÑEZ[1], J. BUXEDA I GARRIGÓS[1], M. MADRID I FERNÁNDEZ[1, 2], J.M. GURT I ESPARRAGUERA[3], J.A. CERDÀ I MELLADO[4]

[1] Cultura Material i Arqueometria UB (ARQ/UB), Facultat de Geografia i Història, Universitat de Barcelona, c/ Montalegre, 6, 08001, Barcelona, Catalonia, Spain

[2] Grup de Recerca d'Arqueologia Clàssica, Protohistòrica i Egípcia (GRAPCE), Facultat de Geografia i Història, Universitat de Barcelona, c/ Montalegre, 6, 08001, Barcelona, Catalonia, Spain

[3] Equip de Recerca Arqueomètrica de la Universitat de Barcelona (ERAUB), Facultat de Geografia i Història, Universitat de Barcelona, c/ Montalegre, 6, 08001, Barcelona, Catalonia, Spain

[4] Associació Catalana de Ceràmica Decorada i Terrissa, c/ St. Honorat, 7, 08002, Barcelona, Catalonia, Spain. (javier_garcia@catalonia.net, mmadrid@ub.edu, jbuxeda@ub.edu, jmgurt@ub.edu, jacme@acatceramica.com)

INTRODUCTION

The purpose of this paper is to present the preliminary findings on Barcelonan majolica production from the Middle Ages and Renaissance. This study forms part of an on-going research project aiming at the archaeometrical characterization of the main production centers of Spain, of which almost all from the Catalan area have been already analyzed (Garcia Iñañez *et al.* in press).

Barcelona was the most important city in the Kingdom of Aragon during Middle Ages and Renaissance, and one of the most significant trade ports all over the Mediterranean Sea. Within this economic and commercial activity, Barcelona developed an important majolica, tin lead glazed ceramic, production. This production is well documented in numerous legal and commercial manuscripts referring to these activities, like the 16th century document *Llibre de les fornades* by Nicolau Reyner (Ainaud de Lasarte 1941), as well as in the names of the streets from the old quarter, like *escudellers st.*, from the word *escudella*, that means porringer, one of the most characteristic shapes of this tableware.

Although small trials had been made before, majolica production of Barcelona can be considered to start during the 14th century following the Mudejar ceramic tradition, using green and manganese colours on a white glaze for their patterns, mainly geometric designs. Renaissance style, due to Italian influence, replaced the geometric patterns, introducing new decorative motifs (human figures, animals or plants among others) and new colours (blue and yellow). These colours, together with metallic luster, became the most characteristic during 16th and 17th centuries (Farwell *et al.* 2003).

The main kiln dump of the Barcelona workshops was discovered outside the walled city, next to the *Drassanes* (shipyard) building, in 1916. On this site, an important deposit of lusterware and majolica ceramics, dated mainly to the 17th century, was documented. As well, there are many other places where this pottery has been found. Amongst others, a set of majolica from the 2nd quarter of the 17th century was recovered from the vaults of the old *Hospital de la Santa Creu* of Barcelona, and another set, but dated to the 14th century, was recovered from the church of *Santa Maria del Pi*.

The aim of the present study is characterization of the tin lead glazed pottery production of Barcelona in order to find out whether chemical and technological differences between 14th (green and black) and 16th-17th (luster, blue, and blue & yellow) pottery production exist. These differences are also investigated in pottery produced in the same period. Finally, the role played by contemporary guilds in majolica provenance studies is considered.

MATERIALS AND METHODS

Ninety-nine individuals were chemically and mineralogically characterized by means of X-Ray Fluorescence (XRF) and X-Ray Diffraction (XRD). Of these, 8 come from the vaults of the old *Hospital de la Santa Creu* (17th century), 15 from the church of *Santa Maria del Pi* (14th century), 27 from the *Drassanes* kiln dump (17th century). In addition, there are several finds of probable Barcelona products from different consumption sites. In order to test the hypothesis, 10 individuals from *Girona* (14th century), and a set of 39 individuals from the consumption centre of *Mataró* were also sampled (Table 1). According to these results and the different types of decoration, a set of selected specimens was studied by means of Scanning Electron Microscope linked to an Energy Dispersive X-Ray Fluorescence and a backscattered electron detector (SEM-BS-EDX).[1]

1 – Experimental analyses by XRF, XRD and SEM-BS-EDX were conducted at the *Serveis Cientificotècnics de la Universitat de Barcelona*. Conditions and analytical routines in Buxeda *et al.* 2001.

Ic	Sta Creu	Sta Maria Pi	Drassanes	Girona	Mataró	Total
Blue		1	1		13	**15**
Luster	8		26		26	**60**
Green & Black		14		10		**24**
Total	**8**	**15**	**27**	**10**	**39**	**99**

Table 1 – Number of individuals per excavations sites.

ANALYTICAL RESULTS

Chemical characterization

Ancient technology of Majolica pottery making is characterized by tin and high lead contents in the glazes, which cause analytical problems because of their diffusion into the clay matrix body. Chemical results have shown that their lead content exceeds our upper XRF regression limit (928 ppm) for semiquantitative data. Besides, other trace elements determination, like Rb, Ga, Th or Y, might also be affected by lead interference. For this reason, those elements will not be taken into account in the statistical data treatment.

For the statistical treatment, the chemical data were logratio transformed (Aitchison 1986; Buxeda 1999). The logratio transformation was performed on the following elements: Fe_2O_3, Al_2O_3, MnO, TiO_2 MgO, CaO, Ba, Nb, Zn, Sr, Ce, V, Zn, Ni and Cr, using SiO_2 as divisor because, according to the variation matrix, it made the smallest contribution to the chemical variability (Buxeda and Kilikoglou 2003). The results can be summarized in the dendrogram resulting from cluster analyses performed on the previous selection, using the Squared Euclidean distance and the centroid agglomerative algorithm, performed by S-plus2000 (MathSoft 1999), without Na_2O and K_2O. These two elements were discarded due to an alteration and contamination process undergone during burial by a large number of individuals. This postdepositional alteration involves the leaching of K_2O and Rb, and the subsequent crystallization of analcime, with an enrichment of Na_2O (Buxeda *et al.* 2001; Buxeda 1999, and references therein).

Study of the dendrogram (Fig. 1) shows a clear structure identifying three clusters: BCN-SC, BCN-PI and BCN-DR, plus seven individuals which do not join any group, remaining unclassified. Chemical differences among the three main groups are due mostly to their

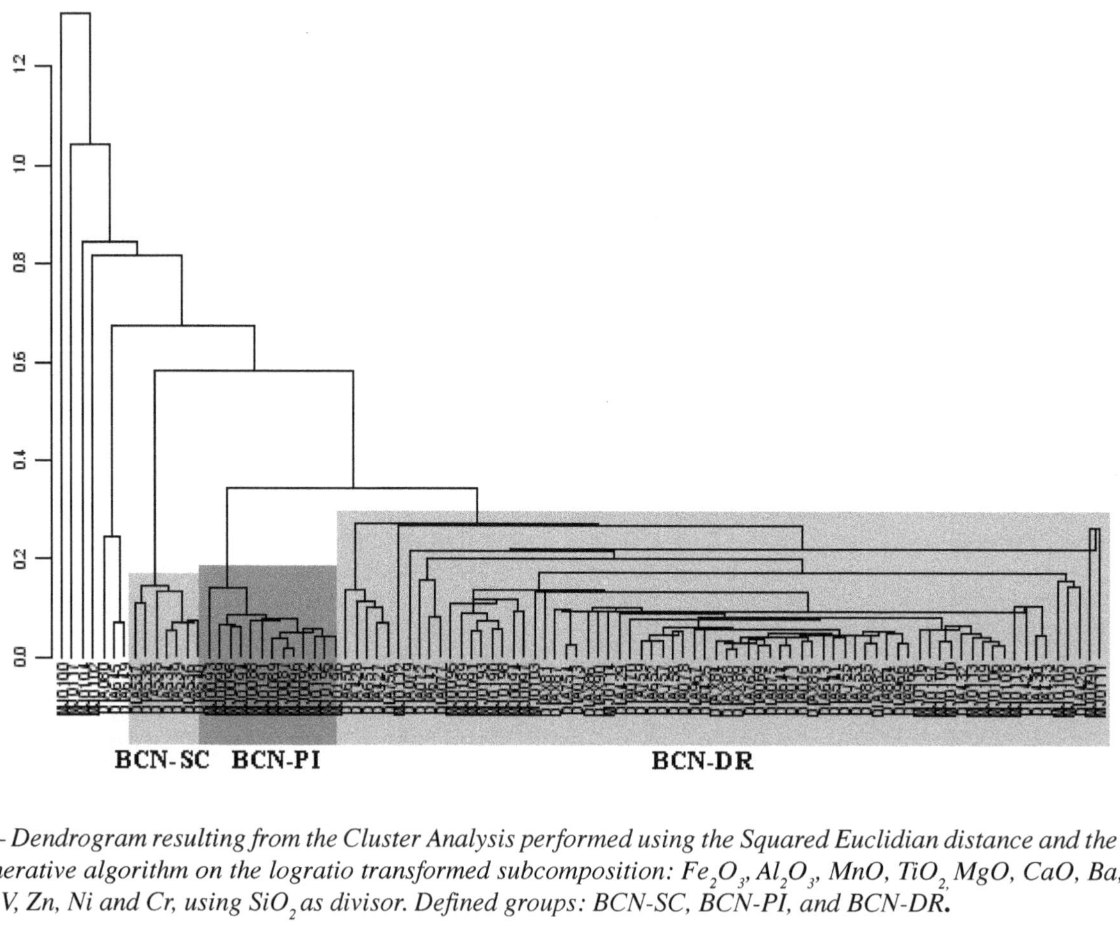

Fig. 1 – Dendrogram resulting from the Cluster Analysis performed using the Squared Euclidian distance and the centroid agglomerative algorithm on the logratio transformed subcomposition: Fe_2O_3, Al_2O_3, MnO, TiO_2 MgO, CaO, Ba, Nb, Zn, Sr, Ce, V, Zn, Ni and Cr, using SiO_2 as divisor. Defined groups: BCN-SC, BCN-PI, and BCN-DR.

different CaO content. Despite this structure, the chemical composition of all the samples presents great homogeneity. This fact could be related to the role played by the guilds. According to legal and commercial manuscripts, the Medieval and Renaissance potters of Barcelona were organized into guilds, which supplied the raw materials to their unionized workshops. These raw materials would be the basis for the paste preparation of majolica pottery, each following their own recipe, probably on the basis of their own tradition. In the present case, the potters of BCN-SC, BCN-PI and BCN-DR might have added different amounts of calcite, making CaO one of the most discriminating elements (Table 2).

Centering our attention on the identified clusters, the biggest one is BCN-DR group, made up of 72 individuals. 24 of them come from *Drassanes* itself and 1 from the *Hospital de la Santa Creu;* all of them exhibit luster decoration and are dated to the 17th century. As well, 6 individuals come from the 14th century site of *Santa Maria del Pi*, 1 with blue decoration and 5 with green and black motifs. With regard to the 17th century samples from the consumption center of *Mataró*, 36 are included in the BCN-DR group, and the 3 remaining individuals must be considered, at this stage, as unclassified. 13 out of the 36 show blue decoration and 23, luster decoration. Finally, 5 green and black decorated individuals from *Girona*, dated to the 14th century, are also included in the BCN-DR group. The next group, BCN-PI, includes 8 individuals from *Santa Maria del Pi* itself and the rest are 5 individuals from *Girona*. The whole set of samples dates from the 14th century and exhibits green and black decoration. Finally, BCN-SC is composed of 7 individuals from *Santa Creu* itself. They show luster decoration and date from the 17th century.

According to these results, it seems reasonable to hypothesize that the three identified productions would have used similar raw materials for the preparation of their pastes. The relatively homogeneous composition for such a long period, with a very high number of potters involved in majolica production, as attested by the written sources, is in total agreement with the guild's role of raw material distribution as explained in the historical documentation.

Technological characterization

Technologically, all the samples considered in this work are calcareous and exhibit a great chemical homogeneity, although BCN-DR, BCN-SC and BCN-PI groups can be differentiated. These high CaO contents, the result of the abundance of calcite ($CaCO_3$), are necessary to achieve a buff light color for the matrix body (Molera *et al.* 1998). This pale color is desirable to avoid a strong signal from the body and to highlight the decoration. Moreover, according to the historical manuscripts, tin oxide, generally included into the glaze as an opacifier, was an expensive material. For this reason, the light buff color can be seen as a way of saving tin oxide in the glaze by the potters.

	BCN-DR (n=72)		**BCN-SC** (n = 7)		**BCN-PI** (n=13)	
	Mean	Stand. Dev.	Mean	Stand. Dev.	Mean	Stand. Dev.
Fe_2O_3 (wt%)	5.34	0.32	4.65	0.21	5.83	0.20
Al_2O_3 (wt %)	15.66	0.45	14.40	0.56	15.81	0.25
MnO (wt%)	0.07	0.01	0.06	0.00	0.08	0.01
P_2O_5 (wt%)	0.74	0.62	0.63	0.80	0.28	0.17
TiO_2 (wt%)	0.70	0.03	0.63	0.03	0.76	0.02
MgO (wt%)	2.22	0.18	1.93	0.15	1.96	0.09
CaO (wt%)	15.88	1.53	25.45	2.54	11.20	1.23
Na_2O (wt%)	1.32 (0.63)*	0.68 (0.40)*	0.54	0.09	0.82 (0.68)*	0.23 (0.02)*
K_2O (wt%)	3.09 (4.34)*	1.01 (0.46)*	2.59	0.30	3.86 (4.11)*	0.38 (0.18)*
SiO_2 (wt%)	54.85	1.16	49.01	1.46	59.25	0.94
Ba (ppm)	439	71	331	27	481	32
Nb (ppm)	13	4	9	1	18	0
Zr (ppm)	178	16	174	10	231	12
Sr (ppm)	171	17	203	19	156	10
Ce (ppm)	62	8	50	6	69	7
V (ppm)	81	5	71	6	83	4
Zn (ppm)	129	17	101	6	118	13
Cu (ppm)	66	52	52	24	104	46
Ni (ppm)	38	3	34	3	36	2
Cr (ppm)	67	6	56	3	78	4

*Table 2 – Mean and standard deviation (normalized data) of the three identified groups from Barcelona. (ppm = parts per million). *: in parenthesis, values calculated using the individuals that do not exhibit analcime by XRD (BCN-DR, n=17; BCN-PI, n=4).*

The evaluation of mineral phases allows estimation of the equivalent firing temperature (EFT). For majolica pottery, the EFT will correspond to the highest temperature achieved in the firing process, which could have consisted of more than one firing. The study of the diffractograms from individuals of the BCN-DR and BCN-SC chemical groups has shown that EFT of majolica is normally in the range of 1000–1050 °C. As can be seen in Fig. 2, the diffractogram shows a total decomposition of the illite-muscovite, while the development, as firing phases, of pyroxene (px) and gehlenite (gh), as well as the crystallization of plagioclase (pg), probably anorthite, and leucite (lct) are evident. As well, an important number of individuals also exhibit analcime (anl) as a secondary phase (Fig. 2), which is related to the relatively higher amounts of Na_2O and lower K_2O contents detected in the chemical composition, as previously discussed. In spite of this uniformity, there are also a few individuals that enable us to estimate an EFT either below or above the previous range. This fact does not mean that these pieces were irregular nor that different firing processes existed. On the contrary, this should really be considered as a normal oscillation within a standardized production, characteristic of pre-industrial pottery making. XRD and SEM-BS-EDX results for individuals of the *Santa Maria del Pi* production are not presented in this article because they are still under study.

Study of the glazes and decorations was undertaken on 6 individuals, 4 of which possess metallic luster and 2, blue and yellow decorations. Examination of the glazes by SEM-BS-EDX shows that their thicknesses vary approximately between 100 to 200/250 μm. It is also evident that the glazes are basically composed of silicon, lead and, in a lesser amount, potassium. In some cases, the presence of aluminium and calcium is also appreciable, which probably diffused from the matrix during firing. Tin oxides particles (SnO_2) dispersed throughout the glaze, as 2 μm maximum size inclusions, are also identified (Fig. 3). These particles, together with intentional or accidental bubbles and quartz and alkali feldspars inclusions, are responsible for the opacity of glazes. This effect is due to the dispersion of the incident light caused by these particles, providing an opaque white colour in the glaze (Mason and Tite 1997; Molera *et al.* 1997). The application of the glaze to the ceramic body could be made before firing, in the leather hard state, or after a first firing, in an already biscuit ceramic. The practice of the later application avoids problems with the glaze, such as bubbling due to gas formation during firing, as well as other difficulties related to shrinkage and porosity.

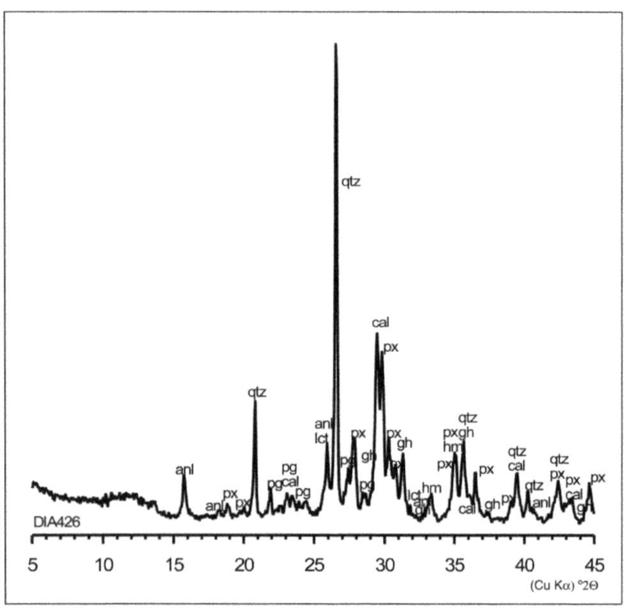

Fig. 2 – Diffractogram of the individual DIA426, from the BCN-DR group. anl: analcime, cal: calcite, gh: gehlenite, hm: hematite, lct: leucite, px: pyroxene, pg: plagioclase, qtz: quartz.

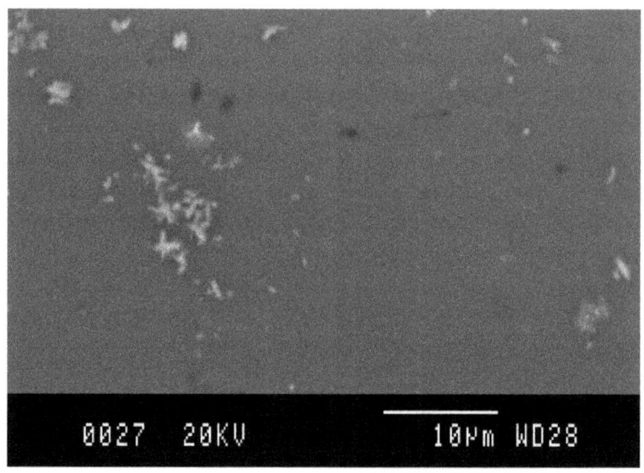

Fig. 3 – Microphotograph by SEM-BS of the individual DIA643 from the BCN-DR group showing tin oxides particles (SnO_2) of 2 μm scattered into the glaze.

In our case, among the blue and blue and yellow decorated pottery, one and two firings have been observed for the same group (*BCN-DR*). In the first case, a great interaction between the glaze and the body occurs, producing strong crystal development in the interface, giving as a result a wide area of contact between the glaze and the matrix. By contrast, in the two firing process, a low interaction between matrix and glaze occurs and a poor growth of crystals in the interface is developed. Consequently, a thin interface between the glaze and the matrix is obtained. That means that the piece has been previously biscuit-fired before the application of the glaze (Molera *et al.* 1997) (Fig. 4). In addition, it must be highlighted that both ceramics show the same EFT (Buxeda *et al.* 2001), supporting the interpretation of two different firing processes. The identification of these two firing processes for the same group could be evidence that we are facing different workshops working with different

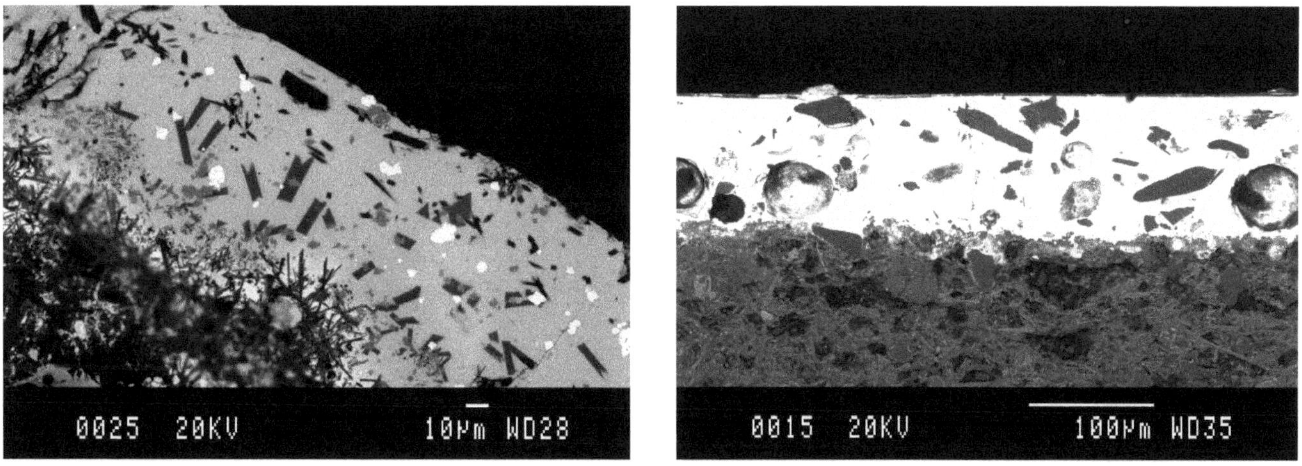

Fig. 4 – Microphotographs by SEM-BS. Left: individual DIA643 from the BCN-DR group showing a great interaction in the interface between glaze and matrix, with strong crystals development. Right: individual DIAX87 from the BCN-DR group exhibiting little interaction in the interface with very few crystals development.

production traditions in the same centre. It is interesting to note that all the metallic luster majolica studied had undergone a biscuit firing before the glaze application, in agreement with other authors (Molera *et al.* 2001; Lazic *et al.* 2003).

Technological studies enable us to observe that pigments are always applied on top of the raw glaze, regardless of the colour (blue or yellow). The identification of cobalt in the microanalyses of blue decoration confirms the use of this element as a pigment, whereas the yellow corresponds to lead antimonate, known in ancient times as "Naples Yellow" (Fig. 5). With regard to the metallic luster decoration, observations by means of SEM-BS-EDX reveal that pigments were always applied after the second firing, developing a differentiated layer from the glaze during a third firing cycle. Generally, this decoration can be considered as based on copper oxide with the addition, sometimes, of silver oxide together with iron or clays.

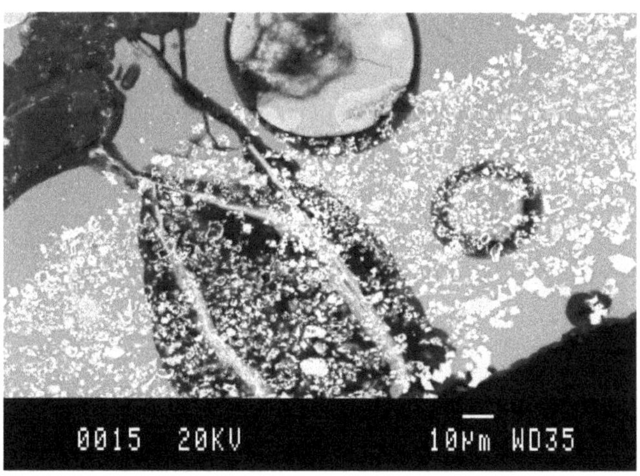

Fig. 5 – Microphotograph by SEM-BS of individual DIA652 from the BCN-DR group showing a lot of yellow lead antimonite particles dispersed into the glaze. Transverse black lines correspond to an alteration of the glaze.

CONCLUSIONS

Archaeometric characterization of majolica from Barcelona has enabled us to identify three different chemical groups corresponding to three differentiated productions related to the city of Barcelona. Likewise, all the assigned individuals from the consumption centre of *Mataró* would have a Barcelonan origin. Similarly, all the samples from Girona have been finally assigned to *BCN-DR* and *BCN-PI* productions, with no evidence of a local production. Furthermore, as determined from their chemical results, *BCN-DR* and *BCN-PI* productions were already making majolica in the 14th century, focused mainly on green and black decoration. In the 17th century, *BCN-DR* would continue its pottery production tradition making luster and blue decorated ceramics, while the production of *BCN-SC* is, by this time, documented as producing only lusterware.

Technologically it has been proved that two different firing processes were used by potters from the same centre for the making of blue and yellow majolica, while for the metallic luster majolica the same double firing process was used by all the potters. These different firing processes result in different glaze thicknesses.

For Medieval and Renaissance pre-industrial tin lead glazed pottery production, it should be considered that most of the raw materials were provided by potter's guilds. In that sense, it has been proved that chemical compositions of the three identified productions are very homogeneous. Therefore, chemical groups can account for "several" workshops with just slight chemical differences. For that reason, diversity of production among workshops in a given production centre may only be traced by their technological differences or similarities.

ACKNOWLEDGEMENTS

The work presented here forms part of the project "Identification, Recovery and Improvement of ancient Mediterranean ceramic manufacturing technologies for the reproduction of ceramic artifacts of archaeological value (CERAMED)", from the Research Directorate General, European Commission, European Community (ICA-3-CT-2002-10018). Javier Garcia Iñañez is also indebted to the support of Departament d'Universitats, Recerca i Societat de la Informació de la Generalitat de Catalunya i Fons Social Europeu. Authors are also indebted to the Museu de la Ceràmica de Barcelona for its collaboration providing the samples.

REFERENCES

Ainaud de Lasarte J. 1941, "Loza dorada y alfarería barcelonesa. Siglos XV-XVI", *Anales y boletín de los museos de arte de Barcelona. Arte Antiguo* 1, p. 89-104.

Aitchison J. 1986, *The statistical analysis of compositional data*, Monographs on Statistics and Applied Probability, Chapman and Hall, London – New York.

Buxeda J. 1999, "Alteration and contamination of archaeological ceramics: the perturbation problem", *Journal of archaeological science* 26, p. 295-313.

Buxeda J., Madrid M., Gurt J.M. 2001, "Provinença i tecnologia de les ceràmiques de pisa i d'obra de Manises del dipòsit de la Plaça Gran de Mataró", *in* J.A. Cerdà (ed.), *La ceràmica catalana del segle XVII trobada a la plaça Gran (Mataró)*, Associació Catalana de Ceràmica Decorada i Terrissa, Barcelona, p. 155-170.

Buxeda J., Kilikoglou V. 2003, "Total variation as a measure of variability in chemical data sets", *in* L. Van Zelst (ed.), *Patterns and process, a Festschrift in honor of Dr. Edward V. Sayre*, Smithsonian Center for Materials Research and Education, Washington, p. 185-198.

Farwell R., Pierce D., Pleguezuelo A. 2003, *Cerámica y cultura. The story of Spanish Mexican Mayólica*, University of New Mexico Press, Albuquerque.

Garcia Iñañez J., Schwedt A., Madrid i Fernández M., Buxeda i Garrigós J., Gurt i Esparraguera J.M. in press, "Caracterización arqueométrica de los principales centros productores catalanes de cerámica mayólica de los siglos XVI y XVII", *in* J. Molera (ed.), *6th CAIG*, Girona.

Lazic V., Colao F., Fantoni R., Palucci A., Spizzichino V., Borgia I., Brunetti B.G., Sgamellotti A. 2003, "Characterisation of lustre and pigment composition in ancient pottery by laser induced fluorescence and breakdown spectroscopy", *Journal of cultural heritage* 4, p. 303-308.

Mason R.B., Tite M.S. 1997, "The beginnings of tin-opacification of pottery glazes", *Archaeometry* 39 (1), p. 41-58.

MathSoft 1999, *S-PLUS 2000, User's Guide*.

Molera J., Vendrell M., García M., Pradell T. 1997, "Technology and colour development of hispano-moresque lead-glazed pottery", *Archaeometry* 39 (1), p. 23-39.

Molera J., Pradell T., Vendrell M. 1998, "The colours of Ca-rich ceramic pastes: origin and characterization", *Applied clay science* 13, p. 187-202.

Molera J., Mesquida M., Pérez-Arantegui J., Pradell T., Vendrell M. 2001, "Lustre recipes from a medieval workshop in Paterna", *Archaeometry* 43, p. 455-460.

COMPOSITIONAL STUDIES ON IZNIK CERAMICS PIGMENTS

ROXANA BUGOI[1], AURELIO CLIMENT-FONT[2], BOGDAN CONSTANTINESCU[1], ALESSANDRA D'ALESSANDRO[3], PAOLO PRATI[3], ALESSANDRO ZUCCHIATTI[3]

[1] "Horia Hulubei" National Institute of Nuclear Physics and Engineering, PO BOX MG-6, Bucharest 077125, Romania

[2] Centro de Micro-Analisis de Materiales, Universidad Autonoma de Madrid, E-28049, Cantoblanco, Madrid, Spain

[3] Istituto Nazionale di Fisica Nucleare (INFN) – Sezione di Genova, Via Dodecaneso 33, I-16146, Genova, Italy

ABSTRACT

The composition of some Iznik glazed ceramics fragments—pieces of vessels, plates for wall decorations—from the National Museum of Romania's History, Bucharest, was investigated by using ED-XRF (Energy Dispersive X-Ray Fluorescence) and PIXE (Proton Induced X-Ray Emission) techniques. The study was intended to shed some light on the painting pigments employed for the decoration of these exquisite objects.

INTRODUCTION

Turkey has a long tradition of manufacturing ceramics and of developing innovative techniques in this field. Its most widely known ceramics are the Ottoman Empire Iznik fritware ceramics, famous for their vibrant colours and beautiful designs, that have been collected ever since their production between the 15th and 17th centuries. Ware such as plates, bowls and ewers were produced as well as tiles for wall revetments for mosques or secular buildings, such as for palaces or fountains. Iznik fritware was the result of a search by the Ottoman court in Istanbul for a recipe to imitate the much-admired and pricey Chinese Yuan and Ming Dynasty blue-and-white porcelain. This required extensive research and experimentation by the Ottoman ceramic workshops which were subsidized by the sultan. The initial copies of the Chinese designs gradually gave way to a uniquely Turkish style which included a broader colour palette.

A detailed discussion on the technology of Iznik ceramics can be found in Tite (1989, p. 115-132) and Paynter (2004, p. 421-437). The ceramic body of traditional Iznik ware was made from a combination of ground quartz, lead-alkali glass frit and a white clay. During firing the frit melted and reacted with the clay, encasing the quartz particles in a glassy matrix. The body is coated with a slip layer made of finer textured glass frit-quartz-clay mixture containing no iron impurities, so it fired completely white. Over this were applied the underglaze mineral-based colorants and the lead-alkali glaze. The introduction of lead as an important component of the body frit, in addition to its presence in the slip and glaze, was an innovation in the technology of Islamic fritware ceramics. Its presence ensured a similar coefficient of expansion and contraction between body, slip and glaze, minimizing crackle in the glaze. In cross-section, the fairly porous traditional Iznik ceramic body, which is a fritware, consists of polyhedric fine quartz particles (~ 25–50 μm) surrounded by an interstitial glassy matrix. The slip layer of the traditional ware, about 300 μm to 1 mm thick, is similar in composition to the body, except that it was prepared from finer quartz particles (~ 25 μm), has less interstitial glass (except where glaze has penetrated the layer), and has a lower iron oxide content (0.5%). The slip was made of 81-90% quartz, 7-11% lead-alkali frit and 3–8% clay. Traditional glazes are 50-200 μm thick. They are of the lead-soda type being composed of about 50% silicon dioxide and about 6% tin oxide. The Na_2O content is much higher than in body and slip and is about 10%, the rest (20–40%) being lead oxide. A small amount of FeO is present.

In this paper, the composition of some Iznik glazed ceramics fragments—pieces of vessels, plates for wall decorations—was investigated (Fig. 1). The curators at the National Museum of Romania's History, Bucharest, from which the ceramics fragments were obtained, had decided on visual grounds that these ceramics sherds were made in the Iznik style, being manufactured during the 17th century. The samples were found at two sites: some of them in the ruins of Prince Vasile Lupu's palace located in Suceava (1600-1640), and the others, at an excavation site at a Danube ford—Piua Pietrii—a renowned commercial center during the Late Middle Age. Two analytical methods—Energy Dispersive X-Ray Fluorescence (ED-XRF) and Proton Induced X-Ray Emission (PIXE) were employed to determine the elemental composition of the analyzed sherds. The study had the purpose of identifying the pigments employed to paint the sherds, with a particular focus on the blue colour.

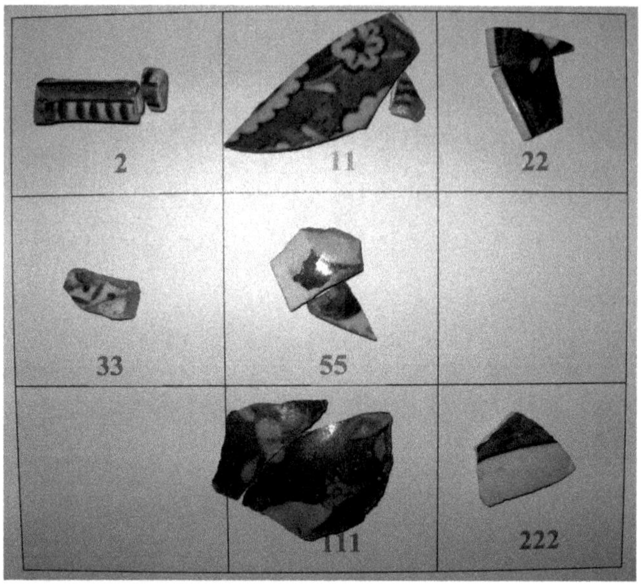

Fig. 1 – Analyzed blue Iznik ceramics samples.

EXPERIMENTAL

The ceramics samples were studied at the National Institute of Nuclear Physics and Engineering "Horia Hulubei", Bucharest, using an Energy-Dispersive X-Ray Fluorescence (ED-XRF) set-up based on a 30 mCi ^{241}Am annular source and a Si(Li) detector (Bugoi 1999, p. 777-781), at Istituto Nazionale di Fisica Nucleare (INFN), Sezione di Genova, Italy using an ED2000 high-resolution ED-XRF spectrometer from Oxford Instruments, based on a 50 kV X-ray tube with a silver anode and a Si(Li) detector, and at Universidad Autonoma de Madrid (UAM), Spain using in-air Particle Induced X-ray Emission (PIXE) with the external micro-beam setup of the 5 MV HVEE type Tandem accelerator.

In principle, both methods—ED-XRF and PIXE—are non-destructive, multielemental and sensitive, thus being appropriate for the elemental composition determination. The advantage of PIXE is that its sensitivity is higher, and that shorter acquisition times are required; it also presents a high lateral resolution. However, ED-XRF technique is much cheaper and more set-ups are available. A rough estimate of the depth which is analyzed through these analytical methods which employ characteristic X-rays, in the case of a silica matrix—the main constituent of ceramics—, is 100 μm.

The first step in the study was the ED-XRF analysis using the Bucharest set-up, in which the excitation of the characteristic X-rays is made with ^{241}Am annular source. The measurements were made to obtain preliminary information on the elemental composition of the different pigments employed to decorate the Iznik ceramics.

In the case of the blue pigments, arsenic is of a special importance, an aspect which will be discussed in more detail in the results and discussions section. Due to the relatively high amount of Pb (more than 25% of PbO) contained in the Iznik glaze, to the overlap of the As K_α and Pb L_α at 10.5 keV, and to the relatively low concentration in As sought, it was quite difficult to find evidence of As in these samples using this set-up. In order to solve the As determination problem, the K_β peak of this element was looked for with a more advanced ED-XRF spectrometer. This was done at INFN Genova, using an ED2000 high-resolution spectrometer from Oxford Instruments. This spectrometer is based on a 50 kV X-ray tube with a silver anode and a Si(Li) detector. The measurements were performed only on the blue regions of the sherds. Particular attention was paid to the focusing of the exciting X-ray beam on the blue spots, since some of the sherds exhibited particularly small areas of these blue spots (of the order of a few square millimeters).

To obtain quantitative results, two sets of measurements under the "fixed conditions" of the spectrometer were performed. All the measurements took place in vacuum (~ 200 mbar), with an 8 mm diameter aluminum collimator in front of the detector and a 2 mm diameter aluminum collimator in front of the X-ray tube window. The measurement time was 120 seconds (live time) for each spectrum, while the current intensity was 1 mA. The first set of measurements was made under the "very light elements" (VLE) arrangement, with no filter in front of the tube window and employing a 5 kV accelerating voltage.

In the second set of the measurements a 35 kV accelerating voltage and a thin silver filter in front of the tube exit window were employed, this time with the purpose of detecting the medium weight elements (ME). The detector resolution was adjusted by changing the counting rate of the electronics: in the first set of measurements the FWHM (Full Width Half Maximum) was 160 eV (at 5.9 keV – Mn K_α line), while in the second set it was 175 eV (at 5.9 keV), allowing a good separation of the peaks of interest (neighboring Z elements, such as in the case of the sought elements Fe, Co, Ni, and also As, and Pb).

To clarify the problem of Co-As-Ni relationships—which is crucial in identifying the Co mineral and information about the technological processing—, further measurements on the same samples were carried out using the micro-PIXE facility of the 5 MV Cockcroft-Walton tandem accelerator of the Centro de Micro-Analisis de Materiales (CMAM) from the Universidad Autonoma de Madrid (UAM), Madrid, Spain, to obtain a more reliable quantification of the results (with lower detection limits for the elements of interest), especially for As, and to check if there are spatial variations in the pigment compositions. For the UAM experiment, the energy of the extracted proton beam was 2995 keV (energy on target), while the detection of the characteristic X-ray was performed using a Gresham Si(Li) detector located at 49° from the target surface normal to the horizontal plane. The sample-detector distance was 83 mm. He was flowed through a cone shaped nose placed in front of the detector. Quantification was made with GUPIX software assuming the metals detected were forming oxides, and considering O to be an invisible element for PIXE. The compounds looked for were: Na_2O, MgO, Al_2O_3, SiO_2, P_2O_5, SO_3, Cl, Ar, K_2O, CaO, TiO_2, V_2O_3, Cr_2O_3, MnO, Fe_2O_3, CoO, NiO, CuO, ZnO, As_2O_3, SrO, SnO_2, Sb_2O_5, BaO, PbO, Bi_2O_3. Glass standards were used to obtain the detector calibration curve. The PIXE results, expressed in oxide weight percentages, as well as the uncertainties and the Minimum Detection Limits (MDL), are summarized in Table 1.

RESULTS AND DISCUSSIONS

The conclusions obtained from the ED-XRF measurements performed in Bucharest with the ED-XRF set-up based on ^{241}Am source excitation are the following:

— The Iznik glaze composition has as main components—apart from SiO_2—a combination of Pb oxide and Sn oxide, with a Pb/Sn ratio of 6/1; these elements were detected in all the examined sherds;

— In the green color pigments, the following elements were detected: Fe, Cu, Cr, Co. Thus, the light green reflects a composition of Cu/Fe=10/1, Cu/Co=1/30, the dark green is given by Cu/Fe=1/12, Co/Fe=1/7, Ni and Co traces while the "normal" green is produced by the following recipe: Cu/Fe=3/1, Cr, with Mn traces. In conclusion, it is most likely that Cu oxides were the main component of green pigments and to obtain specific hues, Co and Cr oxides were added;

— The yellow colour pigment is due to lead antimoniate;

— The dark brown pigment is based on Mn oxides with various amounts of Fe oxides;

— The blue pigments is based on a Co compound.

The case of the blue pigments is definitely one of the most interesting questions when studying the coloring of ancient ceramics. Co minerals are relatively scarce, few sources being exploited during ancient times (Zucchiatti *et al.* 2006). Possible sources for these pigments in the 17th century were either the mining district of Schneeberg in Germany, or the Co ores from the Iranian mines of Qamsar and Anarak (Porter 2000). In Table 2, the main Co minerals from Erzgebirge (Schneeberg) – Saxony are presented (Weber 1986). It is well known (Scott 2001) that

Object	Na_2O	$ErrNa_2O$	MgO	ErrMgO	Al_2O_3	$ErrAl_2O_3$	SiO_2	$ErrSiO_2$	P_2O_5	$ErrP_2O_5$	Cl	ErrCl	K_2O
IZN2	1,45	0,51	< MDL	-	0,67	0,06	62,34	0,29	< MDL	-	0,36	0,03	1,09
IZN222	2,24	0,62	< MDL	-	0,91	0,07	58,48	0,30	1,01	0,07	0,98	0,03	2,05
IZN55	< MDL	-	< MDL	-	0,84	0,06	63,04	0,30	1,49	0,07	0,47	0,03	1,25
IZN111	6,57	0,73	< MDL	-	0,92	0,06	55,60	0,26	< MDL	-	0,16	0,02	1,25
IZN22	6,43	0,72	0,20	0,10	0,94	0,06	54,46	0,26	< MDL	-	0,39	0,03	1,32
IZN33	2,86	0,75	< MDL	-	1,22	0,07	53,52	0,27	0,40	0,06	0,38	0,03	1,72
IZN11	6,74	0,76	< MDL	-	0,49	0,06	53,79	0,26	< MDL	-	0,18	0,03	1,32
MDL	0,9000		0,1000		0,0800		0,0611		0,0376		0,0558		0,0413

Object	$ErrK_2O$	CaO	ErrCaO	TiO_2	$ErrTiO_2$	MnO	ErrMnO	Fe_2O_3	$ErrFe_2O_3$	CoO	ErrCoO	NiO	ErrNiO
IZN2	0,02	1,16	0,02	0,02	0,01	0,02	0,005	0,25	0,01	0,10	0,01	0,03	0,01
IZN222	0,03	1,38	0,03	0,03	0,01	0,01	0,004	0,37	0,01	0,69	0,01	0,31	0,02
IZN55	0,02	1,63	0,03	0,05	0,01	0,02	0,005	0,33	0,01	0,11	0,01	0,05	0,01
IZN111	0,03	1,94	0,03	0,04	0,01	0,01	0,050	0,52	0,01	0,23	0,01	0,12	0,01
IZN22	0,03	1,03	0,03	0,02	0,01	0,87	0,018	0,46	0,01	0,07	0,01	0,06	0,01
IZN33	0,03	1,42	0,04	< MDL	-	0,03	0,006	0,43	0,01	0,07	0,01	0,04	0,01
IZN11	0,03	0,56	0,03	< MDL	-	0,01	0,005	0,22	0,01	0,40	0,01	0,28	0,02
MDL		0,0902		0,0159		0,0069		0,0092		0,0086		0,0101	

Object	CuO	ErrCuO	ZnO	ErrZnO	As_2O_3	$ErrAs_2O_3$	SnO_2	$ErrSnO_2$	PbO	ErrPbO	Bi_2O_3	$ErrBi_2O_3$
IZN2	0,37	0,01	0,01	0,007	0,22	0,09	1,41	0,07	30,19	0,22	0,29	0,07
IZN222	0,12	0,01	0,04	0,007	0,35	0,09	6,95	0,11	23,48	0,19	0,55	0,07
IZN55	0,31	0,01	0,03	0,007	0,16	0,08	2,01	0,08	27,82	0,21	0,14	0,07
IZN111	0,21	0,01	0,02	0,007	< MDL	-	5,10	0,10	27,03	0,20	0,21	0,07
IZN22	0,19	0,01	0,01	0,007	< MDL	-	4,85	0,10	28,34	0,21	0,16	0,07
IZN33	0,11	0,01	0,04	0,007	< MDL	-	6,70	0,12	30,76	0,24	0,13	0,08
IZN11	0,45	0,02	0,03	0,007	< MDL	-	4,74	0,11	30,38	0,23	0,29	0,08
MDL	0,0074		0,0056		0,0676		0,1075		0,0634		0,1020	

Table 1 – The results of the PIXE analysis; the values are oxide concentrations in weight percent. Minimum detection limits (MDL) are also given.

a commercial route existed in the 17th century between Saxony and the Ottoman Empire (Saxony – Breslau – Lemberg – Moldavia – Istanbul). It is therefore expected that the Saxony Co-minerals were intensively traded during that period and used by the Iznik ceramics manufacturers.

Mineral name	Mineral formula	As/Co (atomic)	As/Co (weight oxide)
Skutterudite	$(Co,Ni)As_3$	3.0	3.6
Erythrite	$Co_3(AsO_4)_2{\times}8H_2O$	0.7	0.9
Safflorite	$(Co,Fe)As_2$	2.0	2.6
Cobaltite	(Co,Fe)AsS	1.0	1.3
Shperocobaltite	$CoCO_3$	0.0	0
Siegenite	$(Co,Ni)_3S_4$	0.0	0
Roselite	$Ca_2(Co,Mg)(AsO_4)_2{\times}2H_2O$	2.0	2.6
Linneite	Co_3S_4	0.0	0

Table 2 – Saxony Co minerals used for obtaining Co-blue pigment.

The main purpose of our study was to identify the Co minerals used for the blue pigments, one of the procedure being the determination of different elemental ratios, such as As/Co and Ni/Co. Another issue that has to be verified was the presence or absence of As in the blue pigments. This is due to two different hypotheses that had to be checked. The first one is that As (presence and amount) may be correlated with a certain Co mineral; the second one involves a change in the technology used to obtain the raw Co pigments.

In particular, the way in which the Co blue pigment was obtained during the Middle Age is worthy of mention here. Initially, cobalt was thought to be of no value and to be a contaminant in the metal. In the 15th century, its value as a colorant was discovered, and cobalt blue gained in importance. This circumstance led to the establishment of blue-color mills. A rapid change in the composition of the blue glazes was observed just before 1520 from the correlations between element constituents of the Co minerals. The change from 1520 is marked essentially by the presence of As and Bi and by reduced amounts of Fe and Ni, as compared to the blue cobalt pigment compositions before 1500 (Zucchiatti *et al.* 2000). One can speculate that this was due to procedures introduced to industrialize blue pigment production more than to exploitation of different families of Co minerals (Zucchiatti *et al.* 2006). In the famous book "De re metallica" of Georgius Agricola (Georg Bauer) from 1556 (Georgius Agricola translated by Albert France-Lanord, Gerard Klopp, 1992), cobalt-blue is described as a by-product of silver melting. Agricola describes how after silver extraction in the smelting furnace, solidified juices or "fluores" are present so that "they can make colours that the artists use". In the book "Ars vitralia experimentalis" of Johannes Kunckel (1689) (Rau 1972), a Co-reverberatory oven is described from which "fumi" (coloured gaseous emissions) were obtained (most likely arsenic and sulphur). With this procedure "saffre" (cobalt oxide mixed with silica) could have been made by P. Weidenhammer in Schneeberg starting from 1520. In 1540, C. Schurer mixed the cobalt oxide with sand and potash to produce a blue, finely ground glass, "smalt". However, it is not possible to estimate how much of the As component of the Co mineral was lost as gas—"fumo". It is quite probable that a fraction of As remained attached to the pigment since even in modern processes like the Sevres method (elaborated in 1844) for producing a cobalt blue pigment leads to an As residue: SiO_2 (2.75%), CoO (80.15%), H_3AsO_4 (17.10%).

After data analysis, a relatively high correlation between the Co and Ni elemental concentrations was revealed (Fig. 2), a fact in agreement with previous measurements performed on other samples of Mediterranean origin (Zucchiatti *et al.* 2006). The measurements revealed As only in some of the measured Iznik samples (see Table 1).

Table 3 presents PIXE experimental results in relation to the sought minerals. As can be seen, the glaze composition is compatible with the chemical composition of minerals most abundant in the Erzgebirge (skutterudite, safflorite) and losses of As from 50 to 90%. Roselite also gives compatible results but is a rarer mineral compared to the above two. The absence of As in the last four samples is difficult to interpret. The use of the "fluores"—remnants from silver melting furnaces in Schneeberg region—cannot be excluded although, at the beginning of the 17th century, its use may have been replaced by the "industrial" cobalt blue of Weidenhammer. In Italian Renaissance ceramics (Zucchiatti *et al.* 2006) the absence of As is accompanied by a higher Ni/Co ratio, which seems to be the case also in this work, with the exception of sample IZN22.

Sample number	Ni/Co (weight oxide)	As/Co (weight oxide)	Minerals and As weight loss
IZN222	0.33	2.25	
IZN222	0.42	0.50	
IZN222 average	0.38	1.40	Skutterudite (61%) Safflorite (46%) Roselite (46%) Cobaltite (0%)?
IZN55	0.43	1.50	Skutterudite (58%) Safflorite (42%) Roselite (42%)
IZN111	0.52	0.22	Skutterudite (94%) Safflorite (91%) Roselite (91%) Cobaltite (83%) Erythrite (75%)
IZN22	0.83	1.30	Skutterudite (64%) Safflorite (50%) Roselite (50%) Cobaltite (0%)?
IZN33	0.57	-	As evaporated?
IZN11	0.72	-	As evaporated?

Table 3 – PIXE results analysis.

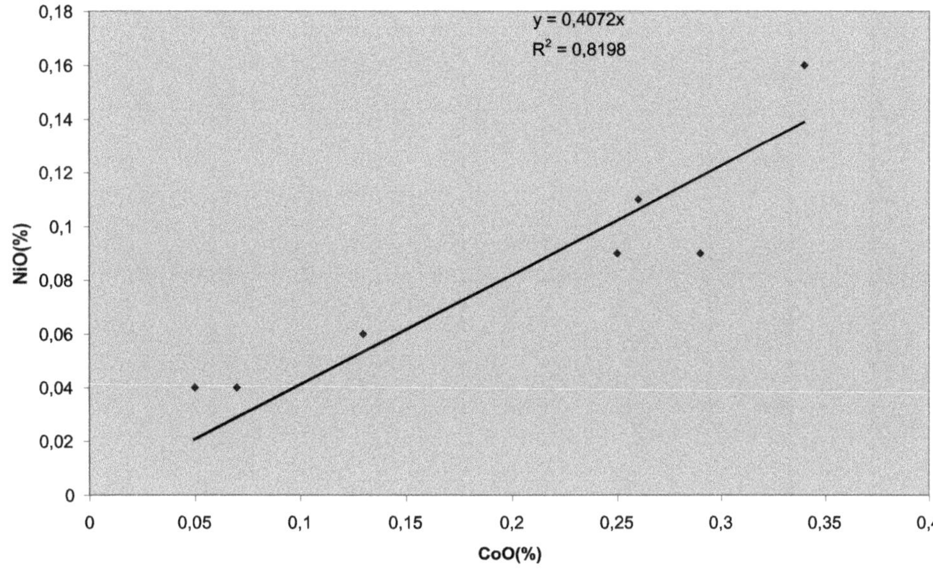

Fig. 2 – CoO-NiO concentrations graph.

The absence of As in some sherds might suggest the hypothesis that the ore might have been the same as the ore of the As-Ni group, but the manufacturing procedure was different. In this case, it would have been a roasted compound of Ni and Co—such as smaltite or erythrite, but roasted at a higher temperature than that in the case of the Co-As-Ni ore. In the first case, As was completely removed, whereas in the second case, most of the As was conserved.

CONCLUSIONS

This paper presents some results obtained from the investigation of some Iznik ceramics sherds using ED-XRD and PIXE techniques. Some conclusions regarding the pigments employed were obtained, especially regarding the Co-blue pigments. The Erzgebirge region from Saxony was the most likely source for the blue color pigment. It is hypothesized that different manufacturing techniques were employed to obtain the Co blue pigment, fact illustrated by the lack of As in some of the analyzed samples.

In the future, we intend to enlarge our archaeological study with Iznik type ceramic sherds from Romanian museums dating from the 16th and 18th centuries, to try to correlate them with the technological evolution of cobalt blue pigment production in Europe.

REFERENCES

Agricola Georgius, *De Re Metallica*, translation by Albert France-Lanord, Gerard Klopp, Thionville, 1992, p. 124-128.

Bugoi R., Constantinescu B., Constantin F., Catană D., Ploştinaru D., Săşianu A. 1999, "PIXE and XRF archaeometrical studies on Greek and Roman silver coins", *Journal of radioanalytical and nuclear chemistry* 242 (3), p. 777-781.

Paynter S., Okyar F., Wolf S., Tite M.S. 2004, "The production technology of Iznik pottery: a reassessment", *Archaeometry* 46 (3), p. 421-437.

Porter Y. 2000, "Le cobalt dans le monde iranien: IX-XVIe siecles", *Taoci* 1, p. 5-14.

Rau H.G. 1972, "Das Glaslaboratorium Johann Kunckels auf der Pfaueninsel", *in Ausgrabungen in Berlin, Band 3*, p. 148-171.

Scott T. 2001, *Society and economy in Germany, 1300-1600*, Palgrave Macmillan.

Tite M.S. 1989, "Iznik pottery: an investigation of the methods of production", *Archaeometry* 31, p. 115-132.

Weber W. 1986, "Der Freiberg Bergbau", *Lapis* 11, p. 13-27.

Zucchiatti A., Bouquillon A., Moignard B., Salomon J., Gaborit J. R. 2000, "Study of Italian Renaissance sculptures using an external beam nuclear microprobe", *Nuclear instruments and methods in physics research B* 161-163, p. 699-703.

Zucchiatti A., Bouquillon A., Katona I., D'Alessandro A. 2006, "The della Robbia Blue: a case study for the use of cobalt pigments during the Italian Renaissance", *Archaeometry* 48 (1), p. 131-152.

TURKISH CERAMICS IN THE CRIMEA ON THE EVE OF THE PORTA INVASION
(PROBLEMS OF CHRONOLOGY OF A CERTAIN GROUP OF VESSELS)

I. TESLENKO

Ukrainian Academy of Sciences, Archeological Institute Crimean Branch,
2 Yaltinskaya str., 95000 Simferopol Crimea, UKRAINE, i_teslenko@mail.ru

An early Ottoman art ceramic of red clay with underglaze paint, known as "Miletus ware" (Sarre 1935, p. 72-75; Lane 1957, p. 251; Erdmann 1963, p. 207; Aslanapa 1965, p. 28-32; 1971a, p. 188-189; 1971b, p. 214-219; Aslanapa *et al.* 1989, p. 25; Hayes 1992, p. 238-243; Özkul Findik 1999, p. 548-550; 2003, p. 106), was widely exported outside the bounds of Anatolia and is often found in excavations in the Pontic region and particularly in the Crimea (Lepper 1913, p. 73-79; Tikhanova, 1953, p. 335-336; Yakobson 1953, p. 399-400, 402; Veimarn, 1963, p. 78-79; Parshyna 1974, p. 76; Mitsyshvili 1976, pl. XII; Tallis 1979, p. 83-85; Baranov 1988, p. 91; Mytz 1988, p. 106; 1991, p. 101-102; 2002, p. 154; Volkov 1993, p. 120; Gin'kut 2001, p. 58; Boguslavski 2002, p. 265-268; Adaksina *et al.* 2003, p. 9, 39, 63; Gertsen and Naumenko 2005, p. 263). The chronology and concentration of "Miletus ware" exports to the peninsula have not yet been identified clearly. This is illustrated by the way in which the researchers took the liberty of dating their ceramics by analogy, within the chronological framework put forward by O. Aslanapa (i.e. by the end of the 14th-15th c.) (Aslanapa 1965, p. 28-32; 1971b, p. 214-219; Aslanapa *et al.* 1989, p. 26), without any attempt to narrow down the chronology by making use of updated information. Nowadays, "Miletus ware" finds are known at practically all late-medieval sites of the Crimea that have been researched. These include the following 15 places: Mangup, Bakla, Inkerman, Balaklava (Chembalo), Chersonese, Simeiz, Oreanda, Gurzuf, Partenit, Ayu-Dag, Alushta (Aluston), Phoona, Sudak, Theodosia (Kaffa), Chaban-Cule (Fig. 1).

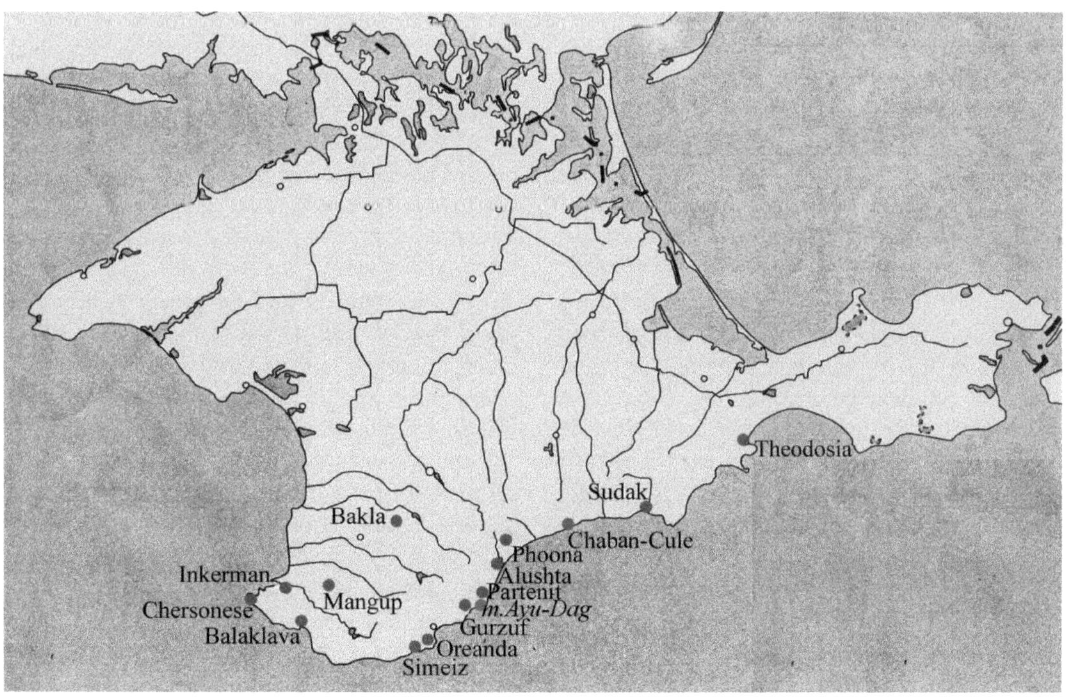

Fig. 1 – Map of finds of "Miletus Ware" in the Crimea.

"Miletus ware" samples found in the Crimea are of a standard nature and differ in no way from the quantity of similar items that have been found overall, for instance in Izmir, Iznik, Istanbul and other places (Fig. 2, Fig. 3). In general, these vessels are of an open form. The pottery is red brown with some slight variations in hue. The original technology used to mould the open vessels is worth a mention: the walls as well as the bottom were drawn up at the same time. At the narrowest point, where the walls curve into the bottom a hole was left. Afterwards, it was covered with a specially manufactured disc, the texture of which was distinguishable by inclusions of a coarse grained sand that was not typical of the whole vessel, made from fine dispersed clay (the contours of the disc are shown in cross section in Fig. 2-1, 3, 5). Such a technological method was evidently connected to the desire to reduce shrinkage of the raw materials during the manufacturing process. This is attested by the fact that closing the hole took place after partial drying of the item. The painting was done on a base of white or light cream colored slip. The decor of "Miletus" vessels is distinguished by the variety of ornamental patterns. Often internal as well as external surfaces of the earthenware were decorated. The colours in use are blue, turquoise, and violet (lilac manganese). The blue color was the dominant one for decorating the internal surfaces of the earthenware. Green and sometimes black colours were used to paint the outlines throughout. Transparent white, or, more seldom, turquoise glazes were used, to cover the internal surface. Slip and glaze covered the external part of the upper portion of the vessel, but did not reach the stem. On the external surface the traces of tripods are usually visible. The most frequent finds reveal three types of dishes and one type of bowls.

In terms of Crimean antiquities, chronological data related to the medieval cultural layers from which "Miletus ware" come are scanty. In scientific published papers on the subject of this group of ceramics, the information has been provided mainly with the idea of showing that it was present at the site. In order to answer questions about early painted Turkish ceramics in the Crimea it is necessary to be familiar with the stratigraphic context of the finds and their relative quantities during the various periods of deposition. The finds from a long-term excavation at the Aluston and the Chembalo fortresses, the castle at Phoona, and Chaban-Cule fortifications as well as the church on the Ay-Constant Glade at Ayu-Dag mountain near Partenit have finds which have been qualitatively analysed, and are highly suitable for data processing. The excavation work was executed by the Mining Crimean Expedition of CB IA NASU under the leadership of Dr. V.L. Mytz, and in Balaklava by the Southern Crimean archeological expedition under the leadership of S.B. Adaksina.

The Aluston fortress excavations yielded two fragmentary bowls which could be partially restored, a bowl and a dish with a full profile, and the fragments of no less than 40-45 vessels of open forms as well as one fragment of a jug that is related to the "Miletus ware" group (Fig. 2). The finds were discovered in the following circumstances: two fragments came from the masonry of a curtain wall that was constructed in the sixties of the15th c. The next group of fragments (6 fragments, 6 vessels) was found in layers which were formed after that, but prior to the capture and destruction of the fortress by Ottomans in 1475 AD. One unbroken cup and 5 remnants of 3 vessels were found in the cinder layer from the fire during the siege of the fortress by the Turks. The finds from the deposits formed after the disastrous events of 1475 constitute the most substantial group. One dish of complete form, two completely and partially restored bowls, and the fragments of no less than 30 various open type vessels were found in the dump at one of the towers of the ruined fortifications (Orta-Culet). The upper date for the deposits was determined by coins, the latest of which are 2 akche from the last quarter of the 15th c. Four fragments were found in a rubbish pit of the 16th c. Three more fragments were discovered in the layers along with coins of the 16th-18th centuries.

The excavations at Phoona revealed one of the best "Miletus" pottery collections in the Crimea (Fig. 3), composed of nine complete forms, another four in fragments and about 150 odd pieces of no more than 90 vessels of open forms. A detailed chronology for periods of construction of the fortification has been worked out (Kirilko 2005, p. 35-81). The chronology enables us to determine the data of the cultural layers, and relate them to the various stages of the fortress and the construction materials to within 16-29 years. The construction periods of the fortifications were defined as follows:
1) 1423: the time when they were already erected and then destroyed as a consequence of a mighty earthquake.
2) 1425-1434 or the middle of the 15th c.: restoration of the fortification and destruction of the advanced post buildings in a fire.
3) 1459-1475: the period when the entire construction was transformed into a castle. Later on it was completely devastated during the Turkish expansion.

As well, constructions from closed contexts which had existed prior to the major fortifications were explored. They had ceased to exist by 1423.

The Turkish ceramic finds were as follows: two fragments of bowls were found in cinder layers of the premises that relate to the second construction period (*i.e.* the second and the early third quarters of the 15th c.). From the contexts and the strata synchronous with the third construction period (building and make-up ground, occupation layers etc.) about 75 fragments from 59 items were uncovered. In the layers of cinder resulting from the fire of 1475 9 complete forms, 4 fragmentary vessels and 75 remnants of no more than 34 items were found. In the deposits from later periods there were three fragments of various vessels.

During excavations of the fortifications at Chaban-Cule (dating from the third quarter of the 15th c.) the remnants of four open form vessels were found (Kirilko and Mytz 2004, p. 212, fig. 22, 7, 8).

While undertaking archeological research at the Christian church at Ayu-Dag Mountain, pieces of a red clay

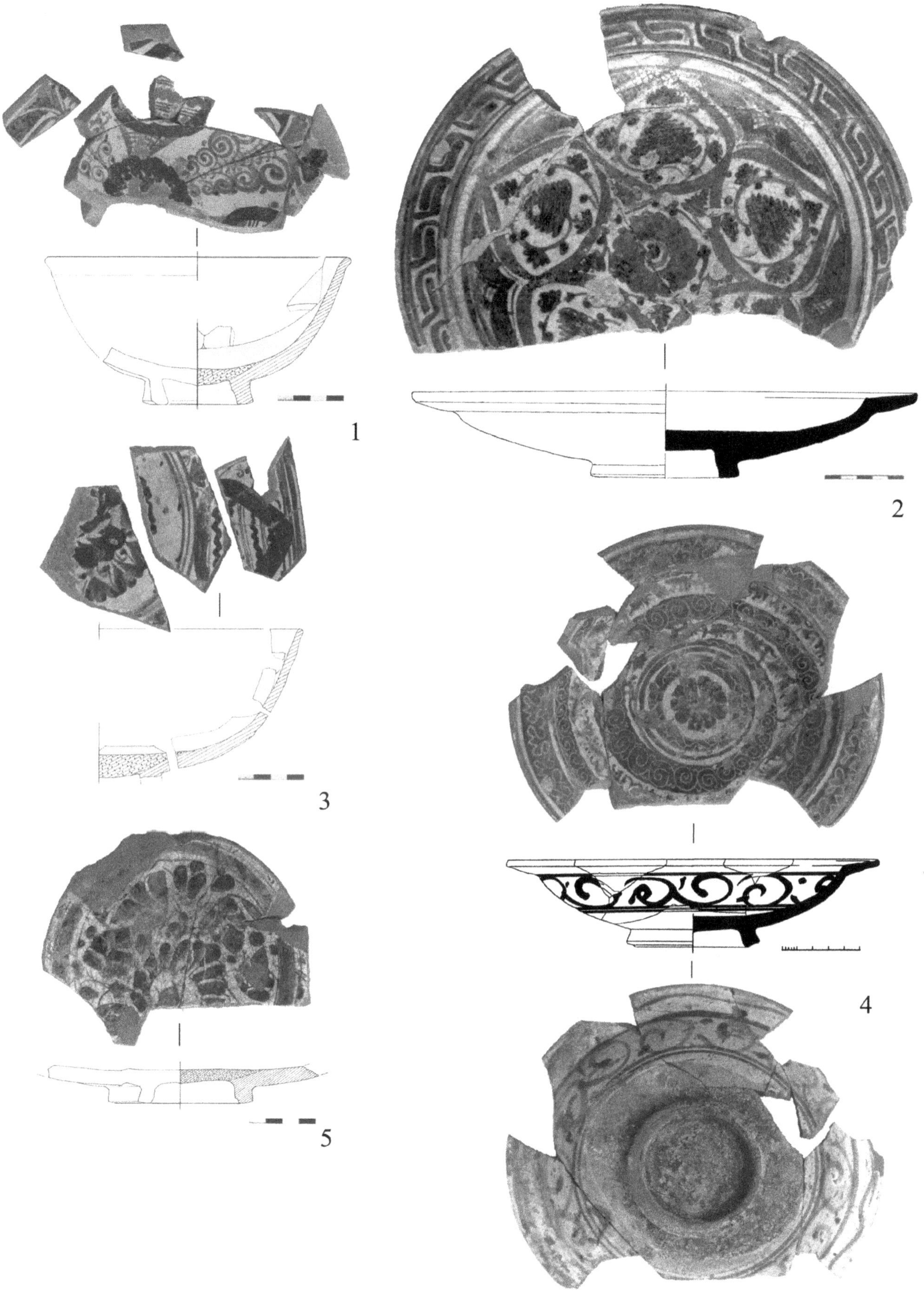

Fig. 2 – "Miletus Ware" from excavations of Aluston fortress.

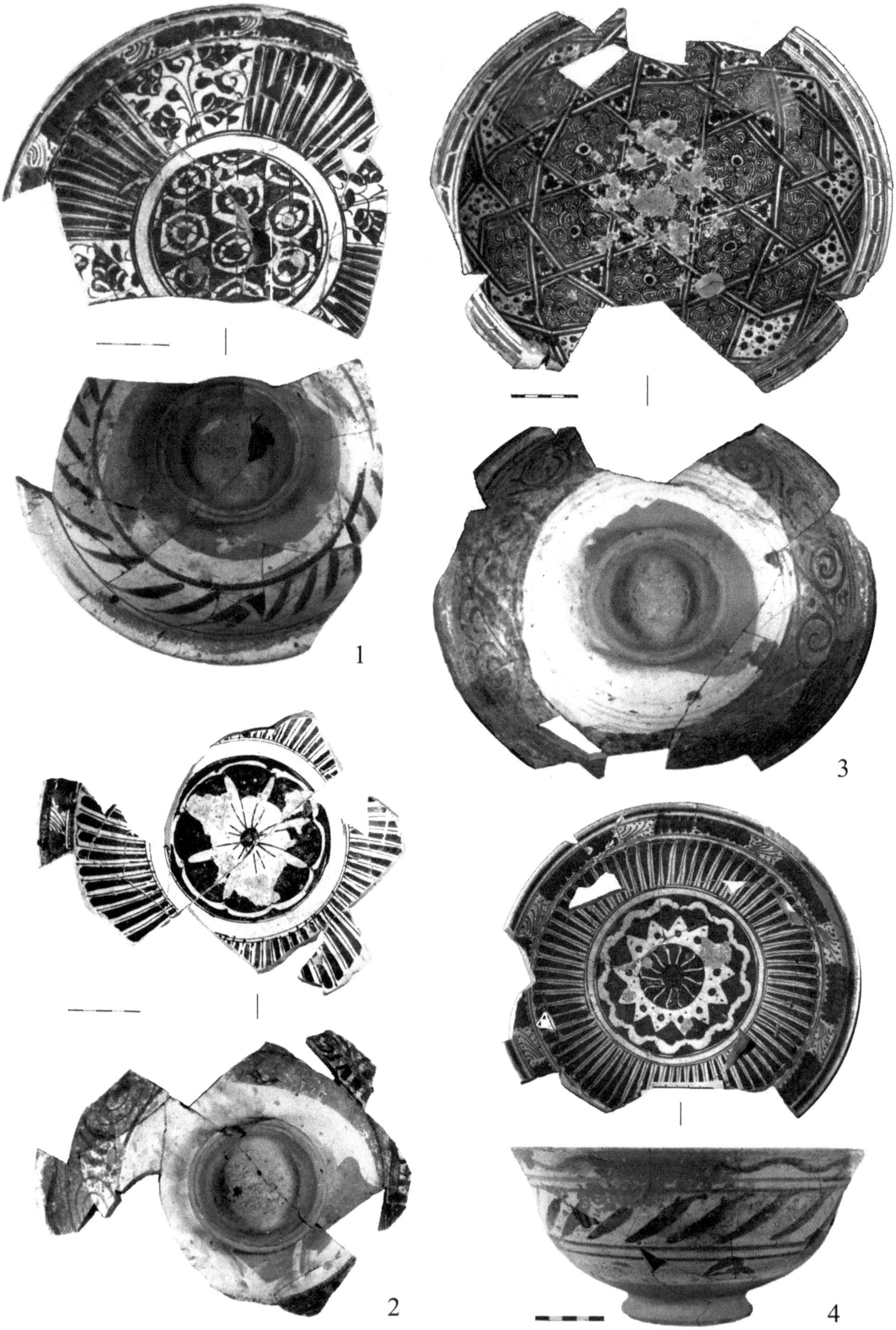

Fig. 3 – "Miletus Ware" from excavations of Phoona castle.

pitcher painted with cobalt were found with a coin dating between 1550 and 1570 AD (Teslenko *et al.* 2001, p. 55).

Thus, one can come across early Turkish art ceramics in various contexts in architectural substructures as well as in diverse archaeological layers. The latter, in terms of their particular circumstances of formation, are subdivided into three categories—closed contexts[1], stratification of long-term formation, embedded in situ and artificially displaced deposits. The information, obtained after analysis of data from these various categories is not equivalent. The first fixes the "horizontal" (single occasion) incursion of a material culture, the second depicts the process of development or replacement of domestic conditions and styles, the third bears testimony to the availability of these artifacts at the time of emplacement of the deposits. All of them can be made use of in order to draw up chronological diagrams, depending upon the presence of the chronological indicators (coins, epigraphic relics and so on) or chronological registration marks (deposits, architectural structures or some definite construction periods, correlated with precisely dated historic events). Data on the frequency of the artifacts in various deposits during definite chronological periods with respect to five historic places are summarised in Table 1.

The most numerous finds of "Miletus" pottery are those that were excavated at the site of the Phoona castle, discovered in the deposits which had formed between 1459 and 1475 AD. At the same time at Aluston, where almost half of the monument area and more than fifty closed contexts with the traces of the fire of 1475 were explored, the finds of Ottoman ceramics in synchronous deposits are rare indeed. Evidently, at the end of the third quarter of the 15th c. that type of ceramic was not widely in use in the Crimea and was not an available luxury item. In the last quarter of 15th c., and at the beginning of 16th c., when the Crimea came under the control of Ottoman Porta, the situation altered. The "Miletus" group pottery came into wider use. The best testimony to that is the large quantity of debris, the fragments of such vessels in the dump of ruins near Orta-Culet Tower at Alushta settlement. The probability of importation of an early art ceramic of Ottoman period into the territory of the Northern Black Sea area, particularly to the Crimea, at the end of 14th c. and even at the beginning of 15th c. is weak, as the finds of it in the layers of an appropriate period are unknown. This view is also confirmed by the observations of I.V. Volkov, referring to the data from Belgorod-Dnestrovsky. According to his point of view it is hardly possible to allow a wider period of usage than 15th c. One of the pieces of evidence for the late origin of "Miletus ware" is a complete lack of the pottery in Golden horde layers in Saraj, Madzhara, Azov, with an upper date of 1396 AD, the devastating campaign under the leadership of Timur (Volkov 1993, p. 120).

Thus, on the basis of the above data, one can define the following chronological boundaries for when "Miletus" pottery reached the Crimea. The first handicraft wares appeared here not earlier than the second quarter—the middle of the 15th century. Taking into account the general political situation in the near Black Sea region, it is logically correct to assume that it had happened by the beginning of the third quarter of the 15th c., when Constantinople was captured by the troops led by Mohamed II "Fatiha" in 1453. The contexts of the finds in the Crimea and particularly those in Istanbul are in no way contradictory to this opinion. In Saraçhane "Miletus" wares are found in quantity in the early Turkish layers, the lower date of which is 1460–77. From the general historic background, J.W. Hayes also concludes that the "Miletus" ceramics appeared here not earlier than in 1453 (Hayes 1992, p. 238). The greatest influx of cobalt painted red clay ceramics to the Crimea was essentially at the end of the 3rd–4th quarters of 15th c., and most likely at the beginning of 16th c. At this time the great bulk of the pottery was discarded in the dumps of even an insignificant settlement (for those times) such as Alushta. The finds of "Miletus" earthenware in an early 16th c. (1509-1516) complex are typical for Azak. We may consider the final falling out of use for "Miletus" earthenware in the Crimea to be the end of the 3rd–the beginning of the 4th quarter of the 16th century but at the same time this does not define the time when production ceased. The researcher of ceramic production kilns in Iznik, Dr. O. Aslanapa considers that final cessation of production of "Miletus" ware can be referred to the end of 15th century (by 1480-90) (Aslanapa 1971b, p. 215; Aslanapa *et al.* 1989, p. 26). J.W. Hayes, on the basis of stratigraphic observations during excavations in Saraçhane, concludes that the final production of this type of ceramics could be 1520, pointing out there was a production boom of "blue-and-white" pottery at that time (Hayes 1992, p. 238). J.M. Rogers stresses that in Iznik from the boundary between the 15th and 16th centuries production of ceramics different in their ornaments from the earthenware of the previous century had commenced (Rogers 1983, p. 9, 95, 100, 113). In any case, production did not extend as far as the first quarter of the 16th century (at a minimum, to the end of 15th century or at the boundary between the 15th and 16th centuries, at a maximum, to the second half of the first quarter of the 16th century). This allows us to determine an average length for its period of currency, since the wares fell out of use in the Crimea as soon as production had ceased in Turkey. According to our calculation the minimum is 30±5 years, and the maximum, 50±5 years.

In conclusion, it has to be admitted that so far research has not solved the question of the initial and final dates of "Miletus" ware production in Turkey. Nevertheless, we have succeeded in defining more clearly the period and intensity of the introduction of this type of earthenware to the Crimea, as well as the period of its circulation after the final cessation of production.

1 – A closed context is an archaeological context formed in a limited area at one time or in a short period of time (for example, a layer resulting from a fire in a building).

Date	Contexts	Number of artifacts							
		Aluston (Alushta)	Phoona	Chembalo (Balaklava)	Chaban-Cule	Ay-Constant	total		
							by context	by period	
1434-1459	1								
	2		2(2)/0-2				2(2)/0-2	2(2)/0-2	1.20%
	3								
	4								
1434-1463	1								
	2								
	3			3(3)/0-3			3(3)/0-3	4(4)/0-4	2.40%
	4			1(1)/0-1			1(1)/0-1		
1450/1460 -1475	1	2(2)/0-2					2(2)/0-2		
	2								
	3	4(4)/0-4	59(75)/0-59		3(3)/0-3		66(82)/0-66	68(84)/0-68	40.70%
	4								
1475	1								
	2	3(5)/1-4	34(75)/13-47				37(80)/14-51	37(80)/14-51	30.50%
	3								
	4								
15th-16th c.	1								
	2	4(4)/0-4					4(4)/0-4	39(59)/3-42	25.20%
	3	30(32)/3-33	3(3)/0-3		1(1)/0-1	1(19)/0-1	35(55)/3-38		
	4								
total by site		43(47)/4-47	98(155)/13-111	4(4)/0-4	4(4)/0-4	1(19)/0-1	150(229)/17-167		
		28.10%	66.50%	2.40%	2.40%	0.60%			100%

Table 1

Contexts:

1 – architectural sub-structures
2 – closed complexes
3 – stratification of a long term formation
4 – artificially displaced sediments

Number of artifacts a(b)/c-d:

a – number of fragmented vessels
b – number of sherds constituting the fragmented vessels
c – number of complete vessels
d – total number of vessels (fragmented + complete)

The number of fragmented vessels is given according to the highest number/indicator.

REFERENCES

Adaksina S.B., Kirilko V.P., Mytz V.L. 2003, *Otchyot ob arkheologicheskih issliedovaniyah srednevekovoi kreposti Chembalo (Balaklava) v 2002 g.*, Sankt-Peterburg, Simferopol.

Aslanapa O. 1965, *Turkische Fliesen und Keramik in Anatolien*, Istanbul.

Aslanapa O. 1971a, "Pottery and kilns from the Iznik excavations between 1963 to 1966", *Proceedings of the 27th International Congress of Orientalists*, p. 188-189.

Aslanapa O. 1971b, "Turkish Ceramic Art", *Archaeology* 24 (3), New York, p. 209-220.

Aslanapa O., Yetkin S., Altun A. 1989, *The Iznik tile kiln excavations: the second round: 1981-1988*, Istanbul.

Baranov I.A. 1988, "Glavnyie vorota srednevekovoi Soldai", *Arkhitekturno-arkheologicheskie issledovaniya v Krymu*, Kiev, p. 81-97.

Boguslavski G.S. 2002, "Polivnaya keramika iz raskopok Belgorod-Tirskoi ekspeditsii (1996-1998)", *Tyras Cetatea Alba/Belhorod-Dnistros'kyj. Sapaturi 1996-1999*, p. 265-284.

Erdmann K. 1963, "Neu Arbeiten zur Türkischen Keramik", *Ars Orientalis* 5, p. 191-219.

Gertsen A.G., Naumenko V.E. 2005, "Polivnaya keramika iz raskopok tsytadeli Mangupa", *in* S. Botcharov and V. Mytz (eds.), *Polivnaya keramika Credizemnomop'ya i Prichernomor'ya X-XVIII vv.*, Kiev, p. 257-287.

Gin'kut N.V. 2001, "Polivnaya keramika XIV-XV vv. iz raskopok 'Konsul'skoi tserkvi' kreposti Chembalo", *Vzaimootnoshenie religioznyh konfessiy v mnogonatsyonal'nom regione*, Sevastopol, p. 53-60.

Hayes J.W. 1992, *Excavations at Saraçhane in Istanbul,* vol. 2, Princeton.

Kirilko V.P. 2005, *Krepostnoi ansambl' Funy 1423-1475,* Kiev.

Kirilko V.P., Mytz V.L. 2004, "Ukreplenie Chaban-Kule", *O drevnostiah Yuzhnogo berega Kryma i gor Tavricheskih*, Kiev, p. 205-245.

Lane A. 1957, " The Ottoman pottery of Isnik", *Ars Orientalis* 2, p. 247-281.

Lepper R.K. 1913, "Arkheologicheskie issledovaniya v Mangupe v 1912 (predvaritel'hoe soobschenie)", *Izvestiya imperatorskoi arkheologicheskoi komissii* 47, p. 73-79.

Mitsyshvili M.N. 1976, *Iz istorii proizvodstva gruzinskoi polivnoi keramiki (XI-XVIII vv.),* Tbilisi.

Mytz V.L. 1988, "Nekotorye itogi izuchenija srednevekovoi kreposti Funa", *Arkhitekturno-arkheologicheskie issledovaniya v Krymu,* Kiev, p. 97-115.

Mytz V.L. 1991, *Ukreplieniya Tavriki X-XV vv.*, Kiev.

Mytz V.L. 2002, "Genuezskaya Lusta I Kapitanstvo Gotii v 50-70 gg. XV v.", *in Alushta I Alushtinskiy region s drevneishyh vremen do nashyh dnei,* Kiev, p. 139-189.

Özkul Findik N. 1999, "Miletus wares found in Iznik Roman theatre excavations", *Turkish art*, p. 547-554.

Özkul Findik N. 2003, "Turkish Glazed Pottery", *in* C.S. Lightfoot (ed.), *Amorium Reports II, Research papers and technical reports,* p. 105-211.

Parshyna E.A. 1974, "Srednievekovaya keramika Yuzhnoi Tavriki (po materialam razvedok 1956-1969)", *Feodal'naya Tavrika,* Kiev, p. 56-94.

Rogers H. 1983, *Islamic Art and Design 1500-1700*, London.

Sarre F. 1935, "Die Keramik der islamischen Zeit von Milet", *Das islamische Milet*, Berlin, p. 72-75;

Tallis D.L. 1979, "Polivnaya keramika Baklinskogo gorodischa", *Sovetskaya arkheologia* 4, p. 63-87.

Teslenko I.B., Siomin S.V., Lysenko A.V. 2001, "Raskopki Khrama XIV-XVI vv. na severo-vostochnom sklone g. Ayu-Dag", *Arkheologichni vidkryttia v Ukraini 1999-2000*, Kiev, p. 53-56.

Tikhanova M.A. 1953, "Bazilika", *Materialy i issledovaniya po arkheologii* 34, p. 334-389.

Veimarn E.V. 1963, "Arkheologichni roboty v raioni Inkermanu", *Arkheologichni pam'yatky Ukrainy* 13, p. 78-79.

Volkov I.P. 1993, "Eschio raz o keramike s rospis'yu kobal'tom iz Belgoroda-Dniestrovskogo", *Drevnosti Pritchernomop'ya. Kratkie soobscheniya Odesskogo arkheologicheskogo obschestva*, Odessa, p. 119-121.

Yakobson A.L. 1953, "Dvorets", *Materialy i issledovaniya po arkheologii* 34, p. 390-418.

PRELIMINARY COMPARATIVE ARCHAEOMETRIC RESULTS ON INKA AND COLONIAL CERAMICS FROM PARIA (ORURO, BOLIVIA)

VERONIKA SZILÁGYI[1], JÁNOS GYARMATI[2], GYÖRGY SZAKMÁNY[1], MÁRIA TÓTH[3]

[1] Eötvös Loránd University, Dept. of Petrology and Geochemistry, Budapest, Hungary

[2] Museum of Ethnography, Budapest, Hungary

[3] Hungarian Academy of Sciences, Research Centre for Earth Sciences, Laboratory of Geochemical Research, Budapest, Hungary

INTRODUCTION

The overlap of native cultures and of colonizing European culture in South America is still not well known. Survival of local traditions in everyday life may be investigated through pottery finds using scientific means. In the settlement of Paria (Bolivia), archaeological evidence of the two consecutive cultures, the Inka and the conquering Spanish, have been found. This paper presents a preliminary archaeometric study of ceramics attributed to these periods.

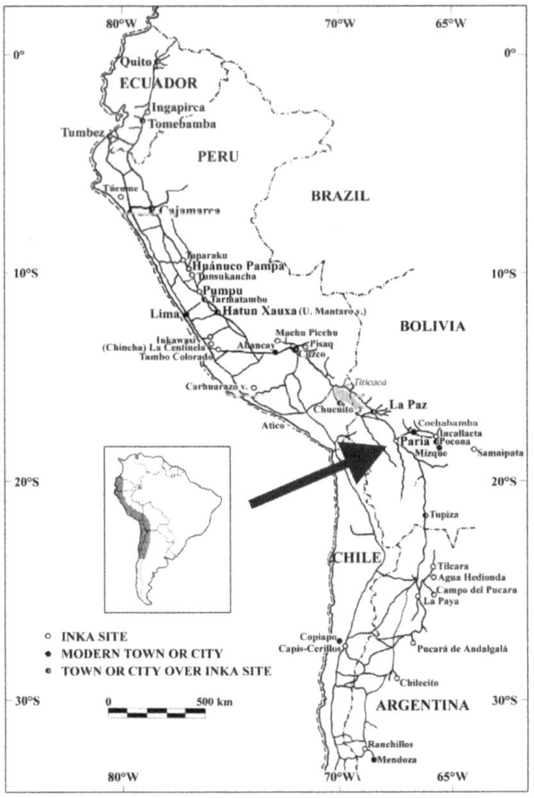

Fig. 1 – The Inka road system with the principal Inka sites. Paria, the site where the ceramics are excavated from signed with the arrow. (Modified after Gyarmati and Varga 1999, p. 2).

INKA PERIOD

The Inka culture was one of the most important in South America before the Spanish colonization. In spite of the many finds we have limited knowledge about the life of the Inkas. It is especially appropriate for archaeometrical examination of the pottery.

There are several facts that suggest a high level of organisation in the Inka Empire which ranged over 900,000 km^2. One example is the unified manufacturing, forming and decorating processes of pottery production.

The subject of our research is Paria, one of the most important provincial centres of the Inka Empire in Bolivia (Fig. 1). The Inka Period in Paria lasted from approximately 1400/1450 to 1550 AD. The settlement was established at the intersection of the principal road from Cuzco that continued along the western side of Lake Titicaca and a branch route running towards the Valley of Cochabamba, the main agricultural zone in the southern part of the empire.

At this first stage in our investigation the observations were carried out on a collection of sporadic surface ceramic finds (54 samples) (Fig. 2a-b). For further comparison, local alluvial sediments (clay-silt-sand) were collected from the vicinity of the archaeological site.

COLONIAL PERIOD

According to historical sources, Paria was the very first Spanish settlement in present Bolivia and was founded at the Inka administrative centre in 1535. Excavations there in recent years have recovered Colonial ceramics of the 16th and 17th centuries in large quantities besides Inka pots. Some forms of the Colonial pottery show a similarity in appearance to the Inka ceramics.

A sample collection excavated by Carola Condarco (the archaeologist at Oruro, Bolivia) contained Colonial (4 samples) and Inka style (5 samples) ceramics in fifty-fifty proportions. Our archaeometric investigation dealt with the Colonial ones (Fig. 2c-d).

Fig. 2 – Ceramic fragments from the Paria archaeological site: (a-b) Inka style and (c-d) Colonial style potteries.

AIM AND METHOD

The object of our research is the archaeometric-petrographic investigation of Inka imperial and Colonial pot sherds excavated in Paria. Our aim is to ascertain if there was any continuity in ceramic making technology between the Inka and the Spanish colonial pottery, whether the colonizers continued their traditions of European (Spanish) culture or followed the practise of the local culture when they made their pots. The main goal is to identify the raw materials used and to characterize the ceramic production technology (raw material preparation and firing processes).

The basic methods for the pottery investigation are macro- and microscopic petrographic observations and X-ray powder diffraction analysis.

PETROGRAPHY

The vessels were usually compact (dense), hard ceramics with homogenous fracture, with an oriented, hiatal (non-continuous grain size distribution) fabric, which has a varied grain size (50–1000 μm) and medial porosity. The basis of the classification of the ceramics was the mineralogical composition of the non-plastic inclusions (Szakmány 1998).

In the case of the pottery fragments coming from the Inka Period layers we distinguished three main petrographic types.

I. "Mineral fragment" type (Fig. 3a): medium-grained (100–250 μm) fabric, subangular or angular grains, dominantly mineral fragments (quartz: mono- and polycrystalline; feldspar: mainly plagioclase and fine-grained biotite) and as an addition to this, fine-grained sedimentary and slightly weathered volcanic rock fragments (consisting of unaltered quartz and biotite with an altered matrix).

II. "Volcanic lithofragments" type: with a varied grain size distribution (50–1000 μm), and rounded grains (the glassy phase forms angular grains). There are three subtypes belonging to this group:

II/A "crystalline" subtype (Fig. 3b): the dominant rock fragments are unaltered acidic-neutral volcanic lithofragments with an isotropic matrix, fine lath-shaped plagioclase, and biotite or amphibole;

II/B "pumiceous" subtype (Fig. 3c): the dominant rock fragments are unaltered acidic-neutral volcanoclastic lithofragments with a pumiceous-glassy matrix containing plagioclase, biotite and amphibole phenocrysts;

II/C "glass shard" subtype (Fig. 3d): the dominant rock fragments are unaltered glassy volcanoclastic lithofragments (glass shards). In addition to these rock grains there are fine- to medium-grained mineral fragments (biotite, quartz, feldspar, amphibole, pyroxene) amongst the non-plastics.

III. "siltstone" type (Fig. 3e): coarse-grained (dominantly 1000–1500 μm), inhomogenous fabric (which suggests poor preparation of the raw material), well rounded grains, dominantly siltstone and sandstone lithofragments (fine-grained oriented quartz and muscovite), and in addition to this, fine-grained mineral fragments (muscovite, quartz, feldspar), volcanic and argillaceous lithofragments.

The petrographic investigation of the Colonial pottery sherds provided partly similar results. We could identify "volcanic lithofragments" type ceramic samples with all rock fragments mentioned above (Fig. 4a-b), and one sherd with carbonised grains and a poorly prepared, mixed matrix (Fig. 4c).

The petrographic composition of the local sediments is dominant siltstone and sandstone fragments (Fig. 5a) with a moderate quantity of slightly weathered volcanic lithofragments with a glassy matrix (and quartz, feldspar, biotite minerals as minor components) (Fig. 5b).

X-RAY POWDER DIFFRACTION ANALYSIS

The results of the X-ray powder diffraction analysis showed that the majority of the Inka pottery fabric consists of quartz, feldspar (mainly plagioclase and rarely potassium feldspar), mica-illite like 10Å phyllosilicate and minor amounts of hematite-goethite, chlorite and amorphous

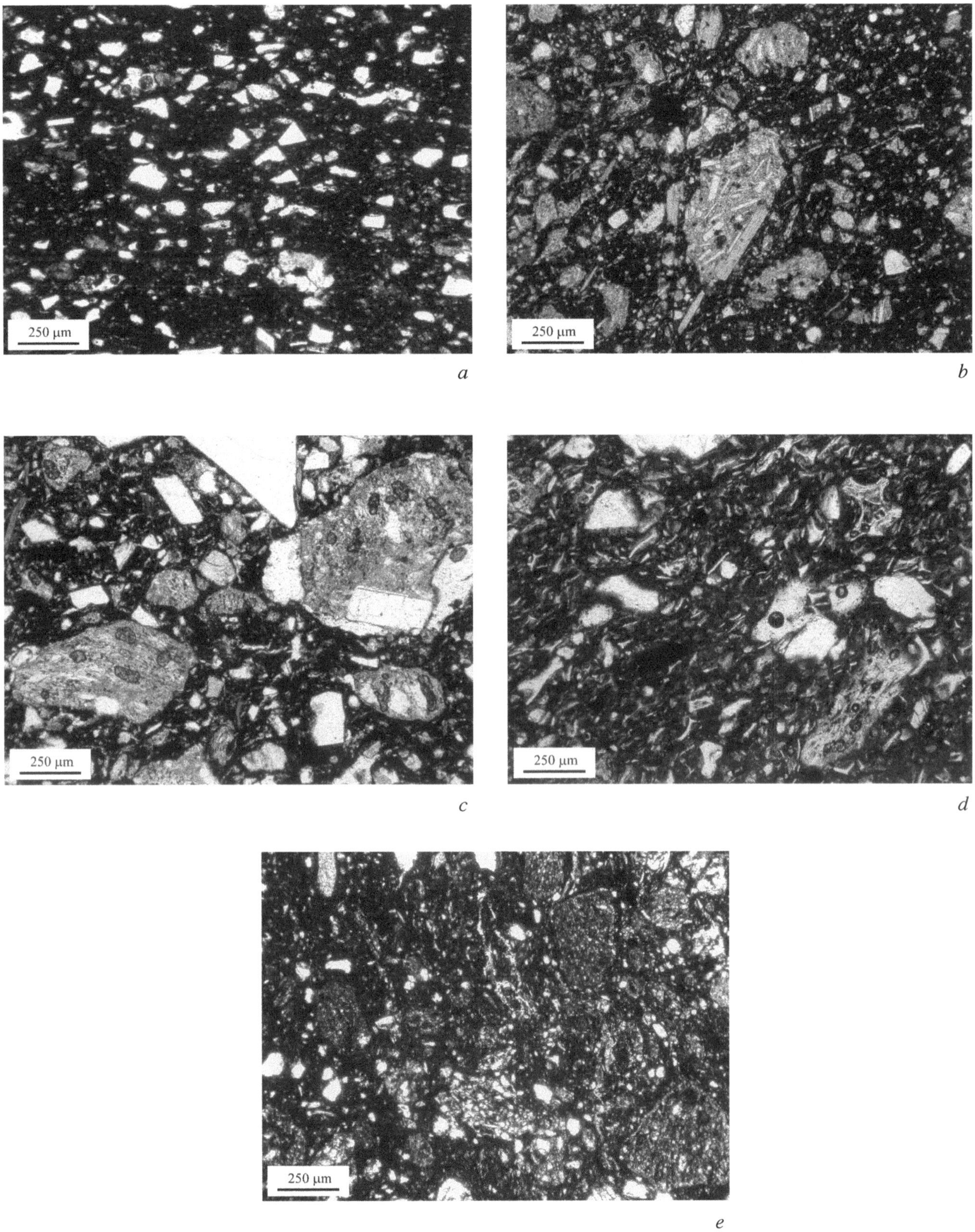

Fig. 3 – Photomicrographs of the petrographic groups of the Inka ceramics: (a) "mineral fragment" type; (b-d) "volcanic lithofragment" type, (b) "crystalline" subtype, (c) "pumiceous" subtype, (d) "glass shard" subtype; (e) "siltstone" type.

phases. The differences between the petrographic types are reflected in the distribution of the above mentioned mineral phases. The presence of chlorite suggests a low (650–700 °C) maximum firing temperature (Peters and Iberg 1978). The elevation of the background indicates the appearance of an amorphous phase which is due to the glassy matrix phase of the volcanic-volcanoclastic rock fragments.

The Colonial ceramics contain quartz, feldspar (mainly plagioclase and rarely potassium feldspar), gehlenite, amorphous phase and minor amounts of hematite-goethite, mica-illite like 10Å phyllosilicate and calcite. The presence of the gehlenite, the increased quantity of the amorphous phase and the absence of chlorite suggests a higher (800–900 °C) maximum firing temperature (Peters and Iberg 1978) for the ceramic samples containing carbonate grains too.

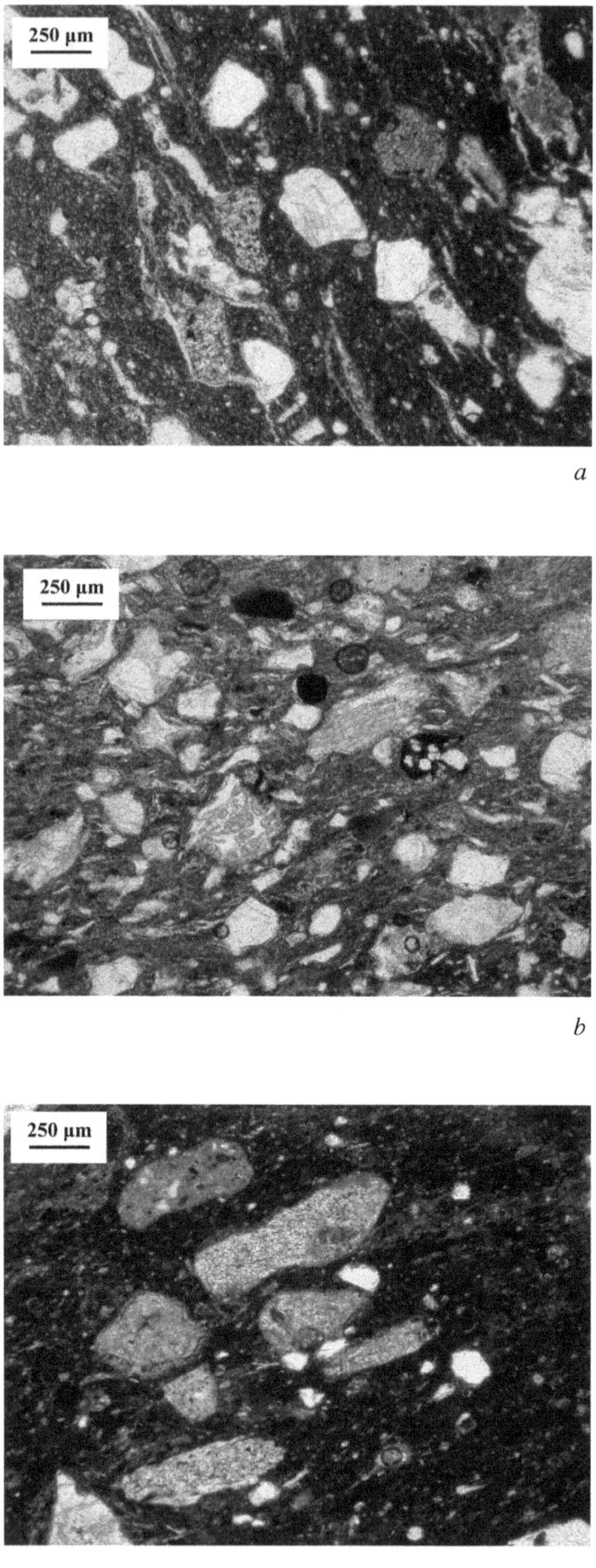

Fig. 4 – Photomicrographs of the Colonial style ceramics: (a-b) samples similar to the "volcanic lithofragment" type Inka pottery; (c) ceramic fragment with non-plastic carbonate grains.

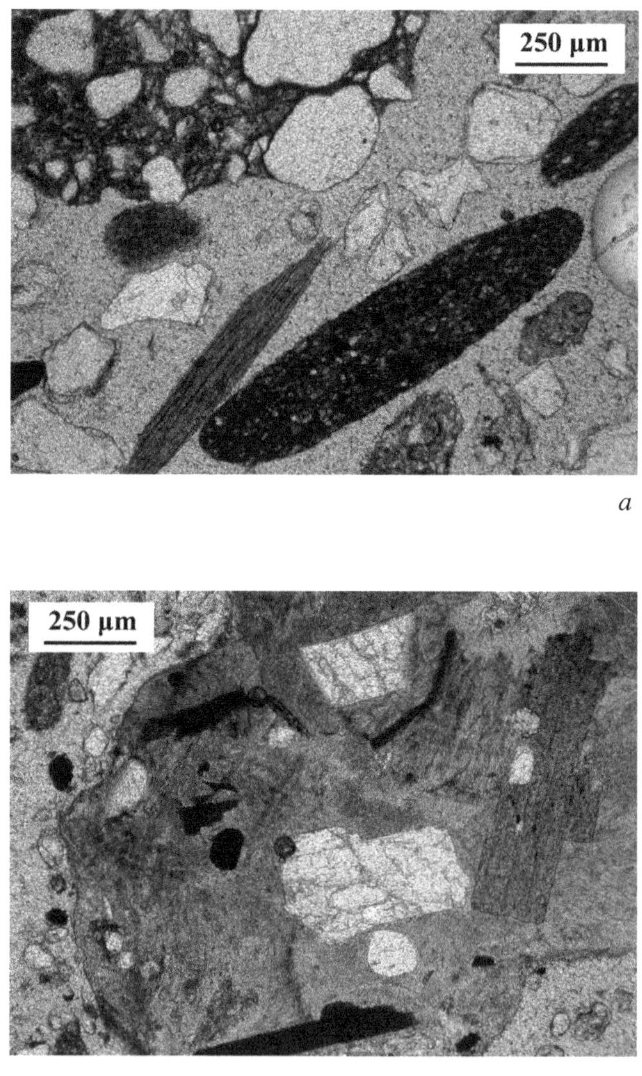

Fig. 5 – Photomicrographs of the local sediments: (a) siltstone and sandstone fragments; (b) slightly weathered volcanic lithofragments.

CONCLUSIONS

According to the above mentioned observations we conclude that:

— the Inka ceramic samples can be classified into three main petrographic types: a "mineral fragment", a "volcanic lithofragments" and a "siltstone" type. The "mineral fragment" and the "volcanic lithofragments" types have non-plastics mainly consisting of acidic-neutral volcanic-volcanoclastic rock fragments and their mineral clasts. The raw materials of these groups may be derived from the same geological setting, but from different parts of it. The relatively homogenous fabric of these ceramics presumes a well prepared or technically elaborated raw material which could be the work of professional potters. The mineralogical compositional and fabric characteristics of the "siltstone" type ceramics suggest that the ceramics here were made by local potters with less elaboration. The non-plastic composition of these ceramics is similar to the mineralogical composition of the local alluvial sediments. These facts make it clear that the "siltstone" type ceramics are local products, while the raw material of the other ceramic types come from the wider vicinity of the site. In the case of all the petrographic types the XRD analysis indicated a low maximum firing temperature (650-700 °C);

— the Colonial ceramic samples consist mainly of acidic-neutral volcanic-volcanoclastic rock fragments and their related mineral fragments. In one case the raw material of a sherd was a poorly prepared clay mix and the non-plastics were carbonate grains. These facts make it clear that the majority of the Colonial samples was made of the same raw material as the Inka samples (from the wider vicinity of the site) but some new materials appeared in the pottery making process related to a carbonate-rich geological setting (the source may be an outcrop from the wider vicinity or the pottery may be an imported vessel), and the local ceramic source disappeared. The XRD analysis indicated a higher maximum firing temperature (800–900 °C).

These data suggest that the raw material usage in pottery making technology did not change to a significant extent in Paria in the transition of the Inka Period to the Colonial Period. The most relevant difference between the subsequent historical periods is the disappearance of local ceramics at the end of the Inka Period, and the appearance of some new raw materials in the Colonial Period. In contrast to the raw material usage, the firing technology appears to have changed to a significant degree: the result of the technological development in the transitional Inka to Colonial Period was a higher firing temperature, which may be related to the Spanish pottery making traditions coming into the area. However, these results, obtained on a small sample, may not be representative of the majority of the archaeological ceramics, and need to be confirmed on the basis of a larger representative sampling.

ACKNOWLEDGEMENTS

We thank Carola Condarco for the Colonial pottery samples. This work was supported by National Research Found of Hungary (OTKA) no. T047048 to János Gyarmati.

REFERENCES

GYARMATI J., VARGA A. 1999, *The Chacaras of War: an Inka State Estate in the Cochabamba Valley, Bolivia*. Museum of Ethnography, Budapest.

PETERS, T., IBERG R. 1978, "Mineralogical changes during firing of calcium-rich brick clays", *American ceramic society bulletin* 57 (5), p. 503-506.

SZAKMÁNY G. 1998, "Insight into the manufacturing technology and the worhshops: evidence from petrographic study of ancient ceramics", *in* L. Költő and L. Bartosiewicz (eds.), *Archaeometrical research in Hungary II*, Budapest-Kaposvár-Veszprém, p. 77-83.

LIST OF PARTICIPANTS

AGOSTINI Lucia
Laboratorio di Acheometria del Museo Fiorentino di Preistoria "Paolo Graziosi"
Via S. Egidio n°21
50122 FIRENZE
ITALY
LABARCHEO@yahoo.it
agostinilucia@libero.it

AMARI Susanna
Superintendence of Catania
Via Vitaliano Brancati 13
95030 NICOLOSI
ITALY
amarisusanna@tiscali.it

BATIGNE-VALLET Cécile
UMR5138 "Archéométrie et Archéologie"
Maison de l'Orient et de la Méditerranée
7 rue Raulin
69365 LYON cedex 7
FRANCE
cecile.batigne@mom.fr

BECHTEL Françoise
Université M. Montaigne-Bordeaux 3
Lab. CRPAA-IRAMAT
Maison de l'Archéologie, Domaine universitaire
33607 PESSAC
FRANCE
francoise.bechtel@u-bordeaux3.fr

BEN AMARA Ayed
Centre de recherche en physique appliquée à l'archéologie
Université de Bordeaux 3 / CNRS / IRAMAT / UMR5060
Maison de l'Archéologie, Esplanade des Antilles
33607 PESSAC
FRANCE
ayed.ben-amara@u-bordeaux3.fr

BEVILACQUA Fabio
c.r.c. Restauri S.r.l.
Via Alfonsine 4/b
40139 BOLOGNA
ITALY
info@crc-restauri.it

BIRÓ Katalin
Hungarian National Museum
Múzeum krt. 14-16
1088 BUDAPEST
HUNGARY
tbk@ace.hu

BLANC Claire
Department of Geosciences, Mineralogy and Petrography, University of Fribourg
Chemin du Musée 6
1700 FRIBOURG
SWITZERLAND
claire.blanc@unifr.ch

BOTTI Alberto
Dip. di Fisica "E.AMALDI"
Universita di Roma tre
Via della Vasca Navale 84
00146 ROMA
ITALY
botti@fis.uniroma3.it

BOURGARIT David
C2RMF
Palais du Louvre, Porte des Lions
14 quai F. Mitterand
75001 PARIS
FRANCE
david.bourgarit@culture.fr

BRUN Céline
BDI / CNRS
31 rue Pasteur
69007 LYON
FRANCE
celine.brun@mom.fr

BRUNI Silvia
Dipartimento di Chimica Inorganica, Metallorganica e Analitica
Università degli Studi di Milano
Via G. Venezian 21
20133 MILANO
ITALY
silvia.bruni@unimi.it

BURRI Elena
Archéologie cantonale
10 place de la Riponne
1014 LAUSANNE
SWITZERLAND
elena.burri@vd.ch

CANTIN Nadia
UMR5138 "Archéométrie et Archéologie"
Maison de l'Orient et de la Méditerranée
7 rue Raulin
69365 LYON cedex 7
FRANCE
nadia.cantin@mom.fr

CAPELLI Claudio
Dipartimento per lo Studio del Territorio e delle sue Risorse (Dip.te.ris.)
Università degli Studi di Genova
Corso Europa 26
16132 GENOVA
ITALY
capelli@dipteris.unige.it

CARPENITO Giulio
Dipartimento di Scienze della Terra dell'Università di Modena e Reggio Emilia
Largo S.Eufemia 19
41100 MODENA
ITALY
carpenito.giulio@unimore.it

CILIANI Serena
c/o Arch. Rita FABBRI
Facultà Architettura Ferrara
via Quartieri 8
44100 FERRARA
ITALY
serena.ciliani@libero.it

CONSTANTINESCU Bogdan
Institute of Atomic Physics
POB MG-6
BUCHAREST
ROMANIA
bconst@ifin.nipne.ro

DECAVALLAS Oreste
C2RMF (CNRS UMR 171) et Équipe de Protohistoire égéenne (CNRS UMR 7041)
Laboratoire de recherche des Musées de France
C2RMF CNRS UMR 171
Palais du Louvre Porte des Lions
14 quai F. Mitterand
75001 PARIS
FRANCE
decavallas@yahoo.fr

DELQUE-KOLIC Emmanuelle
Centre de datation par le Radiocarbone, CNRS UMR5138
Université Claude Bernard-Lyon 1
40 boulevard Niels Bohr
69622 VILLEURBANNE CEDEX
FRANCE
Emmanuelle.Delque@univ-lyon1.fr

DESBAT Armand
UMR5138 "Archéométrie et archéologie"
Maison de l'Orient et de la Méditerranée
7 rue Raulin
69365 LYON cedex 7
FRANCE
a.desbat@laposte.net

DIAS M. Isabel
Instituto Tecnológico e Nuclear
Estrada Nacional 10
2686-953 SACAVÉM
PORTUGAL
isadias@itn.mcies.pt

DURGEAU Sandrine
Univ. Paris 1
Centre Archéologique de Seine-Saint-Denis
1 à 5 route de Saint-Leu
93800 ÉPINAY-SUR-SEINE
FRANCE
sadurgeau@hotmail.com

DYCZEK Piotr
Ośrodek Badań Nad Antykiem Europy Południowo - Wschodniej - Uniwersytet Warszawski
UL. Krakowskie Przedmieście 32
00 927 WARSZAWA
POLAND
novae@uw.edu.pl

ERAMO Giacomo
Département de géosciences
Minéralogie et Pétrographie, Univ. de Fribourg
chemin du Musée, 6
1700 FRIBOURG
SWITZERLAND
giacomo.eramo@unifr.ch

FLEVARI Loucia
University of Crete
Chatzimichali Giannari 23
PC 74100 RETHYMNO CRETE
GREECE
Flevariloucia@hotmail.com

FORMENTI Françoise
CPE Lyon ECTI
FRANCE
formenti@cpe.fr

GARCÍA IÑAÑEZ Javier
Cultura Material i Arqueometria UB (ARQ/UB), Facultat de Geografia i Història, Universitat de Barcelona,
c/ Montalegre, 6
08001, BARCELONA, CATALONIA
SPAIN
javier_garcia@catalonia.net

GLIGOR Mihai
"1 Decembrie 1918" university ;
Systemic archaeology institute
11-13 Nicolae Iorga St.
510009 ALBA IULIA
ROMANIA
m_gligor@yahoo.com

GRAVE Peter
University of New England
Archaeology and Palaeoanthropology, University of New England
NSW, 2351 ARMIDALE
AUSTRALIA
pgrave@metz.une.edu.au

GUALTIERI Sabrina
CNR - Institute of Science and Technology for Ceramics
via granarolo 64
48018 FAENZA
ITALY
sgualtieri@istec.cnr.it

GUIBERT Pierre
CNRS - Institut de Recherche sur les Archéomatériaux, CR2PA
Maison de l'archéologie
33607 PESSAC
FRANCE
guibert@u-bordeaux3.fr

HEIN Anno
Institute of Materials Science, N.C.S.R. "Demokritos",
Aghia Paraskevi
15310 ATHENS
GREECE
hein@ims.demokritos.gr

HENSHAW Christina
Institute of Archaeology, University College London
66c Harestone Hill
CR3 6DH CATERHAM, SURREY
GREAT BRITAIN
PCH@DSL.PIPEX.COM

HEROLD Hajnalka
Department of Prehistory and Early Middle Ages and Vienna Institute of archaeological science, University of Vienna
Franz Klein Gasse 1
A-1190 VIENNA
AUSTRIA
hajnalka.herold@univie.ac.at

HUET Nathalie
ARC'ANTIQUE
26, rue de la Haute Forêt
44300 NANTES
FRANCE
arc.antique@wanadoo.fr

KATONA Ildiko
Centre de Recherche et de Restauration des Musées de France
Palais du Louvre - Porte des Lions
14 quai François Mitterand
75001 PARIS
FRANCE
ildiko.katona@culture.gouv.fr

KEALHOFER Lisa
Santa Clara University
Anthropology/Sociology, Santa Clara University
CA 95053 SANTA CLARA
UNITED STATES
LKealhofer@scu.edu

KILIKOGLOU Vassilis
Institute of Materials Science, N.C.S.R. Demokritos, Aghia Paraskevi,
15310 ATHENS
GREECE
kilikog@ims.demokritos.gr

KORDATZAKI Georgia
University of Crete
Chatzimichali Giannari 23
PC 74100 RETHYMNO CRETE
GREECE
Kordatzaki@students.phl.uoc.gr

KUZMINA Elena
Russian Institute for cultural studies
Bersenevskaya nab, 20
Institute for Cultural Research
119072 MOSCOW
RUSSIA
riku@dol.ru

LADURON Dominique
Unité de Géologie
Université catholique de Louvain
Place Louis Pasteur, 3
1348 LOUVAIN-LA-NEUVE
BELGIUM
laduron@geol.ucl.ac.be

LELU Chloé
UMR7041 : Archéologie et Sciences de l'Antiquité, Maison de l'archéologie et de l'ethnologie
21, allée de l'Université
92023 NANTERRE CEDEX
FRANCE
chloe.lelu@9online.fr

LOPATINA Olga
Institute of archaeology of russian academy of sciences
Molodogvardeyskaya str., 24-1-44
121355 MOSCOW
RUSSIA
olgalop@list.ru

MARTINEAU Rémi
CNRS - Université de Bourgogne, UMR 5594 Archéologie, cultures et sociétés,
UFR des sciences de la terre
6 bd Gabriel
21000 DIJON
FRANCE
remi.martineau@u-bourgogne.fr

MARTÍNEZ FERRERAS Verònica
Equip de Recerca Arqueomètrica de la Universitat de Barcelona (ERAUB), Universitat de Barcelona, Facultat de Geografia i Història, Departament de Prehistòria, Història antiga i arqueologia
c/ Montalegre, 6-8
08001 BARCELONA, CATALONIA
SPAIN
vmartinez@ub.edu

MARTINÓN-TORRES Marcos
Institute of Archaeology, University College London
31-34 Gordon Square
WC1H 0PY LONDON
GREAT BRITAIN
m.martinon-torres@ucl.ac.uk

MIRABAUD Sigrid
Centre de recherche et de Restauration des Musées de France (C2RMF)
UMR 171 CNRS
14 Quai François Mitterrand
75001 PARIS
FRANCE
sigrid.mirabaud@culture.fr

MIRTI Piero
Università di Torino
Dipartimento di chimica analitica
Via Giuria 5
I-10125 TORINO
ITALY
piero.mirti@unito.it

MOMMSEN Hans
Helmholtz-Institut für Strahlen-und Kernphysik, Universität Bonn
Nussallee 14-16
53115 BONN
GERMANY
mommsen@hiskp.uni-bonn.de

MONSIEUR Patrick
University of Ghent
Section Classical Archaeology
Blandijnberg 2
9000 GHENT
BELGIUM
patrick.monsieur@ugent.be

MUNTONI Italo Maria
Facoltà di Scienze, Università degli studi di Bari
Viale Anicio Gallo 3
00174 ROMA
ITALY
imuntoni@katamail.com
Italo.Muntoni@uniroma1.it

NEVORET Joëlle
CREAM Gabriel Chapotat
11 rue du 24 avril 1915
38200 VIENNE
FRANCE
vienne.cream@free.fr

NODAROU Eleni
INSTAP Study Center for East Crete
Pacheia Ammos
72200 IERAPETRA CRETE
GREECE
enodarou@yahoo.gr

OBERWEILER Cécile
Université Paris I-Sorbonne / UMR7041 ArScAn / Protohistoire égéenne
MAE (R. Ginouvès) 21 allée de l'Université
92023 NANTERRE
FRANCE
cecile.oberweiler@mae.u-paris10.fr

ODRIOZOLA LLORET Carlos
Instituto de Ciencias de Materiales de Sevilla (CSIC)
c/ Americo Vespuccio S/N (Isla de la Cartuja)
41092 SEVILLA
SPAIN
carlos@icmse.csic.es

OLLAGNIER Céline
Centre de Recherche en Physique appliquée à l'Archéologie, Université de Bordeaux 3 / CNRS (IRAMAT, UMR5060)
Maison de l'Archéologie
33607 PESSAC
FRANCE
olla_gnier@hotmail.com

PACE Marco
Universita' di Torino
Dipartimento di chimica analitica
Via Giuria 5
I-10125 TORINO
ITALY
marco.pace@unito.it

PACHECO Claire
CRPAA-IRAMAT
Maison de l'Archéologie
Esplanade des Antilles
33607 PESSAC
FRANCE
clairepatch@yahoo.fr

PERARDI Alessandra
Dipartimento Chimica IFM, Universita'di Torino
Via Giuria 7
10125 TORINO
ITALY
alessandra.perardi@unito.it

PÉREZ-ARANTEGUI Josefina
Dep. Analytical chemistry, University of Zaragoza
c/ Pedro Cerbuna 12
50009 ZARAGOZA
SPAIN
jparante@unizar.es

PRICE David Michael
School of Earth and Environmental Sciences
University of Wollongong, northfields avenue
NSW2522 WOLLONGONG
AUSTRALIA
dprice@uow.edu.au

PRUDÊNCIO M. Isabel
Instituto Tecnológico e Nuclear
Estrada Nacional 10
2686-953 SACAVÉM
PORTUGAL
iprudenc@itn.mcies.pt

RAVOIRE Fabienne
UMR 5594 (Dijon) / INRAP Base de Passy
18 rue de la chapelle
89510 VÉRON
FRANCE
f.ravoire@wanadoo.fr
fabienne.ravoire@inrap.fr

RICCIARDI Paola
CNR-ISTEC (Institute of science and technology of ceramics)
via granarolo 64
48018 FAENZA
ITALY
ricciardi_paola@libero.it

RODOT Marie-Angélique
UMR 5594 "Archéologie, Cultures et Sociétés"
Centre des Sciences de la Terre, bâtiment Gabriel
21000 DIJON
FRANCE
Marie-Angelique.Rodot@u-bourgogne.fr

ROUMPOU Maria
University of Bradford, Department of Archaeological Sciences
19 Mesologiou St.
21100 NAFPLIO
GREECE
m.roumpou@bradford.ac.uk

SCHMITT Anne
UMR5138 "Archéométrie et Archéologie"
Maison de l'Orient et de la Méditerranée
7 rue Raulin
69365 LYON cedex 7
FRANCE
anne.schmitt@mom.fr

SCHWEDT Alexander
Equip de Recerca Arqueomètrica de la Universitat de Barcelona (ERAUB), Universitat de Barcelona, Facultat de Geografia i Història, Departament de Prehistòria, Història antiga i Arqueologia
c/ Montalegre, 6-8
08001 BARCELONA, CATALONIA
SPAIN
schwedt@iskp.uni-bonn.de

SCIAU Philippe
CEMES-CNRS (UPR8011)
29 rue J. Marvig BP 94347
31055 TOULOUSE
FRANCE
sciau@cemes.fr

SERNEELS Vincent
Departement Géosciences
Université de Fribourg
PEROLLES
1700 FRIBOURG
SWITZERLAND
vincent.serneels@unifr.ch

SESTIER Claude
UMR 5594 - Archéologie
6 bld Gabriel
21000 DIJON
FRANCE
sestier.flint@infonie.fr

SODO Armida
Dipartimento di Fisica "E. Amaldi"
Universita' Roma Tre
Via della Vasca Navale 84
00146 ROMA
ITALY
sodo@fis.uniroma3.it

SZAKMÁNY György
Eötvös Loránd University, Dept of Petrology and Geochemistry
Pázmány P. Stny. 1/C
1117 BUDAPEST
HUNGARY
gyorgy.szakmany@geology.elte.hu

SZILÁGYI Veronika
Eötvös Loránd University, Dept of Petrology and Geochemistry
Pázmány P. Stny. 1/C
1117 BUDAPEST
HUNGARY
szilagyi.vera@vipmail.hu

TANIMOTO Satoko
Institute of Archaeology
University College London
31-34 Gordon Square
WCIH OPY LONDON
GREAT BRITAIN
s.tanimoto@ucl.ac.uk

TESLENKO Iryna
Ukrainian Academy of Sciences, Archeological Institute, Crimean Branch
2 Yaltinskaya str.
95007 SIMFEROPOL CRIMEA
UKRAINE
i_teslenko@mail.ru

THIERRIN-MICHAEL Gisela
Department of Geosciences, Mineralogy and Petrography, University of Fribourg
Perolles
1700 FRIBOURG
SWITZERLAND
gisela.thierrin-michael@unifr.ch

THIRION-MERLE Valérie
CNRS UMR5138 "Archéométrie et Archéologie"
Maison de l'Orient et de la Méditerranée
7 rue Raulin
69365 LYON cedex 7
FRANCE
valerie.merle@mom.fr

THOMAS Nicolas
INRAP
31 rue Delizy
93698 PANTIN CEDEX
FRANCE
nicolasthomas@noos.fr

TSETLIN Yuri
Institute of Archaeology of Russian Academy of Sciences
Habarovskaya str., 18-1-249
107589 MOSCOW
RUSSIA
yu.tsetlin@mtu-net.ru

TSIRTSONI Zoï
UMR 5138 - CNRS - Archéométrie et Archéologie
Maison de l'Orient et de la Méditerranée
7 rue Raulin
69365 LYON cedex 7
FRANCE
zoi.tsirtsoni@mom.fr

TULUN A. Tulay
Technical university of Istanbul
Faculty of science and letters
Chemistry department
MASLAK 34469
ISTANBUL
TURKEY
tulun@itu.edu.tr

VARVARA Simona
"1 Decembrie 1918" University
Systemic Archaeology Institute
11-13 Nicolae Iorga St.
510009 ALBA IULIA
ROMANIA
s_varvara@bcum.uab.ro
varvarasimona@yahoo.com

VAZ PINTO Inês
Universidade Lusíada de Lisboa
Rua da Junqueira 188 a 198
1349-001 LISBOA
PORTUGAL
invpinto@mail.telepac.pt

VIEILLEVIGNE Emmanuelle
Centre de Recherche en Physique Appliquée à l'Archéologie
Université de Bordeaux 3 / CNRS
(IRAMAT UMR 5060)
Maison de l'Archéologie
Esplanade des Antilles
33607 PESSAC
FRANCE
evieillevign@u-bordeaux3.fr

VILA SOCIAS Llorenç
Equip de Recerca Arqueomètrica de la Universitat de Barcelona (ERAUB), Universitat de Barcelona, Facultat de Geografia i Història, Departament de Prehistòria, Història antiga i arqueologia
c/ Montalegre, 6-8
08001 BARCELONA, CATALONIA
SPAIN
lvila@catalonia.net

WAKSMAN Yona
Laboratoire de Céramologie
CNRS UMR5138 "Archéométrie et Archéologie"
Maison de l'Orient et de la Méditerranée
7 rue Raulin
69365 LYON cedex 7
FRANCE
waksman@mom.fr

ZELENKO Sergiy
Taras Schevchenko Kiev National University
Archaeology and Museum Study Department
Faculty of History
64 Volodymyrska Str.
01033 KIEV
UKRAINE
zel@univ.kiev.ua

ZINK Antoine
C2RMF
UMR 171 CNRS-MCC
Palais du Louvre
Porte des Lions
14 quai F. Mitterrand
75001 PARIS
FRANCE
antoine.zink@culture.gouv.fr

www.ingramcontent.com/pod-product-compliance
Lightning Source LLC
LaVergne TN
LVHW071209210926
848482LV00015B/758